ROCK BLASTING EFFECTS AND OPERATIONS

BALKEMA - Proceedings and Monographs
in Engineering, Water and Earth Sciences

ROCK BLASTING EFFECTS AND OPERATIONS

Pijush Pal Roy
Scientist-in-Charge, Blasting Department
Central Institute of Mining and Fuel Research
(erstwhile CMRI)
Dhanbad, India

CRC Press
Taylor & Francis Group
Boca Raton London New York

CRC Press is an imprint of the
Taylor & Francis Group, an **informa** business
A BALKEMA BOOK

First published 2005 by CRC Press/Balkema

Published 2019 by CRC Press
Taylor & Francis Group
6000 Broken Sound Parkway NW, Suite 300
Boca Raton, FL 33487-2742

ISBN 13: 978-0-415-37230-5 (hbk)

Library of Congress Cataloging-in-Publication Data
Applied for

FOREWORD

In the present scenario of globalization and market driven policy of our country, the mining industry in India is in need of some innovative and cost effective approaches for enhanced coal as well as mineral production. Due to increased proliferation of mining activities towards urban areas, the practising mining engineers are facing severe difficulty in conducting bigger production blasts mainly because of the spectre of ecological disturbance. The burgeoning demand of coal and minerals under varying constraints and conflicts, both locational and rock geologic, call for appropriate technologies and accessories of drilling and blasting operations. In recent times, researchers and practising engineers around the globe are zeroing in on developing rock blasting techniques based on synergistic approaches integrating the knowledge of the rock-explosive interactions, breakage mechanism and geomechanic properties of rock masses with an ultimate aim of optimum benefit.

Utilization of commercial explosives for mining and civil engineering purposes creates several unsavoury situations like formation of flyrock, ground vibration, noise, dust, fumes and air overpressure. Unless such unwanted features are properly and timely warded off there could be every possibility of inviting lawsuits, hindrance of operation, loss of property and cost escalation. Noticeably, the dimensions of such side-effects have widened markedly in recent years due to greater usage of high octane explosives meant for massive production.

Taking stock of the situation, Dr. P. Pal Roy has documented his vast experience and knowledge in the field of rock blasting, spanning over two decades, in the form of an invaluable book which contains fifteen chapters covering fundamental, theoretical and practical aspects of rock blasting technology. Along with many interesting and innovative features, it explains, quite elaborately, some emerging rock blasting techniques that have future prospects. The book undoubtedly will bridge the gap of knowledge in respect of safe and cost-effective contemporary technology on blasting.

I am confident that this book will be of immense use to the mining, civil and explosive industries as well as allied research organizations and academic institutions.

Prof. Dhanus Dhari Misra
Director (Retired)
Central Institute of
Mining and Fuel Research
(erstwhile CMRI)
Dhanbad, India

Preface

In the last two decades there have been many path-breaking advancements in the explosive and blasting techniques around the world. Many revolutionary developments and technical contributions have led to better understanding of the rock-explosive interactions. Innovations like non-electric initiating devices (shock tube systems), detonators of high sensitive primary explosives, emulsion explosives, bulk loading system, and more recently, super accurate electronic detonators, have made noticeable evolution in the development of new explosive and blasting accessories. Emerging blasting techniques like air-decking, remote controlled blasting system, controlled demolition blasting, cast blasting, plasma blasting and vertical crater retreat blasting have already proved themselves of potential advantages and wide acceptability. Several sophisticated software packages for design, prediction, evaluation and assessment of both underground and opencast blasting have made a judicious blend of technology modified to suit mine conditions, desired production and productivity with enhanced safety.

At this juncture of global technological evolution, the idea of writing this book was conceived in perspective of scientific amalgamation of theory and practice of drilling and blasting techniques that, many a time, has been communicated as a science as well as art.

The book contains fifteen extensive chapters covering statistical methods, design parameters, environmental aspects, structural damage, fragmentation, emerging techniques, surface and sub-surface blasting methodologies, safety, explosive characteristics and their accessories. Several case studies and practical issues are discussed so that the readers can make their own judgement reducing the threat of lawsuits, meeting regulatory levels, and maintaining reasonable costs. It also summarizes the results of research carried out by the Blasting Department of Central Mining Research Institute (presently renamed as Central Institute of Mining and Fuel Research or CIMFR) since its inception in the year 1970.

Many people are owed thanks for the structural formation and development of this manuscript. Primary recognition goes to the Blasting Research Group of the Central Mining Research Institute and its erstwhile Heads namely Dr. P.K. Dutta, Dr. R.N. Gupta, Sri A. Bagchi and Sri R.B. Singh. Other colleagues (both present and former) of the Institute Dr. M.M. Bhattacharyya, Dr. A. Sinha, Sri S.K. Gupta, Dr. A.K. Chakraborty, Dr. M.M. Singh, Dr. A.J. Prakash, Dr. P.K. Singh, Sri S.K. Mandal, Sri S.K. Roy, Sri D.P. Thakur, Sri Dinesh Prasad, Sri K.L. Bandyopadhyay and Dr. A.K. Raina have extended all possible help and guidance in the preparation of the manuscript. The valuable research contributions made by several departmental colleagues have also enriched the contents of this book. I am especially indebted to my younger colleague Dr. C. Sawmliana for his tremendous help in tracing most of the important diagrams and contributing many scientific inputs of this book. My other colleagues Sri M.P. Roy, Sri R.K. Singh, Sri N.K. Bhagat have also helped in the preparation of the manuscript.

Dr. Kalyan Sen, formerly Director, CMRI was a source of inspiration for this work, without his encouragement it would not have been possible for me to write this book.

I am thankful to Sri M.O. Sarathy and Sri B. Dash of Gulf Oil Corporation Ltd. (previously IDL Industries Ltd.), Sri V. Bhushan, Sri S. Jana and Sri Mohendra Babu of Orica-IEL, Sri B.K. Paul of Bharat Explosives Ltd. for their cooperation and many fruitful discussions during several field trials. I am indebted to Dr. Sushil Bhandari, Director, Earth Resource Technology Consultants, Jodhpur; Professor A.K. Ghose, formerly Director, Indian School of Mines (presently ISMU), Dhanbad and Professor Bharat B. Dhar, formerly Director, Central Mining Research Institute, Dhanbad for their help, guidance and encouragement throughout the work.

I am confident that this book will prove to be very helpful as a text cum reference book to all those concerned to the field of blasting science and technology. If this happens, I will consider my paramount efforts to be for a noble cause for the understanding of blasting technology for safe and efficient rock and mineral excavations.

Last but not the least, I am thankful to my wife Piyali and son Prameet for encouraging me to write this book and providing me with the mental support and all the necessary help which I needed during the preparation of the manuscript.

Pijush Pal Roy
Blasting Department
Central Institute of
Mining and Fuel Research
(erstwhile CMRI)
Dhanbad, India
e-mail: ppalroy@yahoo.com

Contents

1

Statistical Approaches and Useful Mathematical Formulae

1.1 INTRODUCTION

Various statistical methods are used in the analysis of field data. In addition to some conventional statistical techniques, a few new concepts, which on application may improve predictability, are also included. This chapter covers the following measures:

(i) Characteristic distribution of field data
(ii) Regression analysis for one, two and three independent variables
(iii) Simple correlation coefficient
(iv) Coefficient of determination
(v) Standard error of estimate
(vi) Adjusted coefficient of determination
(vii) Overall *F*-test
(viii) Partial regression and correlation coefficients
(ix) Chi-square distribution
(x) Bar diagram and Pie diagram
(xi) Hemispherical normal distribution

1.2 STATISTICAL ANALYSIS OF FIELD DATA

Over the years, attempts have been made to relate the parameters defining vibratory particle motion with observed structural damages. This has led to the formulation of a number of empirical damage scales relating to observed particle motion in terms of peak particle acceleration, velocities and displacements.

Studies on results of blasting, indicate that the length of collar stemming influences the level of vibration, noise and distance of flying fragments. Increase in collar stemming decreases flyrock and noise, but increases the level of ground vibration. For a rough estimation of flying fragment distance, empirical measures are required which could evolve parameters, such as, ratio of the length of stemming column to burden versus spacing, charge factor etc. Scatter in delay periods of various delay detonators also plays an important role.

Identification of clusters or groupings among rock fracture orientations at a mining site is essential in fragmentation analysis. Subsequently the mean orientation of the fractures within each cluster and the dispersion or scatter among the fracture orientations must be calculated. Points in adjacent patches where the Poisson distribution test shows significant concentrations, are assigned to a single cluster. For a cluster whose fracture orientations follow Arnold's hemispherical normal distribution, a confidence interval for the fracture orientation mean is calculated by applying Fisher's estimates (Mahtab et al., 1972; Denness, 1970; Koch and Link, 1970; Krumbein and Graybill, 1965).

To be able to predict when damage to structures is likely to occur, a statistical analysis of data from many blasts is necessary. The most commonly used statistical techniques include regression analysis based on least-square method, simple correlation coefficient determinations, overall *F*-test for the goodness of fit of the regression equation and chi-square analysis.

The accuracy of a particular model in predicting the vibrations from field data can be assessed by many statistical methods, such as coefficient of determination, standard error of estimates, coefficient of variability and adjusted coefficient of determination. Coefficient of determination and standard error of estimate are very popular methods for determining goodness of fit in the regression analysis. Extensive usage of these techniques is evident in the work of many investigators (Dowding, 1971; Wiss and Linehan, 1978; Siskind et al., 1980).

The adjusted coefficient of determination takes into account the number of independent variables present in the model. It is mainly used for comparing innovative and conventional models (Ghosh, 1983).

1.3 STATISTICAL METHOD AND THEIR PHYSICAL INTERPRETATIONS

1.3.1 Chief Characteristics of Normal Distribution and Normal Probability Curve

(i) The normal probability curve with mean (μ) and standard deviation (σ) is given by the equation:

$$f(x) = \frac{1}{\sigma\sqrt{2\pi}} e^{-(x-\mu)^2/2\sigma^2}, \quad -\infty < x < \infty \tag{1.1}$$

and has the following properties:

(ii) Curve is bell-shaped and symmetrical about the line $x = \mu$

(iii) Mean, median and mode of distribution coincide

(iv) As x increases numerically, $f(x)$ decreases rapidly, the maximum probability occurring at the point $x = \mu$

(v) $\mu_{2r+1} = 0$ (r = 0, 1, 2 ...) and
$\mu_{2r} = 1.3.5. \ldots (2r - 1)\sigma^{2r}$

(vi) Linear combination of independent normal variates is also a normal variate

(vii) x axis is an asymptote to the curve

(viii) Mean deviation about mean is zero

(ix) QD(quartile deviation) = $(Q_3 - Q_1)/2 \cong (2/3)\sigma$

Figures 1.1 and 1.2 show the normal distributions with different standard deviations and means.

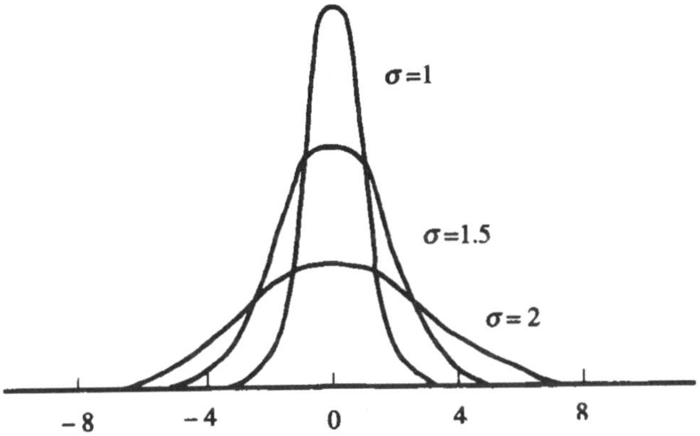

Fig. 1.1 Normal distributions with different SDs but with identical mean $\mu = 0$

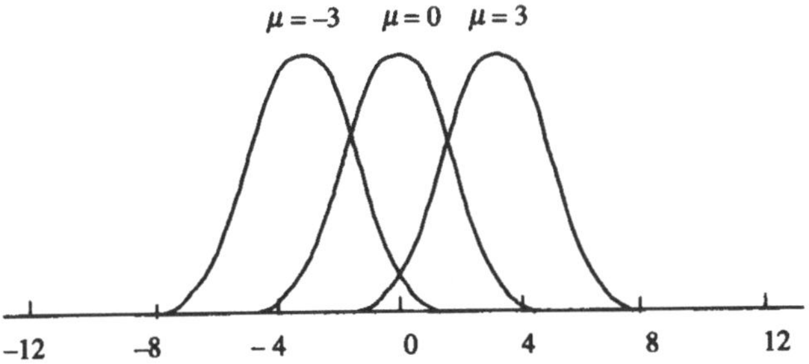

Fig. 1.2 Normal distributions with different means but with identical standard deviations $\sigma = 1.5$

Referring to vibration data analysis, Walter (1981) showed the inherent variability and statistical character of data distribution. He used the term "repeatability", meaning: if a blasting procedure is repeated in an identical manner, keeping the same charge, hole depth, rock conditions, method of initiation etc. then the resultant vibration level will be the same. This logical assumption and expectation does not occur in reality. The vibration level actually differs from blast to blast, sometimes lower, sometimes higher. If the process is repeated a sufficient number of times, the vibration levels will spread out in a statistical distribution that is fairly typical of the variability or lack of repeatability. Figure 1.3 shows the distribution of data from shots of a trenching operation. It reveals a variation approximately 60% above and below the mean vibration level. The distribution characteristics are approximately normal with a spread above and below the mean.

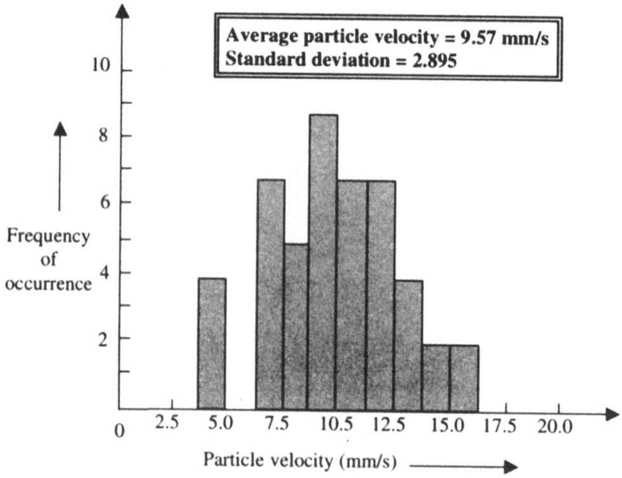

Fig. 1.3 Frequency versus particle velocity distribution (Walter, 1981)

1.3.2 Regression and Correlation Analysis

Many statistical investigations require algebraic relationships between concerned variables to enable prediction of one or more variables in terms of others. Although it is desirable to be able to predict one quantity exactly in terms of the others, this is rarely possible, and in most instances, only averages or expected values can be predicted.

The regression problem considers the frequency distributions of one variable when another is fixed at different levels. The correlation problem considers the joint variation of two measurements, neither of which is restricted by the researcher. A functional relation between two variables implies nothing about cause and effect; nonetheless it enables prediction of the value of the variable on the condition that prior information about the other is available. If two variables are involved, the variable that is the basis of estimation is conventionally called the independent variable and the variable whose value is to be estimated is called the dependent variable. The main use of a regression equation is to predict the most likely measurement of one variable from the known measurement of another.

Correlation methods are considered appropriate when interest is centred on measuring the degree to which two variables are linearly related and when both variables are randomly sampled. The fact that two variables tend to increase or decrease together does not imply that the one has any direct or indirect influence on the other. Both may be influenced by other variables. Such influence can be determined by partial correlation coefficients.

If there are n pairs of observation (x_1, y_1), (x_2, y_2) etc., the problem that frequently arises is the determination of whether an apparent relation between the two variables is significant, and if having shown it to be significant, to determine the best form of representation. In simple regression analysis, the test for the significance is carried out to determine whether the data can be represented by the following linear equation or not.

$$y = a + bx \tag{1.2}$$

By the method of least square, the coefficients a and b are obtained as—

$$b = \frac{n\Sigma xy - \Sigma x \Sigma y}{n\Sigma x^2 - (\Sigma x)^2} \tag{1.3}$$

$$a = \frac{\Sigma y}{n} - b\frac{\Sigma x}{n} \tag{1.4}$$

To test for the significance of an apparently linear relation, the correlation coefficient is determined by—

$$r = \frac{n\Sigma xy - \Sigma x \Sigma y}{\sqrt{[n\Sigma y^2 - (\Sigma y)^2][n\Sigma x^2 - (\Sigma x)^2]}} \tag{1.5}$$

The correlation coefficient is so characterized that if the relationship between the data can be represented exactly by a straight line, then $r = \pm 1$; if, on the other hand, there is no relation at all between the variables, then $r = 0$. However, a high correlation coefficient between two variables does not necessarily indicate a casual relationship. There may be a third variable which may cause the simultaneous change in the first two variables, and produce spuriously high correlation coefficient.

It is important to note that if no assumptions are made about the joint distribution of the concerned random variables, "goodness", of the prediction and "goodness" of the estimates a and b given in equations (1.3) and (1.4) cannot be judged. If the data is considered to follow normal distribution, then r^2 will only determine the strength of the regression line (1.2), and not in other cases.

The percentage of the total variation of the y's accounted for by the relationship with x is 100 r^2. Thus, when $r = 0.9$ then 81% of the variation of the y's is accounted for by the relationship with x. The value of r always lies between 0 and +1 or −1.

Some possible linear equations for curve fitting often used for practical purposes are:

$$y = A + Bx \tag{1.6}$$
$$y = A + B/x \tag{1.7}$$
$$y = Ae^{Bx} \tag{1.8}$$
$$y = Ax^B \tag{1.9}$$
$$y = 1/(A + Bx) \tag{1.10}$$
$$y = x/(A + Bx) \tag{1.11}$$

Obviously, equations (1.6) to (1.11) must be converted beforehand by suitable transformations to linear form.

Regression Analysis of Two Independent Variables

In the case of two independent variables associated in the model one requires regression analysis of multiple variables. For example, in vibration data analysis, the general empirical relationship between peak particle velocity (V), distance (D) and maximum charge per delay (Q) is $V = KD^{-B}Q^A$. The relation takes the following form in a log-log scale.

$$y = a + b_1x_1 + b_2x_2 \tag{1.12}$$

The maximum likelihood estimates of a, b_1 and b_2 are:

$$b_2 = \frac{(AC - nG)(nD - B^2) - (AB - nE)(nF - BC)}{(nF - BC)^2 - (nH - C^2)(nD - B^2)} \tag{1.13a}$$

$$b_1 = \frac{[(C^2 - nH)b_2 - (AC - nG)]}{(nF - CB)} \tag{1.13b}$$

$$a = \frac{(A - b_1B - b_2C)}{n} \tag{1.13c}$$

where n is the number of data in the analysis and

$$A = \Sigma y,\ B = \Sigma x_1,\ C = \Sigma x_2,\ D = \Sigma x_1^2$$
$$E = \Sigma x_1 y,\ F = \Sigma x_1 x_2,\ G = \Sigma x_2 y,\ H = \Sigma x_2^2 \tag{1.14}$$

Calculations for the three independent variables are the same as those for two independent variables. The constants b_1, b_2, b_3 and a can be obtained in a similar manner.

Multiple Regression and Partial Correlation

The simple regression coefficients are given by

$$b_{y1} = \frac{\Sigma' y x_1}{\Sigma' x_1^2} \tag{1.15}$$

$$b_{y2} = \frac{\Sigma' y x_2}{\Sigma' x_2^2} \tag{1.16}$$

where Σ' is the summation of deviations from the means, i.e.,

$$\Sigma' y x_1 = \Sigma(y - \bar{y})(x_1 - \bar{x_1}) \tag{1.17}$$

The correlation coefficient of x_1 and x_2 is—

$$r_{12} = \frac{\Sigma' x_1 x_2}{\sqrt{\Sigma' x_1^2 \, \Sigma' x_2^2}} \tag{1.18}$$

Similarly,

$$r_{y1} = \frac{\Sigma' y x_1}{\sqrt{\Sigma' y^2 \, \Sigma' x_1^2}} \quad \text{and} \quad r_{y2} = \frac{\Sigma' y x_2}{\sqrt{\Sigma' y^2 \, \Sigma' x_2^2}} \tag{1.19}$$

With equations (1.15) and (1.16), when simple regression coefficients are calculated, the effect of the third variable is felt. If for example, $b_{y1} = 3.7$, $b_{y2} = 1.6$, $r_{12} = 0.75$ then, since the higher values for x_1 will tend to be associated with the higher values of x_2 (as $r_{12} > 0$), the values of y for the larger values of x_1 will tend to be larger than they should be, because they are also associated with the larger values of x_2. The partial regression coefficients, however, have the effect of the other variables eliminated. The partial correlation coefficient of y and x_1 when x_2 is kept fixed is—

$$r_{y1.2} = \frac{r_{y1} - r_{y2} \cdot r_{12}}{\sqrt{(1 - r_{y2}^2)(1 - r_{12}^2)}} \tag{1.20}$$

The partial correlation coefficient of y and x_2 when x_1 is kept constant is—

$$r_{y2.1} = \frac{r_{y2} - r_{y1} \cdot r_{21}}{\sqrt{(1 - r_{y1}^2)(1 - r_{21}^2)}} \tag{1.21}$$

The application and importance of equations (1.20) and (1.21) in vibration data analysis were studied by Pal Roy and Gupta (1989). The correlation coefficients were found to be markedly reduced in their transformation from simple to partial coefficients. It was also found that a positive simple correlation coefficient could be transformed into a negative partial correlation coefficient and vice versa.

1.3.3 Coefficient of Determination

The overall accuracy of any predictor equation is determined by the coefficient of determination (R^2). It is the variation in the dependent variable due to the combined linear influence of the independent variables divided by the total variation in the dependent variable. If R^2 has a value of 0.81, then the independent variables $x_1, x_2 \ldots$ in the regression equation explain 81% of the total variation of the dependent variable y. The formula of R^2 is—

$$R^2 = 1 - \frac{\Sigma(y_i - y')^2}{\Sigma(y_i - \bar{y})^2} \tag{1.22}$$

Where $\bar{y}$ is the mean of observed values of y_i and y' is the value calculated from the regression equation.

1.3.4 Standard Error of Estimate (SEE)

The Standard error of estimate (SEE) measures in absolute units the extent of accuracy of the regression equation in predicting the dependent variable y. The formula of (SEE) is

$$SEE = \sqrt{\frac{\Sigma(y - y')^2}{(n - k - 1)}} \tag{1.23}$$

where n is the number of observations and k the number of independent variables present in the model.

When y is normally distributed about the regression line, then 68% of the actual y values will fall within $y \pm 1$ (SEE), 95% of the actual y values will fall within $y \pm 2$ (SEE) and 99.7% of that will fall within $y \pm 3$ (SEE).

1.3.5 Adjusted Coefficient of Determination

The coefficient of determination does not account for the number of independent variables in the model. Seber (1977) has shown that the value of R^2 may increase due to the addition of more independent variables in the model. The formula of adjusted coefficient of determination R_a^2 as proposed by Seber (1977) is—

$$R_a^2 = 1 - \frac{(n-1)}{(n-k-1)}(1 - R^2) = R^2 - \frac{k}{(n-k-1)} \frac{\Sigma(y - y')^2}{\Sigma(y - \bar{y})^2} \tag{1.24}$$

The addition of more independent variables in the model decreases the sum of squares of errors which gives a smaller value of R_a^2.

1.3.6 F-Test for the Goodness of Fit

The F-test is based on the ratio between inter-groups variance and intra-groups variance. The F-distribution is not single curve but a large family of curves, varying with the degrees of freedom. A null hypothesis is proposed that there is no difference between inter-groups variance and intra-groups variance. Then the significance of the *F*-value is determined. The *F*-ratio is defined as—

$$F = \frac{R^2 / k}{(1 - R^2)/(n - k - 1)} \tag{1.25}$$

The degrees of freedom for *F* are *k* and (*n-k-1*). When the calculated value of *F* exceeds the statistical tables critical value for a given significance level, the null hypothesis will be rejected. Otherwise, the observed x_is are not significant at the given significance level. Figure 1.4 shows the nature of *F*-distribution with some specified degrees of freedom.

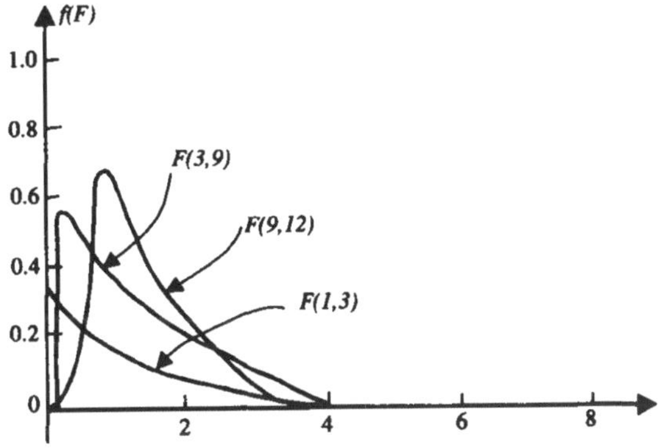

Fig. 1.4 F-distribution with specified degrees of freedom

1.3.7 Chi-square Distribution

Chi-square distribution is a well-known probability distribution. The distribution of χ^2 has very broad applications and ranks close to the normal distribution in general usefulness. An important application of the χ^2 distribution is the test of goodness of fit. Chi-square is actually based on the idea that if the hypothesis upon which the expected frequencies are computed is correct, deviations of actual frequencies from the expected one will be random fluctuations only. In a chi-square distribution the degree of freedom is one less than the number of classifications into which the individuals are stored. The number of restrictions imposed is determined by the total number of observations in the categories. Some of the applications of χ^2 distribution include the test for variance of a single population, the test for goodness of fit and the test of independence. For each value of χ^2, the probability density is given by:

$$f(\chi^2) = K(\chi^2)^{(r/2)-1} e^{-\chi^2/2} \text{ for } \chi^2 > 0 \tag{1.26}$$

where e = 2.71828; r is the number of degrees of freedom and K is a constant depending on r.

Chi-square distributions with 1, 5, 10 and 20 degrees of freedom are shown in figures 1.5 and 1.6.

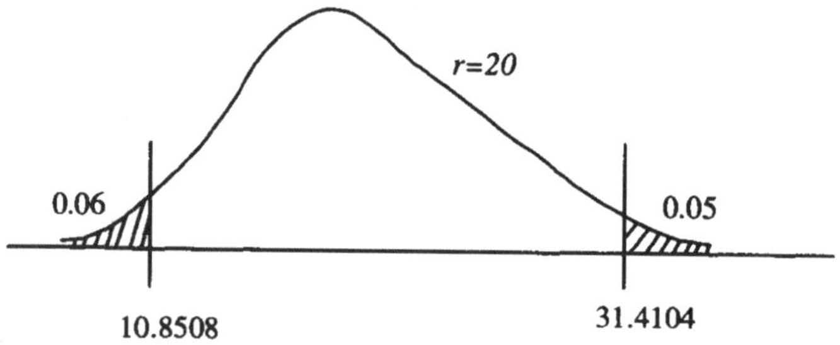

Fig. 1.5 The χ^2 distribution with 20 degrees of freedom

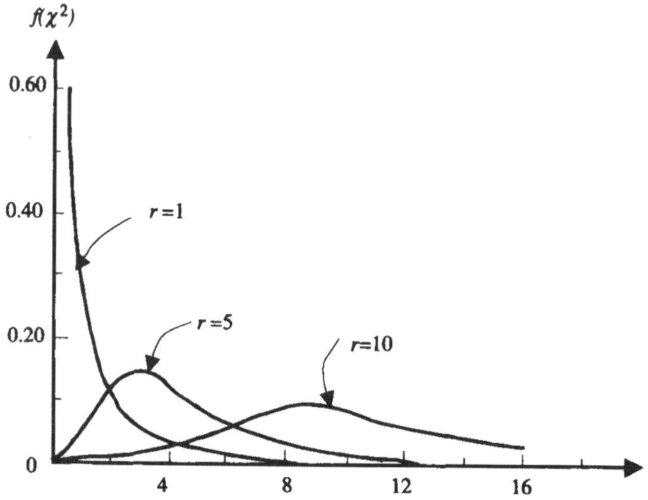

Fig. 1.6 Chi-square distributions with 1, 5 and 10 degrees of freedom

1.3.8 Bar Diagram

Bar diagram consists of a group of equispaced rectangular bars, one for each category (or class) of given statistical data. The bars starting from a common baseline must be of equal width and their lengths represent the values of statistical data. They are shaded or coloured suitably.

There are two types of bar diagrams viz. vertical bar diagram and horizontal bar diagram. Vertical bars are used to represent time series data or data classified by the values of a variable. Horizontal bars are used to depict data classified by attributes only. For each of these types, there are again grouped bar diagram, sub-divided (or component) bar diagram, paired bar diagram, etc. (Das, 1992).

1.3.9 Pie Diagram

Pie diagram is a circle whose area is divided proportionately among the different components by straight lines drawn from the centre to the circumference of the circle. When statistical data are given for a number of categories, and one is interested in the comparison of the various categories or between a part and the whole, such a diagram is very helpful in effectively displaying the data.

For drawing a pie diagram, it is necessary to express the value of each category as percentage of the total. Since the full angle 360° around the centre of the circle represents the whole, i.e. 100%, the percentage figure of each component is multiplied by 3.6° to find the angle of the corresponding sector at the centre of the circle (Das, 1992).

Table-1.1 Calculation of Pie Chart

Name of the Mine	*Number of vibration data recorded*	*Percent of total*	*Angle (in deg.) at the centre of pie chart [Column 3 × 3.6°]*
Chasnalla OCP	35	16.28	58.60
Bermo Mine	25	11.63	41.87
Lingaraj OCP	65	30.23	108.83
Jayant OCP	90	41.86	150.70
TOTAL	215	100	360.00

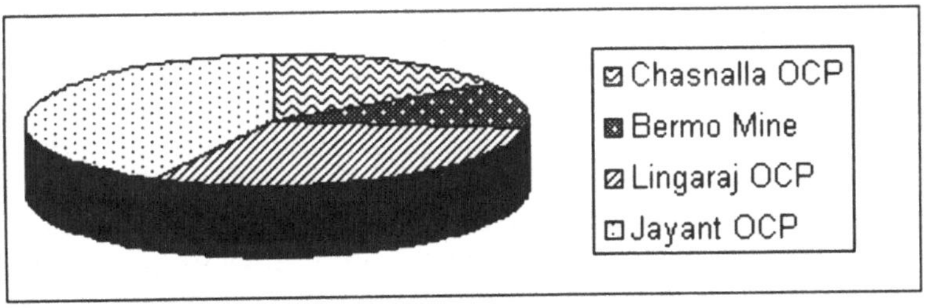

Fig. 1.7 Pie diagram representation of the data of Table-1.1

1.3.10 Histogram

Histogram is the most common form of diagrammatic representation of a grouped frequency distribution. It consists of a set of adjoining rectangles drawn on a horizontal base line, with areas proportional to the class frequencies. The width of the rectangles, one for each class, extends over the class boundaries (not class limits) shown on the horizontal scale. When all classes have equal width, the height of each rectangle will be proportional to the class frequency and it is then customary to take the height numerically equal to the class frequency. If however, the classes are of unequal widths, the rectangles will also be of unequal widths, and therefore the heights must be proportional to the frequency densities. The following relation holds:

Area of each rectangle = width × height
= (width of class) × (frequency density)
= (width of class) × (class frequency / width of class)
= class frequency.

1.3.11 Hemispherical Normal Distribution

Arnold (1941) established the probability density function in two variables corresponding to the co-latitude and longitude coordinates on a hemisphere such that the distribution corresponds, in some manner, to the two-dimensional normal curve for rectangular coordinates on a plane. Models of discontinuous rock containing several joint sets require estimation of a single direction for each set and a measure of precision of this estimate, so that the corresponding precision of this estimate may be computed in the output of the structural analysis that employs these models. Since the distribution characteristic of data points cannot be tested against a known distribution, no estimate of precision of the mean attitude can be formulated.

The attitudes of fractures in a site are very often seen to be non-randomly distributed. Besides non-randomness, it may be possible to group the fractures into sets, such that the elements of each set have a statistically preferred attitude.

In such a problem an efficient scheme would be required for quantifying the preferred (or mean) directions of clusters in a sample of joint orientations. It is necessary to test and illustrate the form of the distribution of points in each cluster and to give an estimate of the population mean in the case of a hemispherical normal distribution.

To compare the data in each cluster with the hemispherical normal distribution, the chi-square goodness-of-fit test is compared by the formula given in equation (1.27).

$$\chi^2 = \sum^{N_C} \frac{(f_0 - f_e)^2}{f_e} \tag{1.27}$$

where f_0 and f_e are the observed and expected frequencies respectively; and N_c is the total number of classes chosen.

It is worth mentioning that in the computer program PATCH, developed by the USBM researchers (Mahatab et al., 1972), the orientations of rock fractures were used quite extensively. PATCH basically identifies the cluster or groupings among the fracture orientations and then calculates (i) the mean orientation of the fractures in each cluster and (ii) the dispersion or scatter among these fracture orientations. It also specifies clusters, which do not follow Arnold's hemispherical normal distribution.

1.4 IMPORTANT MATHEMATICAL FORMULAE

1.4.1 Variation

For variables, *A, B, C* ....., the following rules hold good—

(i) $A \propto B \Rightarrow A = KB$ (*K* is a constant)

(ii) $A \propto 1/B \Rightarrow A = K/B$

(iii) $A, B, C; A \propto BC \Rightarrow A = KBC$, when both *B and C* vary

$\Rightarrow A \propto B$, when C is a constant.

1.4.2 Theory of Indices

(i) $a^m = a.a.a. \ldots\ldots.\ m$ times (*a is a real number and m* is an integer)

(ii) $a^{m/n} = \sqrt[n]{a^m}$ (iii) $a^{-m} = \dfrac{1}{a^m}$, (iv) $a^0 = 1$,

(v) $a^m \,.\, a^n = a^{m+n}$, (vi) $\dfrac{a^m}{a^n} = a^{m-n}$, (vii) $(ab)^m = a^m \,.\, b^m$,

(viii) $(a/b)^m = \dfrac{a^m}{b^m} = a^m \,.\, b^{-m}$, (ix) If $a^x = a^y$, $(a \neq 1) \Rightarrow x = y$

(x) If $a^x = b^x$, $(x \neq 0) \Rightarrow a = b$

1.4.3 Complex Number

(i) $i = \sqrt{-1}$

(ii) $(a + ib)$ and $(a - ib)$ are conjugate complex numbers

(iii) If $(a + ib) = (c + id)$ then $a = c$ and $b = d$; $(a + ib) = 0 \Rightarrow a = b = 0$

1.4.4 Arithmetic Progression

Sum of the following A.P. series are given by;

(i) $a, a + b, a + 2b, a + 3b, \ldots\ldots\ldots\ldots a + (n - 1)b$

$$\Rightarrow S_n = \frac{n}{2}[a + a + (n-1)b] = \frac{n}{2}[2a + (n-1)b]$$

(ii) $$1 + 2 + 3 + \ldots\ldots. + n = \frac{n}{2}[2 + (n-1)] = \frac{n(n+1)}{2}$$

(iii) $1^2 + 2^2 + \ldots\ldots. + n^2 = \dfrac{n(n+1)(2n+1)}{6}$

(iv) $1^3 + 2^3 + 3^3 + \ldots\ldots. + n^3 = \left[\dfrac{n(n+1)}{2}\right]^2$

1.4.5 Geometric Progression

(i) $a, ar, ar^2, \ldots\ldots, ar^{(n-1)} \Rightarrow S_n = \dfrac{a(1-r^n)}{(1-r)}, r<1; \dfrac{a(r^n-1)}{(r-1)}, r>1$

1.4.6 Quadric Equation

(i) $ax^2 + bx + c = 0 \Rightarrow x = \dfrac{-b \pm \sqrt{(b^2 - 4ac)}}{2a}$

(ii) For 2 linear simultaneous equations—
$a_1x + b_1y + c_1 = 0$ and $a_2x + b_2y + c_2 = 0 \Rightarrow$

$$\frac{x}{b_1c_2 - b_2c_1} = \frac{y}{c_1a_2 - c_2a_1} = \frac{1}{a_1b_2 - a_2b_1}$$

(iii) $ax^2 + bx + c = 0 \Rightarrow \alpha, \beta$ are two roots, then

$$\alpha + \beta = \frac{-b}{a} \text{ and } \alpha \cdot \beta = \frac{c}{a}$$

1.4.7 Permutation and Combination

(i) $n! = n(n-1)(n-2)\ldots\ldots 3.2.1 = {}^nP_n$ (!→ is called the factorial)
(ii) ${}^nP_r = n(n-1)\ldots\ldots(n-r+1) =$

$$\frac{n(n-1)(n-2)\ldots(n-r+1)(n-r)\ldots 3\cdot2\cdot1}{(n-r)(n-r-1)\ldots.. 2\cdot1} = \frac{n!}{(n-r)!}$$

(iii) ${}^nC_r = \dfrac{n!}{r!\cdot(n-r)!}, n \le r$

(iv) ${}^nC_1 = n, {}^nC_n = 1, {}^nC_0 = 1, {}^nC_r = {}^nC_{n-r},$

(v) ${}^nC_p = {}^nC_q, p = q$ or $p + q = n$

(vi) ${}^nC_r + {}^nC_{r-1} = {}^{n+1}C_r$

1.4.8 Binomial

(i) $(a+x)^n = a^n + {}^nC_1 a^{n-1}x + {}^nC_2 a^{n-2}x^2 + \ldots\ldots + x^n$

1.4.9 Logarithm

(i) $\log_a (M \cdot N) = \log_a M + \log_a N$

(ii) $\log_a \left(\frac{M}{N}\right) = \log_a M - \log_a N$

(iii) $\log_a M^n = n \log_a M$

(iv) $\log_a M = \log_b M \cdot \log_a b$

1.4.10 Compound Interest

P = Principal, r = % of interest per year, n = number of years

A (Amount) = $P\left(1+\frac{r}{100}\right)^n$;

After n years, the compound interest, $I = P\left[\left(1+\frac{r}{100}\right)^n - 1\right]$

1.4.11 Measurement of Angles

$\pi^R = 180°,\ 1^R = 57°\,17'48.8'',\ \pi = \frac{22}{7} = 3.14,\ 60' = 1°, 60'' = 1'$

1.4.12 Trigonometric Functions

(i) $\tan\theta = \frac{\sin\theta}{\cos\theta}$, $\cot\theta = \frac{\cos\theta}{\sin\theta}$, $\sec\theta = \frac{1}{\cos\theta}$, $\text{cosec}\,\theta = \frac{1}{\sin\theta}$

(ii) $\cos^2\theta + \sin^2\theta = 1$, $1 + \tan^2\theta = \sec^2\theta$, $1 + \cot^2\theta = \text{cosec}^2\theta$

Table-4.1: Functional values of different angles

$\theta \rightarrow$	0°	30°	45°	60°	90°	180°	270°
$\sin\theta$	0	$\frac{1}{2}$	$\frac{1}{\sqrt{2}}$	$\frac{\sqrt{3}}{2}$	1	0	–1
$\cos\theta$	1	$\frac{\sqrt{3}}{2}$	$\frac{1}{\sqrt{2}}$	$\frac{1}{2}$	0	–1	0

1.5 SUMMARY

Field data comprising different parametric values whose proper and systematic utilization would definitely depend on the investigator. Statistical and mathematical analyses of data are only tools for the investigator. Out of the several alternative approaches, only some of the important statistical measures and mathematical formulae, which may be used in the analysis of field data, are discussed in this chapter. The research worker must have an adequate knowledge of the facts, technical and otherwise, of the field in which he is to work. This knowledge is usually insured by the situation. When facts and theories of any particular study are known, the following steps may be undertaken for any plausible outcome: (i) statement of the objectives, (ii) development of an hypothesis, (iii) measurement of the factors and (iv) study and justification of the apparent relationships.

REFERENCES

Arnold, K.J. 1941. On spherical probability distributions, Ph.D. Thesis, M.I.T., MA, USA, p. 42.

Das, N.G. 1992. Statistical methods, Part-I, M. Das and Co., Kolkata, India, p. 504.

Denness, B. 1970. A method of contouring polar diagrams using curvilinear counting cells, Geol. Mag., 107, 1, pp. 61-65.

Dowding, C.H. 1971. Response of buildings to ground vibrations resulting from construction blasting, Ph.D. Thesis, University of Illinois, USA, p. 204.

Ghosh, Amitava, 1983. A new analytical predictor of ground vibrations induced by blasting, M.S. Thesis, The University of Arizona, USA, p. 322.

Koch, Jr. G.S. and Link, R.F. 1970. Statistical analysis of geological data, John Wiley and Sons, Inc., New York, USA, Vol I-II p. 375.

Krumbein, W.C. and Graybill, F.A. 1965. An introduction to statistical models in geology, McGraw-Hill Book Co., Inc., New York, USA, p. 475.

Pal Roy, P. and Gupta, R.N. 1989. Some statistical methods used in the analysis of blast induced vibration data, Jr. Inst. Engg. (India)-MN, Vol. 70, pt. MN2, pp. 54-60.

Mahtab, M.A., Bolstad, D.D., Alldredge, J.R. and Shanley, R.J., 1972. Analysis of fracture orientations for input to structural models of discontinuous rock, USBM-RI 7669, USA, p. 76.

Seber, G.A.F. 1977. Linear regression analysis, John Wiley and Sons, New York, USA, p. 465.

Siskind, D.E., Stagg, M.S., Kopp, J.W. and Dowding, C.H. 1980: Structure response and damage produced by ground vibrations from surface mine blasting, U.S. Bureau of Mines, R.I. 8507, p. 74.

Wiss, J.F. and Linehan, P.W. 1978. Control of vibration and blast noise from surface coal mining, Vol. I-III, Report to U.S. Bureau of Mines, Bureau of Mines OFR (1-3)-74, p. 623.

Walter, J. 1981. Lithologic variation and vibration effects, Proc. 7th Conference on Explosives and Blasting Technique, Int. Society of Explosives Engineers, Phoenix, Arizona, USA, pp. 52-61.

2

Terms and Parameters Influencing Mine and Ground Excavations

2.1 INTRODUCTION

Rock breaking by drilling and blasting is the first phase of the production cycle in most mining operations. Optimization of this operation is very important as the fragmentation obtained thereby affects the cost of the entire gamut of interrelated mining activities, such as drilling, blasting, loading, hauling, crushing, and to some extent grinding. Optimization of rock breaking by drilling and blasting is sometimes understood to mean minimum cost in the implementation of these two individual operations. However, a minimum cost for breaking rock may not be in the best interest of the overall mining system. A little more money spent in the rock-breaking operation can be recovered later from the system and the aim of the coordinator of the mining work should be to achieve a minimum combined cost of drilling, blasting, loading, hauling, crushing and grinding. Only a "balance sheet" of total cost of the full gamut of mining operations vis-à-vis production achieved can establish whether the very first phase—rock breaking—was "optimum" financially; leaving aside factors of human safety.

An optimum blast is also associated with the most efficient utilization of blasting energy in the rock-breaking process, reducing blasting cost through less explosive consumption and less wastage of explosive energy in blasting, less throw of materials, and reduction of blast vibration resulting in greater degrees of safety and stability to the nearby structures.

Important and highly controllable blast parameters include: diameter and length of blastholes; type and configuration of charges; shape, condition and development of effective faces; available expansion volume of broken rock; type and dimensions of the blasthole pattern; initiation sequence and delay timing.

Ground vibration from mine blasting may be expressed by three important characteristics. They are amplitude, frequency and duration of the blast. The variables which influence ground vibration parameters may be divided into two groups (Siskind, 1973):

(i) Non-controllable, and
(ii) Controllable

Controllable variables are those, which can be manipulated or changed by trial and error depending on the characteristics of ground vibration. On the other hand, non-controllable variables are those over which the blasting engineer has no control. Non-controllable variables are:

(i) General surface terrain
(ii) Type and depth of overburden
(iii) Wind

The important controllable variables associated with the characteristics of ground vibration are (Siskind, 1973; Wiss and Linehan, 1978):

(i) Type of explosive
(ii) Charge per delay
(iii) Delay interval
(iv) Direction of blast progression
(v) Burden, spacing and specific charge
(vi) Coupling ratio

Most of these variables are interrelated. A change in one variable in the operating system can change the others. The net change in the magnitude, frequency and duration of ground movement is the combined influence of all variables rather than anyone of them independently. It is of course a very difficult task to quantify the measures or extent of the effect of each variable individually. Basic parameters involved in the process of optimum blasting may be classified as follows (Figure 2.1).

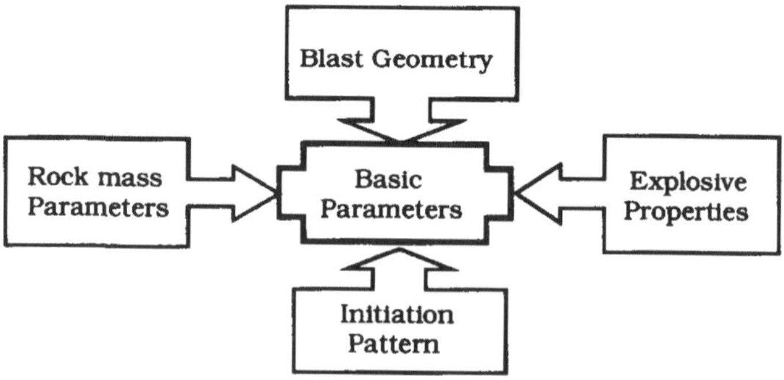

Fig. 2.1 Classification of basic parameters

2.2 PARAMETERS RELATED TO EXPLOSIVE PROPERTIES

Selection of proper explosive in any blasting round is an important aspect of optimum blast design. Basic explosive parameters include:

(i) VOD of explosive (m/s)
(ii) Density (g/cc)
(iii) Characteristic impedance
(iv) Energy output (cal/g)
(v) Explosive type (NG, ANFO, Slurry and Emulsion)

2.2.1 Selection of the Proper Explosive

So many different types of explosives are presently available in the market that it is rather difficult to select the right type for a specific purpose. Some of the explosives extensively used today are:

(1) Special Gelatine
(2) NG-based
(3) Gun powder
(4) Slurry explosives (both cartridge and site-mixed)
(5) ANFO (ammonium nitrate prills mixed with fuel oil)
(6) Emulsion (both cartridge and site-mixed)

Although all of them have relative merits and demerits in utilization, the major factor influencing the mine operator s ultimate choice of the correct explosive is blasting cost. Since this cost includes the cost of drilling plus other fixed expenditures, a more expensive explosive could well mean money saved in the total cost of a blasting operation.

The following simple example illustrates how certain basic considerations could influence the selection of explosives (Dutta et al., 1973).

Let us suppose,

N = Number of boreholes
W = Weight of explosives loaded in each hole (kg)
C = Cost of explosive per kg
D = Cost of drilling and loading of each hole

Then, total cost becomes

$$T = N\,(D + W \cdot C) \tag{2.1}$$

When comparing total cost for two different explosives, A and B, the formula is—

$$T_A - T_B = N_A\,(D + W_A \cdot C_A) - N_B\,(D + W_B \cdot C_B) \tag{2.2}$$

Now, if for the same block, explosive A is such that it requires N_A number of holes to give the same degree of blasting performance as explosive B gives with N_B number of holes, and if S_A, d_A and S_B, d_B are the weight strength and average density of explosives A and B

respectively, then total blasthole power of N_A holes will be the same as the total blasthole power of N_B holes with B explosive, i.e.,

$$N_A \cdot d_A \cdot S_A = N_B \cdot d_B \cdot S_B$$

$$\frac{N_A}{N_B} = \frac{d_B \cdot S_B}{d_A \cdot S_A} \tag{2.3}$$

From the above it can be seen that the number of boreholes needed is inversely proportional to the strength. Also, the weight per borehole is proportional to the density. Therefore,

$$\frac{W_A}{W_B} = \frac{d_A}{d_B} \tag{2.4}$$

Solving equation (2.2), (2.3) and (2.4), we get,

$$T_A - T_B = N_A \cdot D\left[1 - \frac{d_A \cdot S_A}{d_B \cdot S_B}\right] + \left[1 - \frac{S_A \cdot C_B}{S_B \cdot C_A}\right] N_A \cdot W_A \cdot C_A \tag{2.5}$$

The criterion for determining the quality of an explosive for a given job is simply whether the left-hand side of the equation (2.5) is positive or negative. If it is positive, explosive B is better than explosive A. If it is negative, the reverse is true.

It can be seen from the foregoing discussions that while deciding which explosive is best for the job, it is necessary not only to have a correct determination of spacing and burden, but also to take into consideration the density, strength, cost of explosive and cost of drilling.

2.3 PARAMETERS RELATED TO GEOMECHANICAL PROPERTIES OF ROCK MASS

The total geology and rock characteristics have great influence upon proper fragmentation, control of vibration, flyrock, and other safety aspects. They may be divided into the following ten parameters:

(i) *P*-Wave velocity (m/s)
(ii) *S*-Wave velocity (m/s)
(iii) Density of rock mass (g/cc)
(iv) Compressive strength (MPa)
(v) Tensile strength (MPa)
(vi) Characteristic impedance of rock
(vii) *RQD* (Rock quality designation)
(viii) Dip direction (deg) and dip amount (deg) of joints
(ix) *Q* (Rock quality factor)

2.4 PARAMETERS RELATED TO BLAST GEOMETRY

Good fragmentation, displacement and less explosive consumption depend mostly on blast geometry. Basic blast geometry parameters include:

(i) Spacing (m)
(ii) Burden (m)
(iii) Bench height (m)
(iv) Depth of hole (m)
(v) Stemming length (m) and type
(vi) Diameter of hole (mm)
(vii) Loading density (kg/m)
(viii) Charge factor (kg/m^3)
(ix) Length versus width area (m vs. m^2)

2.5 PARAMETERS RELATED TO INITIATION PATTERN

Initiation pattern is important for proper fragmentation, proper throw of blasted materials and fewer blasting hazards. The two most commonly used parameters related to initiation pattern are:

(i) Delay interval (ms)
(ii) Delay pattern or connection

Although it is unlikely that all of the above-listed parameters can be accounted for in detail by means of simple closed-form expressions, it is probable that better predictability can be achieved by explicitly incorporating those parameters, which predominantly govern the blasting operation.

2.6 TERMS AND PARAMETERS ASSOCIATED WITH DRILLING AND BLASTING

A good deal of work in compiling a comprehensive blasting dictionary has been made by Rustan (1998). Other researchers, who have also made contributions in compiling such useful work, are Bhandari (1997), Jimeno et al. (1995), Konya and Walter (1990) and Dowding (1985). Based on such collected documents, a concise compilation work, which will be beneficial to the reader, is presented in this section.

2.6.1 Absolute Temperature

The Kelvin scale quantifies an absolute temperature. The relation is 0°C = 273.15 K.

2.6.2 Absolute Volume Strength (AVS)

It is the energy output of explosive per unit volume (MJ/m^3). Likewise, the Absolute Weight Strength (AWS) is defined as the energy output per unit weight of explosive. While quoting these values, they must be supported by the name of the detonation code used as well as the equation of state and the energy convention factor employed.

2.6.3 Accelerometer

An instrument which measures acceleration (in terms of g) and based on the principle of difference of potential generated by a piezoelectric crystal under force.

2.6.4 Acoustic Impedance, (kg/m^2s, Ns/m^3 or Pas/m)

It is the product of sonic wave velocity and the density of a material which characterizes a material as to its energy transfer properties.

2.6.5 Air blast

It is an atmospheric pressure wave consisting of high frequency sound which is audible and low frequency air-concussion that is sub-audible and cannot be heard.

2.6.6 Air-deck Blasting

A controlled contour blasting method where an explosive column is combined with an air chamber. Wooden spacers and gasbags are generally used for air-decking. It has become an increasingly popular technique in production and presplit blasting (see Chapter-12).

2.6.7 Aluminized ANFO (ALANFO)

Aluminized ANFO or ALANFO is produced by mixing sensitized aluminium powder (AL) with ANFO for higher strength.

2.6.8 Alluvial Mining or Placer Mining

Placer mineral is a secondary deposit formed due to weathering of primary rock and subsequent transport by geological agents like rivers and winds to coastal areas. The exploitation of such alluvial deposits by dredging, hydraulicking, or drift mining is called placer mining.

2.6.9 Ammonium Nitrate (AN)

It is the commonly used colourless oxidizer in explosives and blasting agents. A chemical compound represented by the formula NH_4NO_3. Its properties are: molecular weight 80.04, orthorhombic (–16.3 to +32.3°C), density 1725 kg/m^3 (at +25°C), melting point +169.6°C, soluble in water as well as in ethyl alcohol. Ammonium nitrate is used in crystalline or palletized form in explosives, and as a fertilizer. It becomes a useful explosive when mixed with 5.7-6.0% fuel oil (FO) by weight. ANFO mixed with certain percentage of sawdust or rice-husk, was found effective in metalliferous mines

2.6.10 Backbreak or Overbreak

Rock broken beyond the limits of the last row of holes in a blast is termed as backbreak or overbreak.

2.6.11 Base Charge

Base charge which also occasionally refers to bottom charge, is the detonating component (usually PETN) in a blasting cap, initiated by the priming charge or the initiation element.

2.6.12 Bedding Plane

In sedimentary or stratified rocks, the separation planes of individual layers, beds, or strata are termed as bedding planes.

2.6.13 Black Powder

It is a "low shock energy" and "poor fume quality" explosive consisting of sodium or potassium nitrate, carbon, and sulphur. Black powder which is an essential breaking tool for quarrying of dimensional stones, is generally used with detonating fuse for firing.

2.6.14 Blastability

Blastability can be defined as the blasting characteristics of the rock mass subjected to a specified blast design, site requirement, legislative constraints and explosive characteristics. In other word, blastability indicates how easy it is to blast a rock mass under a specified condition. The most important physical and mechanical rock properties for blastability include—1) Acoustic impedance, 2) Rock structure and friction properties of the discontinuities.

To determine the blastability index, several authors, worldwide, have made different scientific approaches. A few of them are listed for general understanding.

2.6.14.1 Laboratory Based Method

In laboratory-based methods, the blastability is determined by testing of rock and rock mass in the laboratory or by test blasting in the laboratory.

Blastibility is the ratio of compressive strength and tensile strength. This is because; compressive strength represents the maximum rock strength whereas tensile strength represents the minimum rock strength.

Langefors and Kihlstrom (1978) proposed that for practical use, the blastability of rock (c, kg/m^3), can be determined by test blasting with one single vertical hole with 33 mm borehole diameter, hole depth 1.33 m and with that charge which is needed to give a 1 m high vertical bench and 1 m burden, a breakage and throw of maximum 1 m. Such a burden is very close to the critical burden. The rock constant generally comes to 0.4 kg/m^3. However, researchers have shown that it could vary up to 25%.

2.6.14.2 Laboratory Test and Field Study Based Approach

These approaches are based on both laboratories as well as field-testing results and blastability is established by the correlation of these test results. Besides many important contributions made by researchers, worldwide, the empirical model developed by Ashby (1980) can be used as a formidable tool.

Ashby (1980) developed an empirical relationship to describe the powder factor required for adequate blast (in copper mine) based on the fracture frequency and friction angle. The relationship is of the following form:

$$\text{Powder factor} = \frac{1.4 \times \tan(\phi + i)}{\sqrt[3]{\text{fracture/metre}}} \text{ kg/m}^3 \tag{2.6}$$

Where, fracture/metre represents the fracture frequency; ϕ is the friction angle and i is the roughness angle.

2.6.14.3 Empirical Formula Based Method

A Blastibility Index (*BI*), based on rock properties and geological structural parameters, was developed for iron ore pits in Australia, Lilly (1986). Cunningham (1983) subsequently incorporated that concept in the Kuz-Ram fragmentation prediction model. Five parameters were observed in the field and then quantified and summed-up to give a value for the *BI*, as shown in equation (2.7). Historical blast data were used to provide a correlation between *BI* and powder factor, considered valid for large-scale, cable shovel operation having sizeable crushers and using ANFO as the blasting agent.

$$BI = 0.5 \times (RMD + JPS + JPO + SGI + H) \tag{2.7}$$

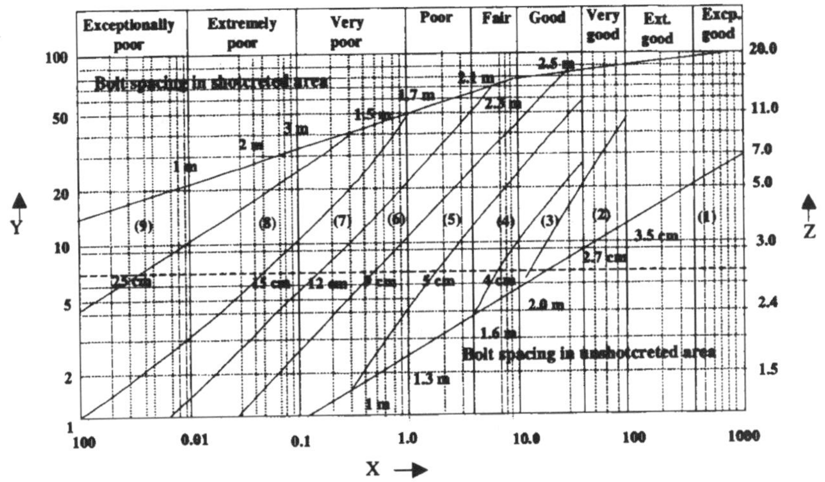

X: Rock Mass Quality, $Q = \dfrac{RQD}{J_n} \times \dfrac{J_r}{J_a} \times \dfrac{J_w}{SRF}$

Y: Equivalent dimension = Span or Height/ESR; Z: Bolt length in m for ESR = 1

Fig. 2.2 Classification of rock, based on Rock Mass Quality factor

Rating for Blastibility Index Parameters

Parameter	*Rating*
1. *Rock Mass Description* (*RMD*)	
1.1: Powdery / Friable	10
1.2: Blocky	20
1.3: Totally Massive	50
2. *Joint Plane Spacing* (*JPS*)	
2.1: Close (< 0.1 m)	10
2.2: Intermediate (0.1 to 1.0 m)	20
2.3: Wide (> 1 m)	50
3. *Joint Plane Orientation* (*JPO*)	
3.1: Horizontal	10
3.2: Dip out of Face	20
3.3: Strike Normal to the Face	30
3.4: Dip into Face	40
4. *Specific Gravity Influence* (*SGI*)	
SGI = 25 x *SG* - 50 (where *SG* is in t/m^3)	
5. *Hardness* (H)	1 to 10

2.6.14.4 Incorporating Blasting Outcome

A good blasting always signifies the overall mining cost and not the only blasting cost. Keeping this idea in view, JKMRC has evolved an approach for the determination of BDI to classify the rock mass according to the properties dependent on blasting performance for coal measure strata. So far, the JKMRC approach has been considered as the best approach to classify the rock mass.

The following properties were considered—

Rock mass ⇒ Strength, density, Young's modulus.
Structure ⇒ Average *in situ* block size, influence of structure.
Design ⇒ Target fragment size, heave desired, confinement provided, scale of operation.
Environment ⇒ Water

Table 2.1 gives an example of analysis of powder factor in a surface coal mine of Hunter Valley, New South Wales using ANFO as the explosive (Dey, 2003).

Table 2.1 Determination of Blastability Index using JKMRC approach

Parameters		*Dragline operation*	*Dragline operation with cast blasting*	*Shovel operation*	*Shovel operation in wet condition*	*Parting FEL*
Rockmass						
Strength (MPa)		60	60	50	50	40
Density (g/cc)		2.51	2.51	2.47	2.47	2.42
Young's modulus [E] (GPa)		12	12	10	10	10
Structure						
Block size (m)		2	2	2	2	0.5
Structure (1-9)		5	5	5	5	3
Design						
Target fragment size (m)		0.5	0.5	0.3	0.3	0.15
Heave (1-9)		5	10	5	5	7
Confine (1-9)		5	5	5	5	7
Scale (1-9)		3	3	5	5	7
Environment						
Water (1-9)		1	1	1	5	1
Indices						
Strength		0.30	0.30	0.25	0.25	0.20
Breakage		0.08	0.08	0.13	0.13	0.06
Heave		0.25	0.51	0.26	0.26	0.36
Modifier		–0.02	0.03	0.00	0.08	0.02
Powder Factor	kg/t	0.18	0.24	0.17	0.21	0.16
	kg/m^3	0.44	0.61	0.42	0.52	0.39

Basic strength and stiffness of the rock material were described by the laboratory-determined values of the compressive strength, density and Young's modulus. Whereas, average *in situ* block size was estimated from the exposed rock surface or from the structural mapping. Influence of structure was considered depending on the favourability of blasting (5 for neutral, 1 for favourable and 9 for strongly difficult). The heave parameter was based on the loading equipment. Confinement was based on the free face availability (for full face, confinement <5 and over-confined >5). The scale factor was considered for blasting in the same material but with separate geometry.

Predicted indices signifies the following—

- Strength Index → Indicates the compressive strength of rock is proportional to powder factor.
- Breakage Index → The degree of required breakage is the ratio of *in situ* block size and target fragment size. This is proportional to Young's modulus.
- Heave Index → The required heave energy is inversely proportional to Young's modulus of rock.
- Modifier Index → This is to adjust powder factor with structure, scale and confinement modifier value. That modifier value of 5 is neutral, >5 is difficult and <5 is easy. Each unit above and below 5 for modifier index respectively increases and decreases the powder factor by 1%.

In water parameter, dry blasting condition is indicated by 1 and each unit above 1 adds 2% to the powder factor.

2.6.14.5 Based on Drilling Indices

There are a few methods developed to correlate powder factor based on the widely popular drilling indices. These methods are purely based on the rock and rock mass properties.

Jimeno et al. (1995) proposed a rock characterization drilling index (D_i) [called I_P drilling Index] based on penetration rate, drilling diameter etc. The equation takes into the following form—

$$D_i = \frac{V_P \cdot D^2}{E \cdot N_r} \tag{2.8}$$

Where,

V_P = Penetration rate in (m/h)

E = Pull down weight on the tricon bit (thousands of pounds)

N_r = Drilling speed in (rpm)

D = Diameter of drilling in (in)

The equation is valid under the following conditions—

⇒ Drilling bit used must be the best for the type of formation

⇒ Air-flow must be sufficient to sweep away drill cuttings

⇒ Only net penetration rate to be counted (not bit changing, positioning time)

The charge factor was correlated with the I_P drilling index (D_i) by the following empirical equation established by the authors (Jimeno et al., 1995)—

$$\text{Charge factor} = CE\ (\text{kg of ANFO/m}^3) = 1.124 \times e^{-0.5727\, D_i} \tag{2.9}$$

2.6.15 Blast Casting

It is a blasting method used for maximum overburden rock displacement and throw in surface coal mining so that with less hauling the underlying

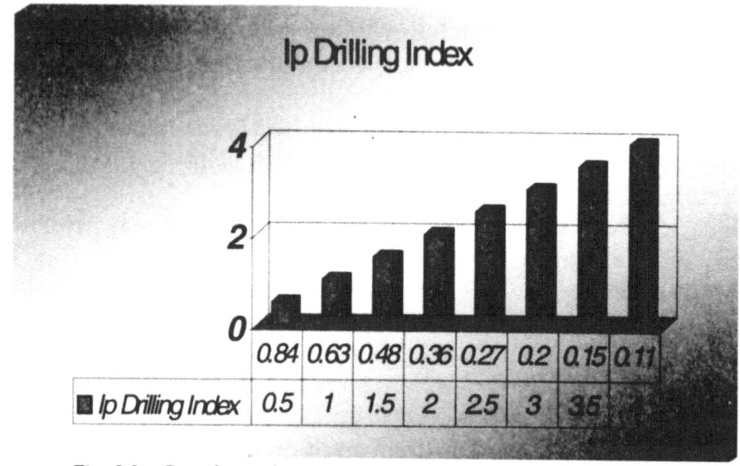

Fig. 2.3 Correlation between I_P drilling index and charge factor

coal is exposed. Cast blasting is commonly used in coal stripping operation since it is often more economical to use the explosive energy to move the overburden instead of the mechanical energy from power shovel, draglines or other excavating equipment. It can increase the profit margin on the coal produced. It can also enable to increase the area of coal reserves by reducing the production cost on deeper reserves (Konya and Walter, 1990). Chiappetta and Borg (1983) proposed the following equation (2.10) as a first approximation of velocity from the data obtained from high-speed photography in dolomite, granite and iron ore mines. The equation is as follows:

$$V_0 = K(B/(Energy)^{1/3})^{-\beta} \tag{2.10}$$

Where,

V_0 = Initial velocity in ft per second
K = Site constant (25 as an example)
B = Burden in ft
$Energy$ = kilocalories per ft of explosive column
β = Site constant (1.17 proposed)

Equation (2.10) does not consider burden stiffness and the ratio of burden to spacing.

2.6.16 Blasting Cap No. 8

It is a specified blasting cap containing 0.40 to 0.45 g of PETN base charge at a density of 1.4 g/cc and primed with standard mass of primer.

2.6.17 Blasting Gelatine

It is a strong (high) and one of the most powerful explosives with gelatinous mixture of 92% nitroglycerin and 8% nitro-cellulose (nitrocotton). It has high-water resistance and is mostly used for open excavation and underground tunnelling work.

2.6.18 Blasting Mat

A covering, usually made of scrap tires, logs, conveyor belts, ropes or wire cables, placed over a blast to prevent flyrock from a blasting face.

2.6.19 Blasting Powder

A powder containing more charcoal and 65 to 75% potassium nitrate in place of potassium salt is known as blasting powder.

2.6.20 Booster

A cap sensitive explosive with high detonation velocity used to initiate and improve the performance of another explosive. The amount of booster required to initiate a column explosive varies between 0.18% (in case of site-mixed slurry and emulsion) and 25% (in case of cartridged form). When the booster is tagged with detonating device, it becomes primer.

2.6.21 Borehole Pressure or Explosion Pressure, (MPa)

When detonated, the pressure exerted by the hot gases of the explosive on the borehole wall before any expansion (of the borehole wall) is termed as borehole pressure or explosion pressure.

2.6.22 Bottom Initiation

Initiation of a charge at the bottom of the blasthole either by Nonel (non-electric initiating devices) or by primer is known as bottom initiation.

2.6.23 Brisance

High explosives such as military explosives or highly sensitive commercial explosives are characterized by an intense shattering effect upon detonation known as brisance. It is an explosive property roughly equivalent to detonation velocity. An explosive with a high detonation velocity has a higher brisance.

2.6.24 Bubble Energy, (J)

In underwater test, the time interval between the initial main shock and the first bubble pulse (i.e. bubble period) is a measure of the energy in the explosion gases. This is termed as bubble energy.

2.6.25 Burden (Actual and Effective)

Burden distance may be defined as the shortest distance to relief at the time the hole is detonated in any blasting operation. It is the most critical parameter. If burden is very little, the rock will be thrown to a considerable distance from the face, air blast level will be high and excessive fines will be produced. Too much burden will produce severe backbreak and shattering on the back wall.

The following different formulae may be used to estimate the burden.

Langefors and Kihlstrom (1978): $$B_m = 0.958\,d\sqrt{\frac{\rho_e \cdot S}{C_o \cdot f\,(S_d / B_d)}} \qquad (2.11)$$

Where,

B_m = Maximum burden (m)
d = Diameter of blasthole (m)
ρ_e = Density of explosive (kg/m^3)
S = Relative weight strength of explosive
C_o = Corrected blastibility factor
C_o = C + 0.05 for B (burden) < 1.4-15 m
C_o = C + 0.07/B for B < 1.4 m
C = Blastibility factor (kg/m^3) (necessary charge factor to break but not to throw the rock)
f = Confinement of the blasthole [free bottom—0.75; fixed slope (2:1)—0.85; fixed slope (3:1)—0.90; fixed vertical—1.0, tunnelling-relief holes—1.25 to 1.50]
S_d = Drilled spacing (m)
B_d = Drilled burden (m)

Konya and Walter (1990): $$B = [2 \times SG_e / SG_r + 1.5] \cdot D_e \qquad (2.12)$$

Where, B = Burden (ft); SG_e = Specific gravity of explosive
SG_r = Specific gravity of rock, D_e = Diameter of explosive (in).

Olofsson (1990): $$B_{max} = K \times \sqrt{q_f} \times R_1 \times R_2 \times R_3 \qquad (2.13)$$

Where, K = Explosive constant (gelatinous explosive=1.47; emulsion=1.45; ANFO=1.36); q_f = Concentration of the bottom charge of the explosive (kg/m); R_1 = Side inclination correction factor; R_2 = Rock type correction factor and R_3 = Bench height correction factor. Table 2.2 may be used for the values of the correction factors.

Table 2.2 Values of correction factors

Inclination	8:1	10:1	5:1	3:1	2:1	1:1
R_1	0.95	0.96	0.98	1.00	1.03	1.10
Rock constant 'C'		0.3	0.4	0.5		
R_2		1.15	1.00	0.90		

Rustan (1990): $B = 18.1 \times D^{0.689}$(+ 52% max. expected value and –37% min. value) (2.14)

Where, B = Practical burden (m) and D = Diameter of blasthole (m).

$$\text{Pal Roy (1995): } B = H \times \frac{D_e}{D_h} \times \frac{5.93}{RQD} + 0.37 \left[\frac{L}{C}\right]^{0.5} \qquad (2.15)$$

Where, B = Burden (m); H = Hole depth (m); D_e = Diameter of explosive (mm); D_h = Diameter of blasthole (mm); RQD = Rock Quality Designation; L = Loading density of explosive (kg/m) and C = Charge factor (kg/m^3).

2.6.26 Cap Sensitivity

It is the sensitivity of an explosive to initiation by a test blasting cap No. 8 (0.40-0.45 g of PETN) or a fraction thereof.

2.6.27 Carrick Delay Detonator

A special designed detonator mainly used in gassy coal mines, where the tube is made of either copper or copper coated steel element, as opposed to the aluminium used in conventional detonators to eliminate any sort of ignition.

2.6.28 Choke Blasting

When a blasting is conducted with insufficient expansion space (may be due to misfire in the same or previous round or uncleared broken rock), it is termed as choke blasting. Researchers have noted that nearly 40 to 50% expansion space produces good fragmentation.

2.6.29 Column Charge

Charge placed between the bottom charge and the stemming in a blasthole is termed as column charge. Usually a non-capsensitive explosive is used in column charge.

2.6.30 Compressional Wave

A seismic wave whose motion is compression-dilatation or push-pull, generated by a rock's resistance to compression. However, the resistance of a material against failure when subjected to compressive stresses is called 'compressive strength (MPa)'.

2.6.31 Controlled Blasting

It is a special blasting technique where safety and environmental needs are met with desired throw and controlled backbreak.

2.6.32 Controlled Contour Blasting

A particular type of controlled blasting method specifically used in tunnels, drifts, raises, open pits and sometimes in production blasts where special care is taken to avoid overbreak and damage of the remaining rock surfaces. It can be smooth blasting, presplitting, cushion blasting or micro-sequential contour blasting.

2.6.33 Coyote Blasting (Shooting)

A method of single round large-scale blasting in which substantial amount of explosive charges are fired in small adits or tunnels driven, at the floor level, in the face of a quarry or slope of an opencast mine or even in small hills. It is used where it is impractical to drill vertically.

2.6.34 Crater Blasting

A blasting method conceptually developed by C.W. Livingston (1956) where the blastholes are drilled perpendicular to the surface being blasted, and the holes are charged with a concentrated charge (length of charge < 6 times the blasthole diameter) near the surface. Livingston determined that a relationship existed between the critical depth D_c, from which the first signs of external action in the form of cracks and fractures are noted, and the weight of the explosive Q, in agreement with the following empirical equation (Jimeno et al., 1995)—

$$D_c = E_t \times Q^{1/3} \text{ (Equation of strain energy)} \qquad (2.16)$$

Where, E_t = Strain-energy factor, which is a characteristic constant in each rock-explosive combination.

2.6.35 Critical Burden

The minimum burden distance with no breakage and displacement is termed as critical burden.

2.6.36 Cushion Blasting

A method of blasting in which a single row of closely spaced holes is fired after the main blast using decoupled charges to produce competent slopes.

2.6.37 Cut-off

When a portion of a column of explosives has failed to detonate because of bridging or to a shifting of the rock formation due to an improper delay system or a break in a path of detonation or initiation caused by extraneous interference, such as flyrock or shifting of ground, is termed as cut-off.

2.6.38 Dead Pressing

Desensitization of an explosive caused by pressurization is called dead pressing. Tiny air bubbles required for sensitivity, are literally squeezed in the mixture. In such cases, an increase in the density of the explosive occurs which when crossed the critical density of the explosive, detonation fails to occur. ANFO has the critical density of 1.2 g/cc (Plate 2.1).

2.6.39 Deck Charging (decking)

When a charge column, of bulk or cartridged explosives in a blasthole is separated by inert material or air cushion, it is called deck charging. The procedure of deck charging is adopted to reduce the maximum charge per delay in a blasting round and also to achieve better fragmentation in bedded deposits of different rock characteristics.

2.6.40 Decoupling

A separation by air between the surface of an explosive charge and the blasthole wall where it is charged is called decoupling. The decoupling ratio is denoted by:

$$D_c = \frac{R_h}{R_e} \tag{2.17}$$

Where, R_e is the radius of explosive (mm) and R_h is the radius of hole (mm). So for decoupled charge, $D_c > 1$ and for fully coupled charge $D_c = 1$.

2.6 41 Delay Blasting

Delay blasting is a technique where delay detonators, relay elements or connectors (through Nonel or sequential blasting machine) are used to separate in time the moment of detonation of a charge with respect to another.

2.6.42 Demolition Blasting

Demolition blasting is a specially designed controlled blasting technique usually carried out for demolition of buildings, towers, bridges and industrial installations. Conceptually, in demolition blasting, destruction occurs by destroying interlocking systems like beams, pillars, plates and steel girders.

2.6.43 Detonating Cord or Detonating Fuse

A flexible cord containing a centre core of high explosives, which when detonated, will have sufficient strength to detonate other cap-sensitive explosives with which it is in contact. Usually it contains 10 g/m of PETN explosive. However, in some special cases, a 6 g/m cord is also used. Low energy detonating cord (LEDC) containing 1 g/m to 2 g/m is unable to initiate the blasting agent. Detonating cord is mainly used to initiate explosive charges. However, it may also be used as an explosive itself, e.g. in controlled contour blasting. The detonating cord is mainly used on surface, but can also be used in underground blasting.

2.6.44 Dewatering of Blastholes

When water is present in the blastholes, it becomes necessary to dewater the blastholes before charging with the help of compressed air operated pumps or electrical submergible impellent pumps so that explosive charges are properly placed at desired locations.

2.6.45 Dip ($0 \leq \phi \leq 90°$)

Dip is the angle of inclination formed by blasthole axis with the vertical; it is also the angle at which the strata, beds, or veins are inclined from the horizontal plane. Regarding the selection of mining methods, the ore bodies are divided into shallow (<15° dip), intermediate (15-45° dip) and steep (>45-90° dip) (Rustan, 1998).

2.6.46 Dragline Excavator

It is a very high capacity mechanical appliance used primarily in surface coal mines for excavating soft to medium hard strata and is suspended from a movable jib. Dragline always operates on high-rise benches (called Dragline face) and during blasting in such faces; care should be taken for good fragmentation. During excavation, the bucket of the machine is dragged by means of a rope to fill it completely. The bucket is then lifted and after a swing, the material is dumped to a new heap.

2.6.47 Dynamite

It is a high explosive, invented by Sir Alfred Nobel of Sweden, in which the sensitizer is nitroglycerine [$C_3H_5(NO_3)$] or similar explosive oil.

Dynamites are classified as high explosives and can be detonated by a blasting cap.

2.6.48 Emulsion Explosive

It is a fourth generation explosive characterized by very intimate mixing between oxidizers and fuels. Oxidizers, which are solutions of ammonium nitrate, calcium nitrate, sodium nitrate, etc., are dispersed in continuous fuel medium comprising oils, waxes and other hydrocarbons. The size of oxidizer droplets is typically less than 5 microns, each surrounded by fuel and stabilized with suitable emulsifier (Figure 2.4).

3500 times expanded Av oxidizer droplet size 5 micron

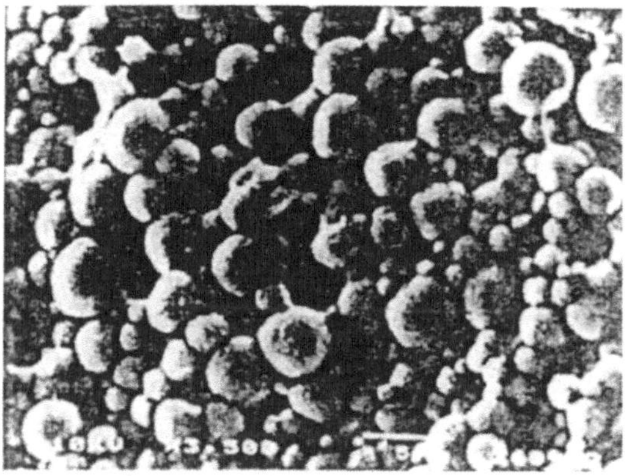

Fig. 2.4 Structure of emulsion under electron microscope

Plate 2.1 Direct loading of ANFO in blastholes

2.6.49 Energy Ratio, (m/s^2Hz)

The ratio of acceleration 'a' (m/s^2) to frequency 'f' (Hz) is termed as energy ratio. It is generally used to classify damage from blasting.

2.6.50 Extraneous Electricity

Flow of electrical current that circulates outside of a normal conductor is termed as extraneous electricity. It can be the result of defective insulation of the equipment, the galvanic action of two different metals etc. In blasting, it is the electrical energy other than the actual firing current, which may be a hazard in connection with electric blasting caps. It includes stray current, static electricity, lightning, radio-frequency energy, and capacitive or inductive coupling (Jimeno et al., 1995; Rustan, 1998).

2.6.51 Fatigue

Degradation through repeated loading of a material is called fatigue. In blasting, due to repeated blast impacts (dynamic pulses), a structure gets weakened, and becomes susceptible to damage even at low intensity of vibration or acceleration.

2.6.52 Fault

A fault is a geological zone of disturbance within a formation such as a fracture or a fracture zone along which there has been shear displacement of the two sides relative to one another and parallel to the fracture. This displacement may be as small as a few centimeters or as large as many kilometers (Rustan, 1998).

2.6.53 Flyrock

Rock that is unpredictably propelled into the air by the force of the explosion from a blasting site is called flyrock. The causes of flyrock are mainly attributed to the poor blast design, inadequate stemming column, geology of the site, or inaccurate drilling etc.

2.6.54 Fragmentation

The extent to which rock is broken into small pieces by blasting is called fragmentation. It is also the act of breaking rock using commercial explosive under certain geology and blast design.

2.6.55 Frequency, (Hz)

Frequency is the number of complete cycles per unit of time.

2.6.56 Gasbag

It is a self-inflating bag with an aerosol can inside containing the inflating gas. The gasbag can be activated by depressing the trigger at the top of the aerosol can. When the trigger is pressed and locked down, it releases the contents of the aerosol can into the gasbag and inflates it to lock the bag in the borehole. Once inflated, each bag is capable of withstanding pressures up to 10 MPa. It is used for air-deck blasting.

2.6.57 Geophone

It is an electromagnetic type transducer which emits an electric current that is proportional to the velocity of the vibrating particle. The electric signal is generated by a moving coil within the field of a stationary magnet. Geophones are designed to measure vibrations within certain frequency ranges (also see Chapter 8).

2.6.58 Ground Vibration (mm/s)

It is the most undesirable side-effects of rock blasting which produce structural annoyance and damage to surroundings. It is measured in terms of peak particle velocity (mm/s).

2.6.59 Haulage

It is the mechanical mode of transportation of mineral, man, machine, supplies and waste both in underground and on the surface.

2.6.60 Heave

Heave is termed to identify the extent to which the broken mass of rock is moved and displaced from its original location (Rustan, 1998).

2.6.61 Hoop Strain

It occurs when compressive or shear waves propagate perpendicularly along the axis (may be of pipe or similar type of things) and cause changes in circumferential stresses (Dowding, 1985).

2.6.62 Ideal Explosive

The boundary between the steady and nonsteady states at the detonation front is referred to as Chapman-Jouget (C-J) plane. At this plane the chemical reaction is complete, assuming an ideal detonation. It is at the

C-J plane that all the thermodynamic properties of pressure, velocity, temperature, internal energy or heat of formation and density are calculated (Dowding and Aimone, 1992). In ideal detonation conditions, an explosive would tend to detonate near its maximum detonation velocity, regardless of confinement and charge diameter. Military explosives are often ideal explosives compared with commercial explosives which are non-ideal.

2.6.63 Joint Frequency

Joints are nothing but the planes of weakness in the rock mass which offer no resistance to separation. Joints occur in sets and the number of joints per metre is termed as the joint frequency.

2.6.64 Kuz-Ram Model

Kuz-Ram model is an empirical formulation with the original Kuznetsov (1973) equation and the modification derived by Cunningham (1987) to determine the mean fragmentation size with any explosive and the index of uniformity.

2.6.65 Misfire

A charge or part of a charge, which failed to fire as planned, is termed as misfired. It may be a single hole or part of a blasting round. Causes for misfire include unskilled charging, defective drilling, defective explosive, detonator or fuse, broken electric circuit and lateral rock movement induced by the firing of preceding holes in the vicinity.

2.6.66 Mud Capping

It is a technique used to conduct secondary blasting where the explosive is placed on the surface of the boulder and covering it by mud. It helps breaking the boulder suitably with reduced air blast impact.

2.6.67 Natural Frequency (f_n), (Hz)

The natural frequency is the number of times in a second the structure or body oscillates back and forth when excited and its damping is proportional to the rate at which the vibration decays with time during free vibration. It is measured in Hertz (Hz).

2.6.68 NG Explosive

Explosives containing any amount of nitroglycerine [NG: $C_3H_5(ONO_2)_3$] or ethyleneglycoldinitrate (EGDN) are collectively referred to as 'NG'. If the NG content is less than 12%, a powder explosive results. For concentrations greater than 25% NG, gel explosives may be formed. Intermediate NG contents form a semi-gel (Rustan, 1998).

2.6.69 Nitrocellulose (NC)

The chemical formula of nitrocellulose is $C_{12}H_{14}N_6O_{22}$. It consists of white fibres bearing density 1.67 g/cc, volume of explosive gases 841 l/kg and heat of explosion 4052 kJ/kg. It is used together with nitroglycerine in dynamites.

2.6.70 Nitropropane

A liquid fuel that can be combined with pulverized ammonium nitrate prills to make dense blasting mixture (Bhandari, 1997).

2.6.71 NONEL

NONEL is the non-electric initiating devices (sometimes called shock tube system). It consists of a plastic 'bunch block' connector attached to a No. 8 strength NONEL delay detonator at the end of the tube. The delay periods vary, depending on the particular manufacturer's product. The relay type of each TLD (Trunk-line Delay), functions unidirectional, ensuring true hole-by-hole initiation with correct sequencing.

2.6.72 Overbreak or Backbreak

Excessive breakage of rock beyond the desired excavation limit is called overbreak or backbreak. In underground rock excavation, the amount of overbreak depends on the blasting method, the strength and structure of the rock, the size of the opening and delay sequence. Some correlation has been found between overbreak and rock mass quality designated as Barton Q-factor.

2.6.73 Pentolite

A mixture of PETN and TNT in the ratio of 1:1 which- when cast- is used as a cast primer or shaped charges. The density and detonation velocity are approximately 1.65 g/cc and 7400 m/s respectively.

2.6.74 Percussion drilling

It is a drilling method in which hammer blows are transmitted by drill rods to a drill bit. In some occasions, it also employs a string of drilling tools that are raised and lowered by a cable. An in-the-hole hammer is also based on percussion drilling (Rustan, 1998).

2.6.75 Permitted Explosive

An explosive which is approved by the competent authority (in India, The Directorate General of Mines Safety or DGMS) for use in underground gassy coal mines of degrees–I, II and III and is characterized by a low flame temperature that does not ignite certain specified methane/air or coal dust/air mixtures.

2.6.76 Plasma Blasting

It is a non-chemical explosive means of rock fragmentation in which an electric probe, situated in a borehole and surrounded by a highly conductive and gelled electrolytic solution, is fed with a very high capacitance discharge. The plasma transforms from the electrolyte solution in the borehole reaches an energy level of 280 kJ by electrical input of 8.5 kV and 200 A, and this rapidly released energy in the borehole develops a shock wave, which in turn produces a stress field that fractures the rock without producing excessive dust and flyrock. The range of plasma strength varies with chemical additives.

2.6.77 Pocket Charge

In a production blasting face, sometimes, a small explosive charge is placed in the stemming portion in blastholes for breaking strong surface material in surface blasting. This practice is termed as pocket charging.

2.6.78 Presplitting

It is a form of controlled blasting in which a series of closely spaced holes (generally 10 to 20 times of the borehole diameter), at the periphery of the excavation, is fired prior to the main blast to create a crack along the contour.

2.6.79 Pyrotechnic Delay

Pyrotechnic is a phenomenon in which the burning rate of delay element in a detonator is maintained. However, while manufacturing, it is quite impossible to maintain very high precision and thereby some scatter in timing may occur which increases with the delay number. In cases of supreme high precision, it is always recommended to use electronic detonators.

2.6.80 Quasi-static Pressure

It is a time-varying pressure condition which does not require fully dynamic analysis.

2.6.81 Rayleigh Wave or R-wave

Rayleigh (R) waves are usually generated in rock blasts and are important in view of the damage to constructions from blast vibrations on surface. These are characterized by elliptical particle orbit, usually a motion that is contrary to the propagation direction of the wave. An example of Rayleigh wave is the ocean wave.

2.6.82 Relief Hole

In blasting, relief holes are used in various situations for relieving or removing part of the burden for the charge to be fired in the main blast or to get rid of misfired holes which are a subject of handling hazards.

2.6.83 Ring Blasting

It is a specific blasting method in which a blasting round is carried out by a number of holes drilled in a ring pattern, usually used in conventional sublevel open stoping of underground metal mining and Blasting Gallery method of coal extraction (for thick seam) from underground.

2.6.84 Rosin-Rammler-Sperrling (RRS) Distribution

It is a mathematical distribution function to describe the amount of rock of different sizes as obtained after blasting. The *RRS*-distribution is a special case of the more general Weibull distribution and it is the most commonly used distribution in blasting and comminution fields (Rustan, 1998).

2.6.85 Safety Fuse

It is a cord used to initiate plain detonator in surface blasting. It contains a core of black powder covered by textile and waterproof material and does not explode.

2.6.86 Scabbing

Scabbing is nothing but the spalling of rock by the high amplitude tensile wave which is created due to the reflection of the compressive wave at a rock/air interface (Rustan, 1998).

2.6.87 Scaled Distance (SD)

Scaled distance is generally used to predict ground vibration and air overpressure levels generated due to blasting. It is obtained by dividing the distance (D, in m) from the blasting source to the point of observation by a root (usually square root for cylindrical charge symmetry or cube root for spherical charge symmetry) of the explosive mass per delay (Q) in kg. In some specific situations, some authors used 'total charge' in place of maximum charge per delay to establish the vibration predictor equation.

2.6.88 Secondary Blasting or Boulder Blasting

Secondary blasting is used to break oversize boulders which appear after a primary blast. It is achieved either by "pop shooting" or by "plaster shooting".

2.6.88.1 Pop Shooting

In this boulder blasting technique, the explosive charge is placed in a drill hole, and confined with clay up to the hole collar. The hole length may be slightly more than the half of the thickness of the boulders. For larger boulders, several holes may be drilled (with jackhammer) keeping an approximate spacing of 0.6 m. For small boulders, the hole may be drilled at the centre of the boulder. The average charge may be one full cartridge i.e. 130 gm of S.G. 80% for every 2-2.5 m^3 of boulder size. Pop shooting generally creates flyrock and thus it is advisable to keep a minimum distance of nearly 200 m from the nearest structure.

2.6.88.2 Plaster Shooting

Plaster shooting (mud capping) is a commonly used method for boulder blasting. In this method, a thin layer of mud is placed on the boulder and the explosive cartridge is firmly placed into the mud. The charge is then covered hemispherically with some handful of mud to make a wave shaper. The average charge ratio is 1.0-1.5 kg of gelatinous explosive for every cubic metre of boulder size. Plaster shooting creates a lot of noise which may lead to human psychological effect and may be heard over a periphery of 1 km.

2.6.89 Seismograph

Seismograph is an instrument which measures and supplies a permanent record of earth-borne vibrations induced by blasting and/or earthquake. Very sophisticated microcomputer based seismographs are available today for blast monitoring. These are all supplied with compliance software for detailed blast analysis and FFT calculations.

2.6.90 Shaped Charge

Shaped charge is an explosive charge of special shape. In this type of charge, the shock wave is concentrated along the axis of an explosive column if a conical depression is made at the end of the charge. This process gives explosives a locally higher penetration power. Moreover, if the cone is lined with metal, a different and much more pronounced result is obtained. As the metal liner of the cone collapses during the detonation process, a jet, followed by a 'slug', is projected forward at high velocity, producing destructive effects to greater depths (Sen, 1995). The three-stage detonation process of a shaped charge is shown in figure 2.5.

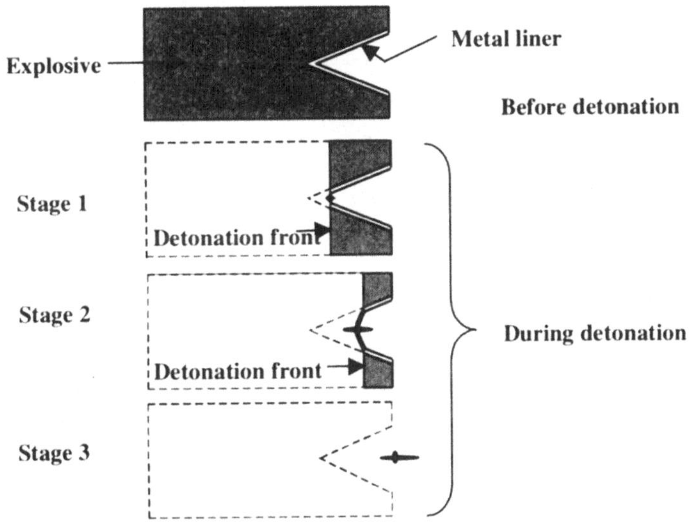

Fig. 2.5 Formation of 'slug' in a shaped charge (Sen, 1995)

2.6.91 Shot Firing Cable

A cable that is used to fire a blasting round (or several rounds) with the help of an exploder, is called a shot firing cable. The cable consists of a pair of insulated copper or iron conductors which joins the exploder (or sequential blasting machine) to the blasting circuit through a detonator.

Iron is used where copper cannot be tolerated in the muckpile, as it can be removed by magnets. It may consist of a twin core where both conductors are contained in the one cable or two separate conductors twisted around each other.

2.6.92 Slab Blasting

Slab blasting consists of blastholes which are open at both ends and it aims to create slice of rock or ore from the parent body.

2.6.93 Slurry Explosive

Slurries are blasting agents based on saturated aqueous solutions of AN, often with other oxidizers such as sodium nitrate and/or calcium nitrate, in which the fuels, sensitizers, gallants and cross-linking agents are dispersed to avoid the segregation of the solids (Jimeno et al., 1995). It is the third generation explosive.

2.6.94 Sound Pressure (level), (dB)

The sound pressure is expressed in decibels (dB), and is defined by the following equation—

$$dB = 20.\log_{10}(P/P_0) \tag{2.18}$$

where,

dB = Sound level in decibels

P = Overpressure in psi

P_0 = Overpressure of the lowest sound that can be heard = 3×10^{-9} psi.

2.6.95 Spacing (S), (m)

It is the distance between boreholes or charges in the same row. Spacing (S in m) should be determined in relation to burden (B in m), joints of rocks and hole depth. It is always advisable to understand the difference between spacing in drilling (S_d) and spacing in blasting (S_b). The following formula may be used, as a broad guideline, for the determination of spacing—

$$S = (1.2 - 1.5)B \tag{2.19}$$

In some opencast mines, it has been noticed that S/B ratio up to 3 also worked well.

S < B is used in cast blasting, smooth blasting and cushion blasting and in places where the specific requirement is to get bigger-size boulders with clean wall cut rather smaller fragments of blasted rock. Very small spacing may cause excessive crushing between charges and superficial crater breakage, large blocks in front of the blastholes would occur, and toe problems cannot be overruled.

A few other formulae developed by a number of researchers are given below—

Jimeno et al. (1995) suggested for staggered drilling pattern: $S = 1.15\ B$ for vertical blastholes and $S = 1.15\ B \times \cos\theta$ for inclined holes, where θ is the angle of inclination with respect to the vertical.

Pal Roy (1995): $S = 1.30 \times B - 4.0 \times [L/C]^{0.5} \times [1/RQD]$ (2.20)

All parameters in equation (2.20) are already defined while describing equation (2.15). Comparisons of predicted and optimum values of spacing are depicted in figure 2.6.

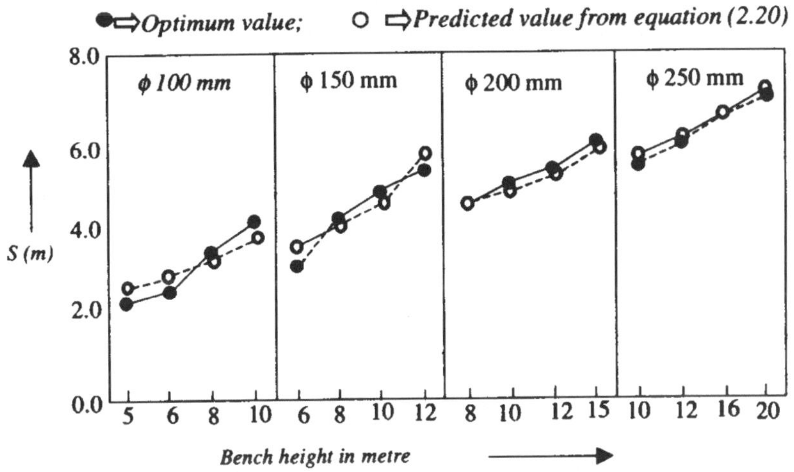

Fig. 2.6 Comparison of predicted and optimum values of spacing (*S*, in metre)

2.6.96 Spalling

Spalling in rock occurs under various situations like: a) Longitudinal splitting in uniaxial compression, b) Breaking-off of plate-like pieces (spalls) from a free rock surface due to high tensile stresses caused by the reflection of a compressive wave at a free surface (Rustan, 1998). c) In underground roof and pillars due to nearby surface and underground blasting.

2.6.97 Specific Charge, (kg/m³) or (kg/t)

Amount of explosive consumption (planned or actual) per ton or m^3 of blasted rock is called specific charge. The specific charge varies for different kinds of rock from 0.3 to 0.9 kg/m^3. According to Pal Roy and Dhar (1993), the specific charge can be correlated to the *RQD*.

2.6.98 Stemming

The inert material of dense consistency, such as drill cuttings (free from pebbles and gravel), sand, clay or water in plastic bags, which is inserted in the collar of the drillhole after charging and stemmed to seal the hole temporarily in order to prevent venting of gas, increase of blasting efficiency, reduce air shock waves or dampen any open flames. The stemming length (S_t) varies from 12 times of the borehole diameter for hard competent rock to 30 times of the borehole diameter for soft competent rock (Rustan, 1998). The following relations between stemming length and burden are found to be most suitable for fruitful blasting operations—

$$S_t / B > 0.6 \text{ (for controlling flyrock)} \qquad (2.21)$$

$$S_t = B \text{ (for symmetrical stress distribution)} \qquad (2.22)$$

2.6.99 Stress or Strain Energy

When an explosion takes place in a rock, the transmitted energy, generated by the shock wave is termed as stress or strain energy. However, shear strain is defined as the change due to deformation of the angle between two line segments.

2.6.100 Stray Current, (A)

The current flowing outside its normal conductor is known as stray current. As a result of defective insulation, it may come from electrical equipment, electrified fences, electric railways or similar items. Such current flow is usually facilitated by conductive paths like pipelines, wet grounds and wet materials (Bhandari, 1997). This undesired electric current may enter or leave a blasting circuit, due to inadequate insulation, resulting in misfires of detonators or premature detonation.

2.6.101 Stress, (MPa)

It is a tensor quantity, which in a uniaxial case, is defined as the force per unit area obtained within the limits of the force acting on an infinitesimal area element at a point.

Let us draw three planes through the general point P(x) parallel to the coordinate axes (Figure 2.7). We assume the positive side of the plane x_1 = constant is the side from which x_1 increases. The stress vector at P for the plan x_1 = constant is σ_{1j}. Similarly we have the stress vector at P for the plane x_2 as σ_{2j} and for x_3=constant, σ_{3j}. Hence at P we have 9 components of stress σ_{ij}; σ_{11}, σ_{22} and σ_{33} are the normal stresses, and others like σ_{12}, σ_{23}, etc., the shear stresses (Milklowitz, 1978).

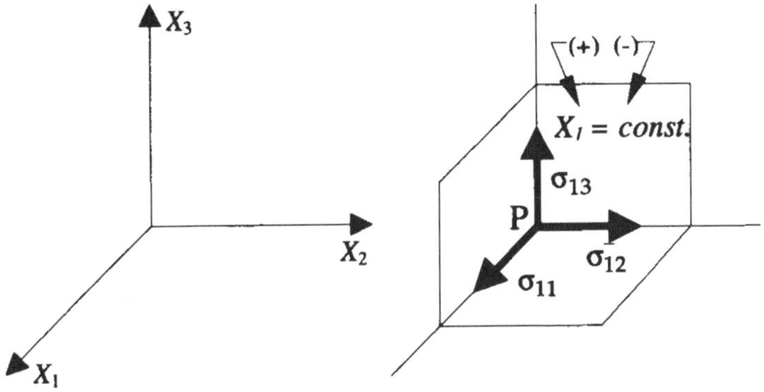

Fig. 2.7 Stress components of the stress vector σ_{ij} at P for the plane X_1=constant

2.6.102 Subdrilling

Subdrilling means to drill blastholes beyond the planned grade lines or below floor level. When subdrilling is used, there will be a large zone of maximum tension and it will occur close to the floor level which must be sheared for good breakage. Subdrilling may be used as per the following relation—

$$J(m) = 0.1 \text{ to } 0.25 \text{ times of the burden (m)} \tag{2.23}$$

2.6.103 Sublevel Caving

It is an underground extraction method used in metal mining where longitudinal drifts or transverse crosscuts are driven through the ore body on successive levels. From the crosscuts, the ore just above is fan drilled. The ore is then blasted and moved by gravity to the drift, where it is loaded, transported to the footwall and dumped into shafts (Rustan, 1998).

2.6.104 Swell Factor

Swelling is the increase in the volume of rock after breakage. The ratio between the total volume of broken rock and the volume of intact rock before blast is termed as swell factor. It can be used for a quantitative assessment of fragmentation. In blasting, the swell factor varies between 1.0 and 1.7. The normal value of swell factor in a multiple row blasting (generally 3 to 5 rows), is normally 1.4. For more than 10 rows, the swelling will be considerably reduced to lie between 1.1 and 1.2.

2.6.105 Tracer Blasting

It is a special blasting technique adopted by Singh (1994) where reduction of the linear charge concentration of a blasthole was made with the help of a low strength detonating cord placed along the explosive column. The detonating cord deflagrates the explosive (nearly 10 to 15%) and there are deleterious effects of detonating cord downline the borehole depending on the explosive composition and coreload of the cord (Bhushan et al., 1986). The detonating cord is used for the initiation of the primer at the bottom of the hole and is detonated first. The compressive wave desensitizes the explosive by increasing the pressure (increase in explosive density) such that when the primer is initiated, the explosive detonates with a lower detonation velocity. The method is generally used in controlled contour blasting.

2.6.106 Velocity of Detonation, (VOD), (m/s)

The velocity at which the detonation wave travels through a column or mass of explosive is known as velocity of detonation. It may be measured in confined or unconfined conditions. For low explosives, VOD ranges between 1500 and 2500 m/s while for high explosives, it ranges between 2500 and 7000 m/s. Emulsion explosive has the VOD range of 3200-4500 m/s.

2.6.107 Young's Modulus (E), (GPa)

Under given loading conditions, the ratio of stress to strain induced in a solid material, which is numerically equal to the slope (tangent) of the linear part of the stress-strain curve is known as Young's modulus.

REFERENCES

Ashby, J.P. 1980. Production blasting and the development of open pit slopes, Proc. 6th Conference on Explosives and Blasting Techniques, Intern. Society of Explosives Engineers (ISEE), Solon, OH, USA, pp. 291–311.

Bhushan, V., Konya, C.J. and Lukovic, S. 1986. Effects of detonating cord downline on explosive energy release, Proc. 2nd Mini-Symposium on Explosives and Blasting Techniques, International Society of Explosives Engineers (ISEE), Montville, OH, pp. 41-55.

Bhandari, S. 1997. Engineering rock blasting operations, A.A. Balkema, Rotterdam, Netherlands, p. 375.

Chiappetta, R.F. and Borg, D.G. 1983. Increasing productivity through field controls and high-speed photography, Intern. Symposium on Rock Fragmentation by Blasting, Lulea, Sweden, pp. 301-331.

Cunningham, C. 1983. The Kuz-Ram model for prediction of fragmentation by blasting, Proc. I Intern. Symposium on Rock Fragmentation by Blasting, Lulea, Sweden, pp. 439-453.

Dey, Kaushik, 2003. Personal communication, India.

Dowding, C.M. and Aimone, C.T. 1992. From chapter 9.2 of SME Mining Engineering Handbook: *http://books, smenet.org*.

Dowding, C.H. 1985. Blasting vibration monitoring and control, Prentice-Hall, Inc., Englewood, Cliffs, NJ, USA, p. 297.

Dutta, P.K., Barman, B.K. and Moitra, B.C. 1973. On some problems of opencast drilling and blasting, Proc. Seminar on Limestone and Dolomite Industry in India, Rourkela, p. 150.

Jimeno, L.C., Jimeno, L.E. and Carcedo, A.J.F. 1995. Drilling and blasting of rocks, A.A. Balkema, Rotterdam, Netherlands, p. 391.

Konya, C.J. and Walter, E.J. 1990. Surface blast design, Prentice Hall Int. Inc., USA, p. 303.

Kuznetsov, V.M. 1973. The mean diameter of the fragments formed by blasting rock, Soviet Mining Science, 9(2): 144-148.

Langefors, U. and Kihlstrom, B. 1978. The modern technology of rock blasting, John Wiley and Sons Inc, New York, USA, p. 438.

Lilly, P.A. 1986. An empirical method of assessing rock mass blastibility, Proc. Large Open Pit Conference, August, IMM, Australia, pp. 89-92.

Livingston, C.W. 1956. Fundamentals of rock failure, Quarterly of the Colorado School of Mines, 51(3), July, USA, pp. 1-14.

Miklowitz, Julius. 1978. Elastic waves and wave guides, North-Holland Publishing Company, Vol. 22, Oxford, UK, p. 618.

Olofsson, S.O. 1990. Applied explosive technology for construction and mining, Nora boktryckeri AB, Sweden, p. 303.

Pal Roy, P. 1995. Mathematical modelling to evolve explicit relationships in the study of energy balance in rock blasting, EMR Report of Investigations, December, CMRI, Dhanbad, India, p. 117.

Pal Roy, P. and Dhar, B.B., 1993. Rock fragmentation due to blasting—a scientific survey, The Indian Mining and Engineering Journal, Vol. 32, No. 9, pp. 27-32 (Also given in the unpubl. Ph.D. thesis of M.M. Singh submitted to IT, BHU).

Rustan, P.A. 1990. Burden, spacing and borehole diameter at rock blasting, The Australasian Institute of Mining and Metallurgy, Proc. 3rd Int. Symposium on Rock Fragmentation by Blasting, Brisbane, 26-31 August, Australia, pp. 303-309.

Rustan Agne 1998. Rock Blasting terms and symbols, A.A. Balkema, Rotterdam, Netherlands p. 193.

Sen, G.C. 1995. Blasting technology for mining and civil engineers, UNSW Press, Sydney, Australia, p. 146.

Singh, S.P. 1994. Blast damage control during underground mining, Proc. 20th Annual Conference on Explosives and Blasting Technique, Inter. Society of Explosives Engineers, January 30 to February 3, USA, pp. 329-341.

Siskind, D.E. 1973. Ground and air vibrations from blasting, Chapter 11.8 of SME Mining Engineering Handbook, Vol. 1, Society of Mining Engineers, New York, USA, pp. 11-99 to 11-111.

Wiss, J.F. and Linehan, P.W. 1978. Control of vibration and blast noise from surface coal mining, Vol. III, Bureau of Mines, US Department of the Interior, Washington, DC, USA, Report No. OFR 103(3)-79, December, p. 623.

3

Detonation Principle and Rock Breakage Mechanism

3.1 INTRODUCTION

When commercial explosives are detonated to break the rock, its energy is partitioned in different components (Figure 3.1). Out of the energy released some is useful energy and the rest is waste energy. While useful energy is capable of doing work, the waste energy cannot be used in a constructive way and it creates some environmental problems and annoyance. When high explosive detonates, two basic types of useful energy components are produced, one is called shock energy and the second gas energy. The partitioning of this gas and shock energy depends on the selection of the explosive and its suitability towards a particular application.

In the case of secondary blasting where explosives are used to break large size of boulders produced in the primary blast or for structure

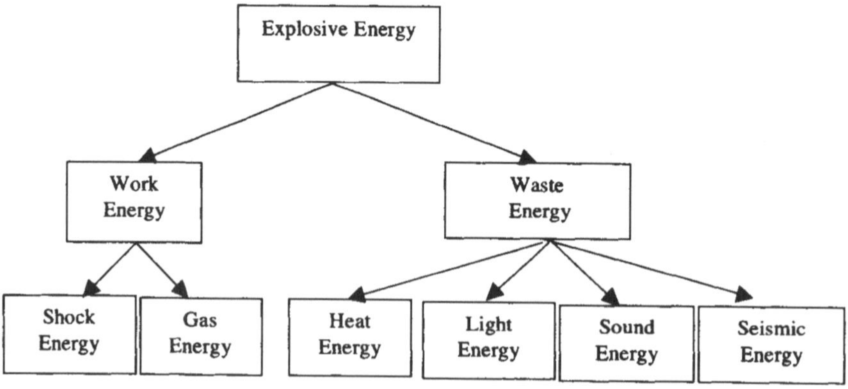

Fig. 3.1 Energy partitioning of detonating explosives

demolition, an explosive with high shock energy is preferred. Explosives with high gas energy are used in boreholes in mines and quarries and are confined by proper stemming materials.

This chapter gives a fundamental concept of rock breakage mechanism and principle of breakage. It also discusses strain wave propagation and the resulting effect of initial compression which expands the rock in concentric cylinders surrounding the blasthole, and ultimately visualizes the role of gas pressure in the blasthole for rapid destressing resulting in rock failure.

3.2 BASIC INTRODUCTION OF HIGH AND LOW EXPLOSIVES

Low explosives are those which deflagrate or burn very rapidly. These explosives usually have reaction velocities of 600 to 1200 metre per second. Such explosives produce no shock energy. They produce energy only from gas expansion. A typical example of low explosive is black powder. High explosives, on the other hand, on detonation, produce both gas and shock pressures. Figure 3.2 shows detonation fronts of low and high explosives (Konya and Walter, 1990).

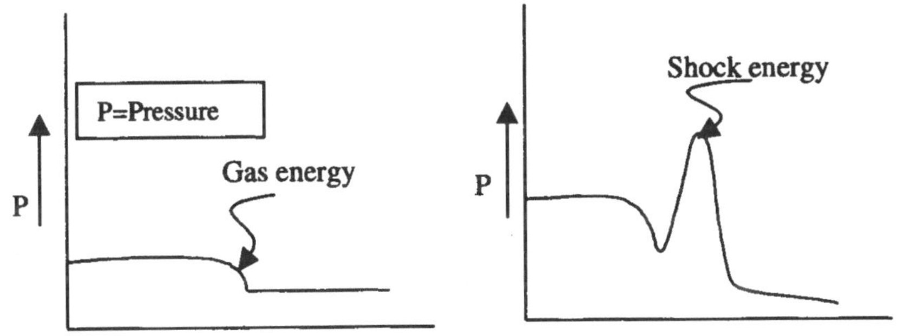

(a) Low explosive (b) High Explosive

Fig. 3.2 Detonation of low and high explosives

A high explosive detonation exhibits shock pressure at the reaction front which travels through the explosive before the gas energy is released (Figure 3.3). Shock energy generally produces higher pressure than the gas energy does. After the shock energy passes, gas energy is released. The shock pressure of high explosive is a transient pressure that travels at the explosive's rate of detonation. Some authors have adequately explained through model studies that approximately 15% of the total available useful work energy is spent in shock pressure. The gas pressure accounts for approximately 85% of the useful work energy. The gas energy produces a force that is constantly maintained until the borehole ruptures.

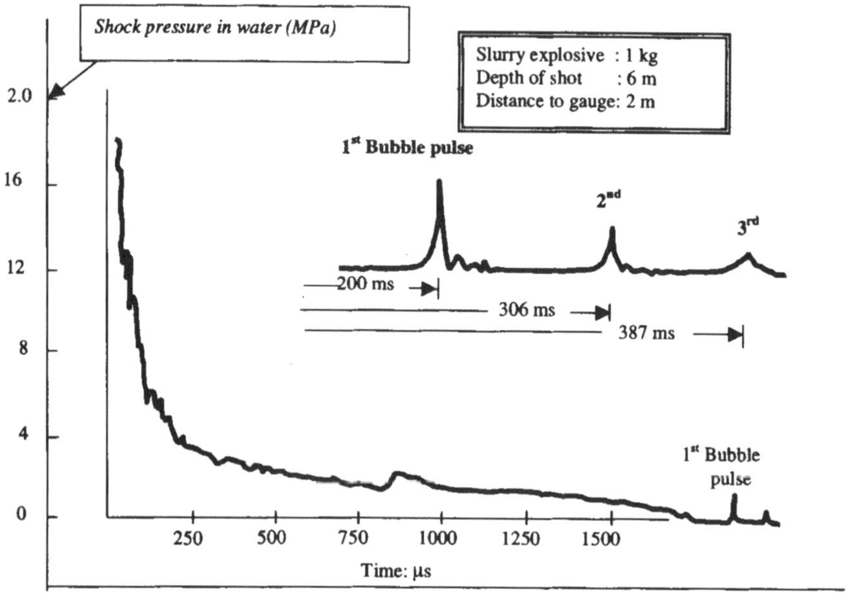

Fig. 3.3 Underwater test record sample

3.3 THE PRINCIPLE OF DETONATION AND BREAKAGE

When the exploder is switched on, the firing current pulse from the exploder flows through the bridge wire of the detonator fuse head. Within 4 ms the fuse head fires, igniting the detonator's igniting charge which in turn detonates the base charge. The detonator bursts and the detonation is passed, by the shock and impact of incandescent particles, to the primer cartridge. The explosive is compressed and hot spot is formed due to friction and the adiabatic compression of small gas bubbles which ultimately initiates the detonation reaction. The gaseous products are projected in the direction of detonation propagation, while the pressure and heat from these further accelerate the detonation reaction.

The initiation may be of low order with the reaction zone moving initially with a velocity of about 600 m/s but the rate of reaction rapidly accelerates and, within about 0.0001 ms from the bursting of detonator, fully developed detonation is achieved and the velocity of detonation leaps to the characteristic value of the explosive (Jeffrey, 1990—Personal communication).

The pressure in the blasthole rises rapidly as the detonation process progresses. At the reaction zone the pressure is maximum and transient in nature. This pressure is known as explosion pressure P_e. The explosion pressure rapidly attenuates as the reaction products disperse into the borehole and fracture the rock. When it is loaded to the borehole, it becomes the borehole pressure P_b. The borehole pressure spontaneously loads the surrounding rock and generates a compressive shock pulse that moves rapidly away from the borehole at a velocity that may initially exceed the sonic velocity of the rock.

Sonic velocity is a function of Young's modulus which is a measure of the material's elasticity. It is also characteristic of the material's inherent integrity in terms of large, small and hairline cracks, grain size, structural geological features, moisture content and porosity.

The following Table 3.1 shows the typical sonic wave velocity of some common Indian rock masses (Dhar et al., 1993).

Table 3.1 P-wave velocity of different rock masses

Name of rockmass	*Rock type*	*P-Wave velocity (m/s)*
Dunite		7000
Diabase		6500
Gabbro	Compact	5600
Dolomite		5500
Granite		5000
Limestone		4000
Slate & Shale	Less Compact	4000
Sandstone		3000
Alluvium		1000
Loam	Unconsolidated	1000
Sand		1000
Loess		500

3.4 ROCK BREAKAGE

There are two principal explosive mechanisms namely—

a) Shock (or brisance)

and b) Gas pressure

Chapman and Jouguet proposed the first theory of detonation in 1899. This was modified by Zeldovich in 1943 and has gone through many improvements since then. Modern computers have made it easy to apply the theories to explosive compositions and calculate the expected detonation parameters. Calculated energy values based on thermochemical and hydrodynamic theories are commonly quoted by explosive manufacturers.

Different explosives exhibit different degrees of gas and shock pressures. The extent to which an explosive possesses these properties is described by the Aquarium test for bubble energy and shock energy.

On detonation, explosive energy is converted into mechanical energy, heat and incomplete reaction products. The initial detonation pressure P_d is quickly reduced to the explosion pressure P_e and as the gaseous reaction products expands to fill all available voids, it reduces to blasthole pressure P_d. This pressure is maintained until venting occurs. The relationship between P_d and the explosion pressure P_e depends on the type of explosive used. In general the relationship between them is [$P_d = n \times P_e$] where n lies between 1 and 2. The blasthole pressure will depend on the extent to which the hole is filled.

The energy transferred from the explosive charge to the rock will depend on the extent to which the blasthole is coupled with the explosive charge. Any air-gap will cushion the blast and will reduce the efficiency of the energy transfer. This effect is called the decoupling effect.

Decoupling factor, D = Diameter of borehole/diameter of charge. D is always greater than or equal to 1. D is '1' when the explosive charge is fully coupled with the borehole wall.

The smaller the decoupling factor, the greater will be the blasthole pressure and efficiency of the energy transfer. With this point in view, slitting of explosive cartridges, as practiced in opencast mines, is done to fill the entire diameter of the blasthole thereby reducing the value of the decoupling factor.

The shock pulse transmitted to the rock creates stresses resulting in radial compressive strains which tend to crush the rock close to the borehole, and also tensile hoop stresses which are concentric to the borehole and cause local radial cracking. If the transmitted shock pulse is strong in relation to the confined strength of rock, then considerable initial crushing and fracturing near the explosive charge is bound to occur (Figure 3.4).

Beyond the fracture zone, the pulse travels as an elastic wave until it reaches the free face, where the wave is reflected as a tensile wave. The amount of energy transferred to a given rock is a linear function of the product of density and the rate of detonation. This term is called the characteristic impedance of explosive. An explosive, which has the larger characteristic impedance or close to the characteristic impedance of rock, transfers more energy to the rock mass.

The explosive must be selected to match with the operational aspects of blasting so that maximum energy could be transferred to the surrounding rock mass. The ideal one being that which breaks the rock in tension. If the compressive strength of the rock is exceeded by the shock pulse then annular rock around the borehole will be crushed to a fine size, and this powder will prohibit the propagation of the strain pulse. In addition, increased amount of dust will be formed.

Out of the energy available in the explosive, it is estimated that only 3 to 20% is transmitted into stress waves. The major portion comes out as heat, which is not wasted but expands the gas products of detonation.

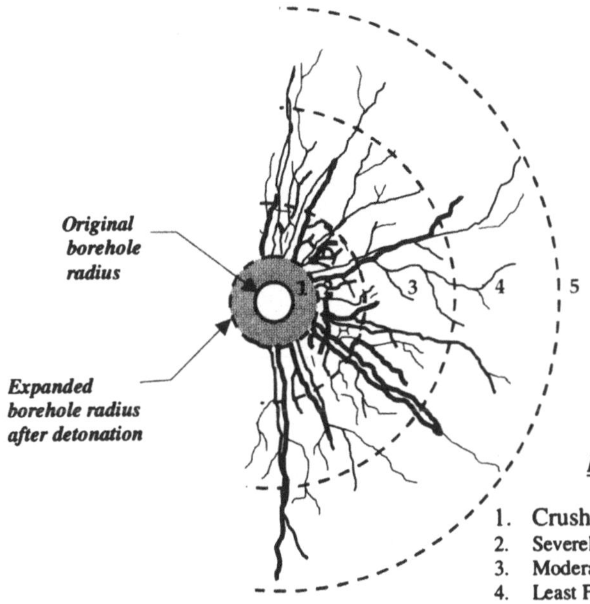

Fig. 3.4 Process of rock breakage

3.5 FORMATION OF SEISMIC WAVES

When the seismic waves, produced due to detonation, reach the surface of the earth, get reflected and refracted thereby produce both vertical and horizontal motion in the earth's surface. Seismic waves of different types would propagate at their respective sonic velocities. Two groups of seismic waves are seen to be generated by detonation of explosive charges in drill holes. They are classified as body waves and surface waves. Body waves travel within a medium while surface waves are restricted to travel along free interface, such as the ground surface.

Body waves comprise of two discrete components—compression or P-wave and shear or S-wave. Two types of surface waves are usually produced from normal mine blasting—Rayleigh (R) waves and Love (L) waves. The P-wave is the fastest one, followed by the S-wave. The velocity of the Love wave remains less than the shear wave velocity of the overlying layer. Rayleigh waves propagate at a velocity slightly less than the S-wave velocity. Unfortunately most explosives are detonated as a series of smaller explosions which are delayed by milliseconds, and differences in travel paths and delay times result in overlapped arrival of both wave fronts and wave types (Pal Roy, 1995).

3.6 SHOCK ENERGY

The shock energy transmitted to the rock depends on the detonation pressure (P_d) of the explosive. The detonation pressure is a function of the explosive density times the detonation velocity squared. It is a form of kinetic energy. There are a number of different computer codes written to calculate this pressure. The most commonly used equations are as follows:

$$P_d = 4.18 \times 10^{-7} \times SG_e \times V_e^2 / (1 + 0.8\ SG_e) \tag{3.1}$$

Where, P_d is in kilobars (1 kilobar = 14504 psi),
SG_e = the specific gravity of the explosive, and
V_e = the detonation velocity of the explosive in ft per sec.

Another simplified form of equation (3.1) is—

$$P_d\ (\mathrm{N/m^2}) = \frac{V_e^2\ (\mathrm{m/s}) \times d_e (\mathrm{kg/m^3})}{3.8} \tag{3.2}$$

where, d_e is the density of the explosive.

The detonation pressure is maximum in the direction of shock wave travel. It is thus maximum in the explosive cartridge at the opposite end from where initiation occurred. The detonation pressure on the sides of the cartridge is nearly zero since the detonation wave does not extend to the edge of the cartridge. To get maximum detonation pressure effects from an explosive, it is necessary to place the explosive on the material to be broken and initiate it from the opposite end to that in contact with the material. To maximize the use of detonation pressure, maximum contact area between the material and explosive is required. This technique is widely used in mud capping or plaster shooting of boulders.

3.7 GAS ENERGY

When the charges are detonated in boreholes, then the expanding gas energy causes majority of the rock to break. The gas pressure is the pressure which is exerted on the borehole walls by the expanding gases after the chemical reaction is completed. The amount of explosion pressure is related to the volume of gas liberated per unit weight of explosive and the amount of heat liberated during the reaction. The higher the reaction temperature at a constant gas volume, the higher the gas pressure. If more gas volume is liberated at a constant temperature, the pressure would also increase. For an easy calculation, it is often assumed that the explosion pressure is approximately one-half of the detonation pressure. This is not always the case, as there may be some conditions where explosion pressure can exceed the detonation pressure. Explosion pressures nowadays are calculated either from computer codes or measured in underwater tests. The gas products formed during detonation can provide useful information on reaction, efficiency and useful energy release. The gas energy produces a force that is constantly maintained until the borehole ruptures. The action of gas pressure is important in loosening, expanding and throwing the mineral.

3.8 FRACTURING BY RELEASE OF LOAD

For the blasthole charge, it is estimated that less than 5% of the explosive energy is carried into the rock by the initial strain wave. It is however estimated that by the time the strain wave reaches the free face, the total energy transferred to the burden by the initial compression is around 65% of the blast energy. The effect of initial compression is to expand rock in concentric cylinders surrounding the blasthole and develop hoop strains, resulting in tension cracks formed tangential to the cylinders (Jeffrey, 1990; Personal communication).

After passing of the compression wave, the gas pressure in the blasthole remains in equilibrium with the strain in the rock, while venting causes the strain to be released, rapidly destressing the rock. Such rapid destressing results in failure of the rock.

It follows that failure first occurs close to the blasthole and the burden tends to be detached almost completely, moving forward with a low velocity. As the load release progresses towards the face, subsequent layers of rock detach from the moving burden and the velocity of the face increases to a higher value.

In large blast, consisting of a line of blastholes, initiated in sequence, the compressive strain waves and gas expansion effects from adjacent holes superimpose. An undesirable effect of this is the creation of vertical overbreak cracks parallel to and well behind the face and consequent mine bench instability.

3.9 GAS EXPANSION

After the formation of radial cracks by the strain wave, the gases expand and fill these cracks, exerting pressure on the crack surfaces. The radial cracks extend due to stress concentration at the crack tips.

The stress concentration at the crack tips extends the crack length and thus formed long cracks are most unstable. It follows that there is a higher stress concentration around a blasthole with pressure cracks than around the original pressurized blasthole.

Extension of explosion-produced cracks may be masked by the preexisting fractures. Preexisting cracks may be widened and extended and the formation of new cracks inhibited.

3.10 REFLECTION BREAKAGE

When the compressive strain waves strike the free face, two reflected waves are generated—one in tension and the other in shear. As the tensile strength of rock is far less than the compressive strength, the reflected tensile strain may have considerable breaking effect, theoretically causing progressive spalling from the free face. But high-speed photographs indicate that, in practice, rock blasting burden spalling does not occur (Figure 3.5).

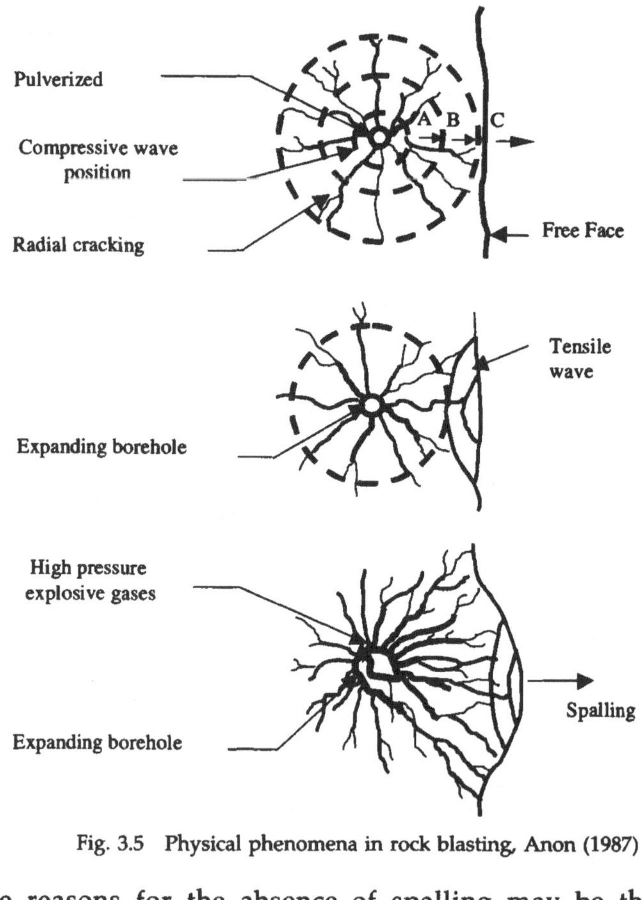

Fig. 3.5 Physical phenomena in rock blasting, Anon (1987)

The reasons for the absence of spalling may be that internal discontinuities reflect the compressive strain waves, before they reach the free face. Reflected tensile waves, too weak to initiate new fractures, may still interact with radial cracks and cause them to extend, thus contributing to failure.

Spalling is generally an unimportant aspect of breakage and may result from overcharging. Reflected energy may, however, be important in interacting with the direct stress pulse.

REFERENCES

Anon. 1987. Explosive and rock blasting, Atlas Power Company, Dallas, USA, p. 662.

Dhar, B.B., Pal Roy, P. and Singh, R.B. 1993. Optimum blasting for Indian geomining conditions—suggested standards and guidelines, CMRI Publication, Dhanbad, India, March, p. 40.

Jeffrey, G.S. 1990. Personal communication, Trent Polytechnic, Nottingham, U.K.

Konya, C.J. and Walter, E.J. 1990. Surface blast design, Prentice Hall Int. Inc., USA, p. 303.

Pal Roy, P., 1995: Mathematical modelling to evolve explicit relationships in the study of energy balance in rock blasting, EMR Report of Investigations, CMRI, Dhanbad, India, December, p. 117.

4

Strength of Explosives—Theoretical Derivation and Laboratory Determinations

4.1 INTRODUCTION

In earlier years, the strength of explosives was simply related to the percentage composition of NG in dynamite. However, with the development of more complex formulations, this method of classification was abandoned in favour of easily measured physical parameters. A number of different computer programmes are now available for calculating explosive energy and other relevant properties. These are based on mathematical models of thermo-hydro-dynamics of the explosion phenomenon.

4.2 STRENGTH OF EXPLOSIVES

There are several ways of measuring strength of commercial explosives. Bulk strength is the energy per unit volume of explosive (kJ/m^3). When the bulk strength is referred as related to the bulk strength of a standard explosive such as ANFO, then it is called relative bulk strength (*RBS*). ANFO is the normal reference explosive at densities between 0.75 and 0.87 g/cc, while Blasting Gelatine (BG) is often quoted at a density of 1.55 g/cc (Bhushan and Srihari, 1991). Weight strength is the calculated energy per unit mass of explosive (MJ/kg). This is also called absolute weight strength (*AWS*). Relative weight strength (*RWS*) is the weight strength of one explosive related to standard explosive like ANFO. Relative weight and bulk strengths of an explosive can be measured by the following equations.

$$RWS = (AWS \times 100)/(AWS \text{ of ANFO}) \tag{4.1}$$

Relative bulk strength (*RBS*) compared to ANFO is given by:

$$RBS = (RWS \times \text{Density})/(\text{Density of ANFO}) \qquad (4.2)$$

4.3 Laboratory Determination of Explosive Energy

A considerable amount of work is underway at several institutions for developing computer codes to arrive at reliable energy estimates by allowing for partial reaction and incomplete heat transfer between gas and condensed reaction products. There are many drawbacks in comparing explosives solely on the basis of calculated energy or strength values. A wide gap always exists between theoretical energy and actual performance of an explosive. The calculated energy does not depend on factors like particle size, degree of mixing, sensitivity and priming or confinement. Thus it is always better to measure the energy released by an explosion through suitable laboratory experiments like Ballistic Mortar, Trauzl Lead Block and Underwater tests. Absolute volume and weight strengths of some commercial explosives are given in Table 4.1.

Table 4.1 Properties of some commonly used explosives (Data collected from various manufacturers)

Explosive Brand	*Density (g/cc)*	*Velocity of Detonation (m/s)*	*Absolute Vol. Strength*	*Absolute Wt. Strength*
ANFO	0.90	4200	1139	911
Toeblast	1.15-1.25	3500-4500	1400-1500	x
Supergel	1.15-1.25	3200-4200	1250-1350	x
Energel	1.12-1.22	3000-4000	1000-1100	x
Powergel-I	1.20	4500	1100	850
Powergel-II	1.25	4800	1235	950
Kelvex Extra	1.18-1.25	4300-4800	960-1200	800-975
Apexgel-80	1.25	4000	1260-1298	1010-1100
Ecoblast	1.20-1.25	4000-4500	1450	1175
Indoprime	1.11	3900	1083	800
Ecogel-I	1.15-1.20	3000-4000	1050	925
Ecogel-II	1.20-1.25	3000-3500	1300	1025
Apex-80	0.80	3000	720	900
Aquadyne	1.15-1.25	3400-4400	1125-1225	x
OCG	1.40	6000	1600	1130
Superdyne	1.15-1.25	3000-4000	1400-1500	x
Permadyne	1.07-1.17	2800-3800	925-1000	x
Pentadyne	1.05-1.15	2800-3800	850-950	x
Powerflow-I	1.20	3800	1000	830
Kelvex-700	1.18-1.25	4300-4800	960-1200	800-973
Pentex	1.60	7000	2090	1310
Bharat Toe	1.15-1.28	3900-4100	960	880
Bharat Prime	1.15-1.28	3800-3900	880	760

4.3.1 Ballistic Mortar Test

The system for the test consists of a steel mortar, weighing about 360 kg, suspended in an aluminium frame of 3 m height. The mortar contains a firing chamber (295 cm^3) and a steel projectile weighing 18 kg (Figure 4.1). In the firing chamber 10 gm of test explosive is detonated with a standard blasting cap and the projectile is ejected. The recoil deflection of the ballistic pendulum is a measure of energy in the explosive; this is usually expressed relative to a standard explosive. Generally Blasting Gelatine (BG), a mixture of 92% NG and 8% Nitrocotton, is used as the standard explosive. Strength of the test explosive as a per cent of Blasting Gelatine (BG) is given by the following equation :

$$RWS = 100 \times (1 - \cos A)/(1 - \cos B) \qquad (4.3)$$

Where A is the angle of deflection with the test explosive and B is the angle of deflection with BG.

The principal drawbacks of the ballistic mortar test are the lack of confinement of the explosion during its initial phase, over-rating of explosives with high gas volumes because of the propulsive nature of the test, and the requirement of very small charge quantity.

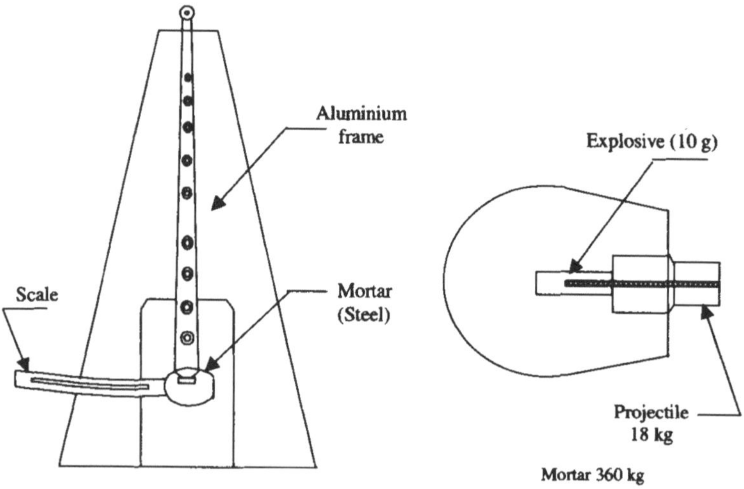

Fig. 4.1 Ballistic mortar mechanism

4.3.2 Trauzl Lead Block Test

Trauzl proposed the lead block test in 1903 for the determination of relative weight strength of explosives. In this test, the explosive is placed in an axial hole in a cast cylindrical block of pure lead. The block is of 200 mm diameter, 200 mm long and the hole is of 25 mm diameter, 125 mm deep. The test charge of 10 g primed with a detonator is placed in the hole and stemmed with dry sand. When fired, the chamber in the block is expanded. The change of volume of the cavity after detonation of the test explosive is a measure of explosive energy. For most sensitive explosives, an adequate correlation is found between ratings of explosives from this test and those of the ballistic mortar. Good correlation between results from Trauzl lead block test and the 'total' energy by underwater test of a few selected cap-sensitive explosives have been reported by several authors (Mohanty, 1981; Satyavratan and Vedam, 1980). Usually, relative weight strength is calculated with reference to a standard explosive like PETN or picric acid. Limitations of lead block test depend on two factors; namely, firing of small charge and high cost (75 kg lead block has to be cast preciously for each test).

4.3.3 Underwater Test

Underwater testing of explosives is well established and its applications as well as principles are well documented. In a typical underwater test, a spherical or short cylindrical charge weighing up to 5 kg is detonated at a relatively shallow depth (<10 m). The detonation front meets the explosive-water interface; this causes the surrounding water to be highly compressed and a high-intensity shock wave propagates outward. In the process, the water in the immediate vicinity is heated up very rapidly up to a few hundred degrees, and because of the lesser density and 'stiffness' of water, the amplitude of the shock pressure drops off very rapidly within a few charge radii.

While the pressure pulse propagates in an expanding spherical shell, the remaining high-temperature high-pressure explosion gas continues to expand at a relatively slower rate, resulting in a "bubble" of diminishing internal pressure and an outward flow of water behind the shock wave. Ultimately the internal pressure of the gas bubble falls below the equilibrium hydrostatic pressure due to inertial effect. At this point the process is reversed and the gas bubble begins to collapse. This process sets up an oscillation of the gas bubble, resulting in the release of a series of shock pressure pulses of diminishing amplitude. The time interval between the initial main shock and the first bubble pulse (i.e. 'bubble period') is a measure of the energy in the explosion gases; as it is for the maximum size of the bubble (Figure 4.2).

If detonation takes place underwater, the shock wave is transmitted to water and moves as a pressure discontinuity. This shock pressure pulse can be recorded by a suitable transducer by placing it at some distance away. Energy in the pressure pulse is given by the following equation.

$$E_S = \frac{(12.5 \times R^2)}{(D_o \times C_o \times W)} \int_o^{t_i} P^2 (t) dt \qquad (4.4)$$

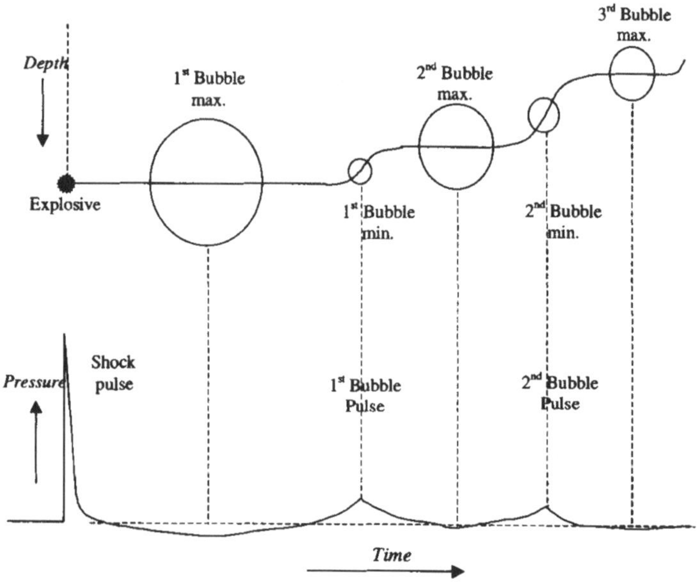

Fig. 4.2 Underwater explosion. Pressure-Time History

Where, R = distance between the shot and the transducer (m)
W = charge weight (kg)
D_o = density of water (g/cc)
C_o = velocity of sound in water (m/s)
$P(t)$ = pressure as a function of time
t_i = time constant of the falling pressure pulse (s)

t_i is calculated by dividing the peak pressure by $e = 2.718$ because up to one time constant, pressure falls exponentially.

The integral term can be calculated from the pressure-time profile recorded by the transducer. E_s calculated from equation (4.4) determines the value of shock energy away from the charge. Close to the charge, there is some loss of shock energy. The loss of shock energy depends on VOD and density of the explosive. The shock loss factor 'S_{lf}' can be calculated as follows.

$$S_{lf} = 1.542 + 6.552 \times 10^{-9} \times De \times V^2 \tag{4.5}$$

Where, De = Density of the explosive in g/cc and V = VOD in m/s.
When 'S_{lf}' is multiplied to E_s of equation (4.4), then the appropriate shock energy (E_{sh}) released by the explosive is obtained. Thus:

$$E_{sh} = S_{lf} \times E_s \tag{4.6}$$

Energy in the bubble or the explosion gases (E_b) can be calculated from the period of first oscillation, by the following equation:

$$E_b = (2.5 \times T^3 \times P_0^{2.5})/W \tag{4.7}$$

Where, T = period of first bubble oscillation (in seconds) and

P_o = atmospheric and hydrostatic pressure at the charge location (atm)

If a booster explosive having bubble energy Q_B and weight W_B is used to initiate the test explosive then:

$$E_{bl} = \frac{(2.5 \times T^3 \times P_0^{2.5}) - Q_B \times W_B}{W} \tag{4.8}$$

Where, W = weight of the test charge alone (kg).

Total energy released by the explosive is a simple simulation of E_{bl} and E_{sh} if the charge shale is spherical ($L/D < 2$). Table 4.2 summarizes the gas energy (Cal/g) of some commercial explosives tested at CMRI explosive laboratory. Separate calculations of gas and shock energies are shown in Tables 4.3 and 4.4 (Jagdish et al., 1993; Kogarko et al., 1975).

Table 4.2 Gas energy of commercial explosives

Explosive name	T_b *(ms)*	S_d	*W (g)*	E_{bl} *(Cal/g)*
NG based	128	2.449	100	498.82
	130			
	125			
	130			
	125			
NG based (P_5)	108	10.307	100	211.83
	100			
	105			
	90			
	80			
Slurry; Non-permitted	120	3.741	100	448.83
	125			
	130			
	125			
	120			
Slurry (P_1)	108	4.833	150	298
	110			
	115			
	110			
	108			
	100			
Slurry; Non-permitted	122	2.244	100	430.93
	120			
	125			
	120			
	125			
Slurry Al; Non-permitted	123		100	443
	125			
	125			
	125			
	120			

Where, T_b = Pulsation period in (ms) i.e. the time between first expansion and compression of bubble; S_d = Standard deviation; W = Weight of explosive in (g) and E_{bl} = Bubble energy (Cal/g).

Table 4.3 Gas and shock energies of some commercial explosives

Name of the explosive	E_{bl} *(Cal/g)*	E_{sh} *(Cal/g)*	*Total energy (Cal/g)*
Pentolite (50% TNT + 50% PETN)	490	234	724
TNT	505	215	720
Pentolite	475	469	944
OCG	495	408	903
Emulsion	460	370	830
ANFO	500	340	840
Heavy ANFO	470	360	830

Table 4.4 Shock and bubble energies of some explosives

Explosive Name	*Shock Energy, Joule/g*	*Gas Energy, Joule/g*
Pentolite	—	1840
MMAN	860	1840
TNT	1030	2060
Slurry (Al)	730	1560
NG-Based	—	1780

In order to achieve a good performance from any explosive, it is essential to match explosives for a particular rock. Division of explosive energy between shock and heave components is decided by the rock properties. From the principle of mechanics, if a large plate having a hole is pressurized from inside, then the pressure on the hole walls is given by the following equation (4.9).

$$P = \frac{0.5 \times E \times (V/V_0 - 1)}{(1 + \nu)} \tag{4.9}$$

Where, P = Pressure on the hole walls, GPa

V = Volume of the expanded hole, m^3

V_0 = Volume of the original hole, m^3

E = Young's modulus of the plate material, GPa

ν = Poisson's ratio

When it is the case of rock blasting, equation (4.9) represents the equation for confined rock. It's curve is a straight line in a *P-V* plot. When it is plotted on the curve of expanding explosion gases, the point of intersection of equation (4.9) with the explosion gas expansion curve will define the division of energy between shock and heave. Theoretically, work done by the explosive up to the point of intersection is called the shock energy. Work done by the gases in expanding beyond this point is called the heave or the gas energy. Depending on the desired results, we require more shock and hence choose explosives which would maximize the desired results.

4.4 THEORETICAL DETERMINATION OF ENERGY RELEASE

In energy release calculations, it is assumed that the explosive reactions are neither losing nor gaining any energy from outside. The heat energy is supposed to heat the explosive products. The reaction is assumed to be complete or ideal. The pressures, due to explosive detonation, are a direct function of the number of molecules and the temperature of the gases. Thus the explosive work potential is directly related to the amount of heat released (Q_e).

The heat released is the difference between the total heat of formation of the reaction products (Q_p) and the total heat of formation of the reactants (Q_r). Thus it can be said:

$$Q_e = Q_p - Q_r \tag{4.10}$$

The respective heat of formation of various compounds is given in Table 4.5 (Konya and Walter, 1990).

It is assumed that no heat is required to form any pure element and thus 'C', 'N', 'H', 'O', 'Al' etc., have a zero heat of formation. If heat is released during reaction, the compound is said to have a negative heat of formation (exothermic). If heat is added to the reaction to cause it to occur, the compound is said to have a positive heat of formation (endothermic). All explosive reactions must be exothermic.

A temperature of 298°K and a pressure of 760 mm (STP) are considered when listing heats of formation in data tables.

Table 4.5 Heats of formation for selected chemical compounds

Compound	*Formula*	*Mol. Weight*	Q_p *or* Q_r *(Kcal/Mole)*
Corundum	Al_2O_3	102.0	–399.1
Fuel Oil	CH_2	14.0	–7.0
Nitromethane	CH_3O_2N	61.0	–21.3
Nitroglycerine	$C_3H_5O_9N_3$	227.1	–82.7
PETN	$C_5H_8O_{12}N_4$	316.1	–123.0
TNT	$C_7H_5O_6N_3$	227.1	–13.0
Carbon monoxide	CO	28.0	–26.4
Carbon dioxide	CO_2	44.0	–94.1
Water	H_2O	18.0	–57.8
Ammonium nitrate	$N_2H_4O_3$	80.1	–87.3
Aluminium	AL	27.0	0.0
Carbon	C	12.0	0.0
Nitrogen	N	14.0	0.0
Nitrogen oxide	NO	30.0	+21.6
Nitrogen dioxide	NO_2	46.0	+ 8.1

4.5 CALCULATION OF HEAT OF EXPLOSION

Three different examples are cited from the work of Konya and Walter (1990) to show the methods of calculating heat of explosions.

4.5.1 Ammonium Nitrate and Fuel Oil

$3\ AN + 1\ FO \rightarrow CO_2 + 7\ H_2O + 3\ N_2$

$3\ AN + 1\ FO =$ Reactants

3 (–87.3 Kcal) + (–7 Kcal) = Heat of formation of reactants (Q_r)

$Q_r = -268.9$ Kcal

$$\text{Heat of explosion} = Q_e = Q_p - Q_r$$
$$= -498.7 \text{ Kcal} - (-268.9 \text{ Kcal})$$
$$= -229.8 \text{ Kcal}$$

Using molecular weight of compounds from Table 4.5,

$$3\ (AN) + 1\ (FO) = 3\ (80.1\ g) + 1\ (14\ g) = 254.3\ g$$

The following calculations give, Kilocalories per kilogram (Kcal/kg), which is:

$$(-229.8 \text{ Kcal}/254.3g) \times (1000g/kg) = -903.7 \text{ Kcal/kg}$$

Therefore, the heat of explosion = -903.7 Kcal/kg

4.5.2 Aluminized ANFO

$$AN + FO + AL \rightarrow CO_2 + H_2O + N_2 + AL_2O_3$$

Balanced Reaction:

$$4.5\ N_2H_4O_3 + CH_2 + AL \rightarrow 4.5\ N_2 + 10\ H_2O + CO_2 + 0.5\ AL_2O_3$$
$$4.5\ AN + FO + AL = \text{Reactants}$$

Heat of formation of reactants (Q_r)

$$= 4.5\ (-87.3 \text{ Kcal}) + (-7 \text{ Kcal}) + (0 \text{ Kcal})$$
$$= -399.85 \text{ Kcal}$$

Products = $4.5\ N_2 + 10\ H_2O + CO_2 + 0.5\ Al_2O_3$

Heat of formation of products (Q_r) = 4.5 (0 Kcal) + 10 (-57.8 Kcal) + (–94.1 Kcal) + 0.5 (–399.1 Kcal)

$$= -871.65 \text{ Kcal}$$

$$\text{Heat of explosion} = Q_e = Q_p - Q_r$$
$$= -871.65 \text{ Kcal} - (-399.85 \text{ Kcal})$$
$$= -471.8 \text{ Kcal}$$

Molecular weight of compound from Table 4.5:

$4.5\ (AN) + (CH_2) + (AL) = 4.5\ (80.1\ g) + 14\ g + 27\ g = 401.45\ g$

To calculate kilocalories per kilogram (Kcal/kg)—

$$(-471.8 \text{ Kcal}/401.45\text{g}) \times (1000 \text{ g/kg}) = -1175.24 \text{ Kcal/kg}$$

Therefore, heat of explosion = –1175.24 Kcal/kg

4.5.3 Underfuelled ANFO

$$6 \text{ AN} + \text{FO} \rightarrow CO_2 + H_2O + N_2 + NO + NO_2$$

$$6 \text{ } N_2H_4O_3 + CH_2 \rightarrow CO_2 + 13 \text{ } H_2O + 5 \text{ } N_2 + NO + NO_2$$

$$6 \text{ AN} + \text{FO} = \text{Reactants}$$

Heat of formation of reactants (Q_r) = 6 (–87.3 Kcal) + (–7 Kcal)
= – 530.8 Kcal

Products = $CO_2 + H_2O + N_2 + NO + NO_2$

Thus,

Q_p = Heat of formation of products
= –94.1 Kcal + 13 (–57.8 Kcal) + 5 (0 Kcal) + 21.6 Kcal + 8.1 Kcal
= – 815.8 Kcal

$Q_e = Q_p - Q_r$ = – 815.8 Kcal – (–530.8 Kcal) = – 285 Kcal

Molecular weight of compound = 6 (80.1 g) + 14 g = 494.6 g

Heat of explosion in Kilocalories per Kilogram (Kcal/kg)
= (–285 Kcal/494.6 g) × (1000 g/kg) = –576.2 Kcal/kg

Heat of explosion = –576.2 Kcal/kg.

The example cited in (4.5.2) considers the case of aluminized ANFO which is formed by adding a small amount of aluminium to an ammonium nitrate/fuel oil mixture. Powdered metals produce a high value for heat release and form solid oxides rather than gaseous products. The resulting gas pressure cannot be linearly related to heat release.

The example cited in (4.5.3) illustrates the reaction of ammonium nitrate and insufficient quantities of fuel oil, which would produce a positive oxygen balance.

4.6 IDEAL MIXTURE OF ANFO AND FUEL OIL

Proper and improper explosive energy utilization may be visualized in field by observing the nature of gases released just after detonation. When light-grey coloured steam is released, oxygen balance is near ideal and maximum energy is released. When gases are either yellow or rust coloured, they indicate an inefficient reaction, which may be due to an oxygen positive mixture. An oxygen negative mixture produces dark grey gases and often leaves carbon on borehole walls.

Whenever coloured fumes are seen just after detonation, it indicates that the explosive reaction was inefficient and some useful energy was lost. In this situation the blasthole may not have sufficient energy to break the rock properly in front of the hole. Blastholes are therefore underloaded for the amount of rock present. Such a situation may lead to airblast and flyrock and the blasthole may not break up to the desired depth.

At Sharda OCP of Sohagpur Area, SECL (South Eastern Coalfields Ltd.) the management had a long-standing problem of bigger-size boulders appearing from overburden blasting of the shovel face. The rock formation of the face was mixed with hard and soft sandstone but highly jointed. Geological mapping of the site confirmed that there were vertically oriented two major joint sets. The boulders were coming from the topmost 3 m of the deposit. To avoid such boulders, the management was giving pocket charges of nearly 3 kg at a depth of 1.5 m from the hole face. But there was no improvement. As a solution to this problem, CMRI conducted 10 trial blasts, one after another, at the said face and critically observed the post-blast fumes through zoom lens at a close distance.

Yellow and also dark-grey coloured gases were observed just after detonation in all such blasts, which indicated an inefficient explosive reaction. The reason was due to pocket charging which initiated detonation of explosive column from the top, resulting in premature opening of cracks on the top. The gas pressure energy was thus less effective causing formation of bigger size boulders. The problem was tackled by simply eliminating pocket charges and putting uniform explosive distribution in each hole (Pal Roy, 1995).

In the mixture of ammonium nitrate and fuel oil, it has been calculated that approximately 6% fuel oil mixture makes the optimum balance. Table 4.6 shows the theoretical energy and gas energy release in different combinations of ammonium nitrate and fuel oil mixtures. The calculation has been made using 'CHEF' software of the Precision Blasting Services, USA. It may be observed from Table 4.6 that as the oil content decreases from the desired 6% there is rapid energy loss. The extent and the rapidity of energy loss is more in less fuel oil mixture than that in the excess fuel oil mixture (Figure 4.3).

The strength of the explosive and the energy delivered in the field are not linearly related with the heat released and therefore, using the heat released chart alone as an indicator of strength or energy loss is misleading.

4.7 PARTITION OF EXPLOSIVE ENERGY

Recent researchers have provided evidence to the fact that there is a definite interaction between the explosive and the rock in the fragmentation process. Notable among them is the work of Udy and Lownds (1990). They analysed this interaction by modelling the partition of explosive energy to predict how various explosives, with ideal and non-ideal reaction behaviours, interact with various rock types (Figure 4.4).

Table 4.6 Theoretical and gas energy calculations for different mixture of ammonium nitrate and fuel oil

Ammonium Nitrate (%)	*Fuel Oil (%)*	*Theoretical energy (%)*	*Gas energy (%)*
99	1	58.23	57.45
98	2	71.93	64.27
97	3	80.66	73.37
96	4	88.16	83.04
95	5	95.15	92.94
94	6	99.74	99.87
93	7	98.86	99.58
92	8	97.15	98.67
91	9	94.61	95.80
90	10	91.24	86.72

Figure 4.4 shows the concept of energy partitioning. The Pressure-Volume (*P-V*) curve is based on the simplified assumption that the explosion in the rock is an adiabatic reaction. The kinetic component of the shock energy is considered essentially lost as 'useful work' in the form of crushed rock immediately around the borehole and as seismic waves propagating into the ground. The strain component of the shock energy produces the pressure pulse which causes fracture and cracking. The gas energy produces effective work in propagating the cracks and fragmenting the rock between pressures P_4 and P_5 on the *P-V* curve. The major part of the fragmentation work is done by the sum of energies in areas 2 and 3 below the *P-V* curve. For the non-ideal explosive, the velocity of detonation (VOD) is reduced which reduces the detonation pressure, which in turn affects the magnitude of P_3 and produces a different partition of the energy between shock and gas energies. Point P_4 corresponds to the equilibrium of borehole state. As the rock is compressed and crushed, the volume of the borehole increases with a corresponding decrease in pressure until the stress in the rock just balances the pressure. A simplified model developed by Udy and Lownds (1990) is used to simulate the explosive-rock interaction. Fragmentation and heave energies can also be estimated from the same model for various rock stiffness (having Young's modulus varying from 5 to 150 GPa), with ideal and non-ideal explosives. It is concluded that fragmentation energy decreases to a greater extent as rock stiffness is increased with a non-ideal explosive. Heave velocity is systematically higher for the non-ideal explosive; with a greater improvement rate as rock stiffness is increased.

The proposed theory on the partition of explosive energy adds support to the increased dependency of fragmentation efficiency on the properties of rock. Furthermore, this dependency will be, in all likelihood, more pronounced for non-homogeneous rock masses with discontinuities and flaws. Also, the proposed theory demonstrates the limited use of total available energy or available absolute strength values or relative weight strength, to provide a clear measure of an explosive for the fragmentation of a particular rock mass.

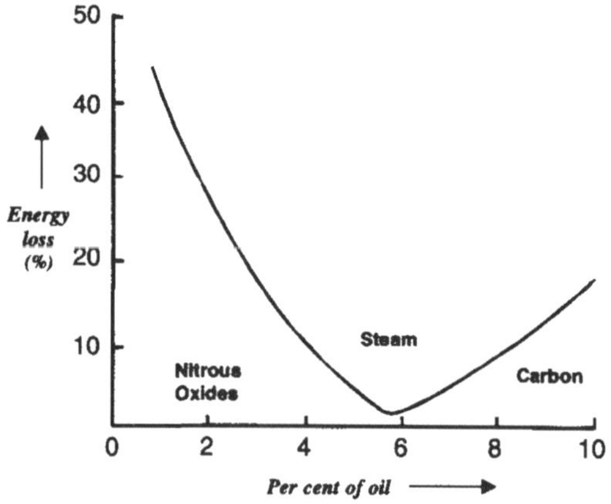

Fig. 4.3 Energy loss in ANFO and Fuel oil mixture

4.7.1 Comminution Theory

Fragmentation prediction models based on comminution theory and other empirical fragmentation models have achieved various degrees of success (Lizotte and Scoble, 1994). Da Gama (1983, 1984) formulated two basic models. The work is based on Bond's third theory of comminution, whereby the work spent to fragment a rock mass is proportional to the length of the new cracks created.

BOND's equation: $W = 10\ W_i\ [1/\text{sqrt.}\ (P_{80}) - 1/\text{sqrt.}\ (F_{80})]$ (4.11)

Where, W = Energy required for fragmentation (kWh/t)

W_i = Bond's Work Index

P_{80} = 80% passing size, Product, microns

F_{80} = 80% passing size, Feed, microns

Da Gama (1983): $P_s = a\ W^b\ (s/B)^c$ (4.12)

Where, P_s = Per cent cumulative undersize of the fraction size 's'

$$W = 10\ W_i\ [1/\text{sqrt.}(P_{80})] \quad (4.13)$$

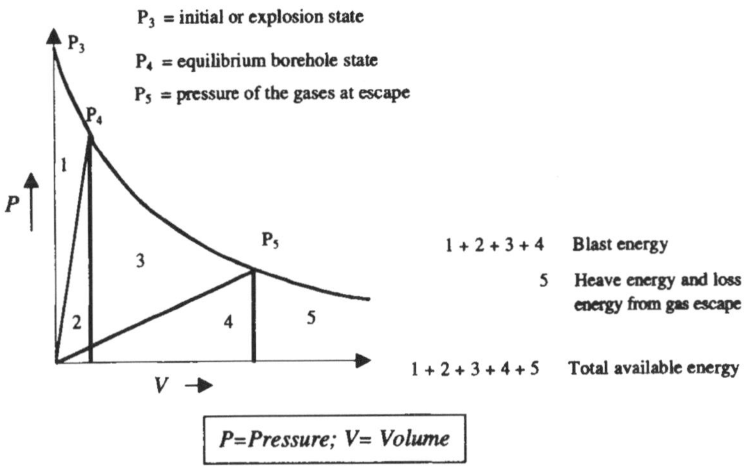

Fig. 4.4 The partition of explosive energy, Udy and Lownds (1990)

B = burden

a, b, c = Site specific empirical constants.

$$\text{Da Gama (1984): } Ps = a\, W^b\, (s/B)^c\, (1/F_{50}^d) \tag{4.14}$$

Where, F_{50} = Average joint spacing or inherent block size

d = Site-specific empirical constant.

REFERENCES

Bhushan, V. and Srihari, H. N. 1991. Explosive performance indices, IM and EJ, India, October, pp. 19–24.

Da Gama, C. D. 1983. Use of comminution theory to predict fragmentation of jointed rock masses subjected to blasting, Proc. 1st Intern. Symposium on Rock Fragmentation by Blasting, Lulea, Sweden, pp. 565–579.

Da Gama, C. D. 1984. Microcomputer simulation of rock blasting to predict fragmentation, Proc. 25th US Symposium on Rock Mechanics, AIME, USA, pp. 1018–1030.

Jagdish, Nabiullah, Md., Pingua, B. M. P. and Bagchi, A. 1993. Studies on shock and gas energy of commercial explosives using underwater blasting method, Proc. National Symposium on Advances in Drilling and Blasting V. R. Sastry (ed.), 15–16 October, pp. 110–115.

Kogarko, S. M., Popov, D. E. and Novikov, A. S. 1975. Combustion, explosion and shock waves, Translated from Russian, Vol. 11, pp. 648–657.

Konya, C. J. and Walter, E. J. 1990. Surface blast design, Prentice-Hall Int. Inc., USA, p. 303.

Lizotte, Yves C. and Scoble Malcom J. 1994. Geological control over fragmentation, CIM Bulletin, September, Canada, pp. 57–71.

Mohanty, B. 1981. Energy, strength and performance, and their implications in rating commercial explosives, Proc. 7th Conference on Explosives and Blasting Technique, Phoenix, Arizona, Intern. Society of Explosives Engineers, pp. 293–306.

Pal Roy, P. 1995. Mathematical modelling to evolve explicit relationships in the study of energy balance in rock blasting, EMR Report of Investigation, CMRI, Dhanbad, India, December, p. 117.

Satyavratan, P. V. and Vedam, R. 1980. Some aspects of underwater testing method, Propellants and Explosives, Vol. 5, pp. 62–66.

Udy, L. L. and Lownds, M. C. 1990. The partition of energy in blasting with non-ideal explosives, Proc. 3rd Intern. Symposium on Rock Fragmentation by Blasting, Brisbane, Australia, pp. 37–43.

5

Prediction, Control and Damage Thresholds of Ground Vibration from Opencast Blasting

5.1 INTRODUCTION

One of the undesirable side effects of rock blasting, i.e. producing structural damage and annoyance, is due to ground vibrations. Ever-increasing usage of commercial explosives in the mining and quarrying industry, coupled with the sprawl of urbanization and, in part, man's increased sensitivity to environmental disturbances are dimensions looming large in recent years. With expanded mining activity, the increase in open-pit mining and the necessity of large blast sites demand a balancing approach to ensure cost-effective rock excavation techniques.

Vibration, noise and flyrock due to blasting create great socioeconomic problems for the mine management as well as for the people residing in the vicinity of the mine. Pre-operational planning is the key to efficiency, economy and environmental safety. Modern opencast mines are in general larger and deeper than the earlier ones, giving rise to a higher percentage of overburden removal and a greater density of blasting in any area. Solutions to these problems call for an in-depth understanding of the basic parameters involved in the process of blasting, to ascertain the basic features of generation, propagation and prediction of ground vibration from opencast blasting.

When a certain amount of explosive detonates at a specific depth below the Earth's surface, approximately 20 to 30% of its energy is utilized in fragmenting the rock or other materials around. However, as an explosion is an imperfect use of energy, there is a loss of energy transmitted through the Earth in the form of pulsating waves or vibrations. A part of the energy is also dissipated in the air, which produces noise. Almost 90% of the earth-borne energy reaches the surface at a few metres from the total displacement area and acts as surface waves. The explosion-generated seismic waves are of two types, namely, longitudinal and shear waves, travelling at different velocities. When the seismic waves reach the surface of the Earth, they are reflected thereby exhibiting both a vertical and horizontal motions of the Earth's surface. The vertical vibration of the earth's surface produces an equivalent particle velocity in the air. The particle velocity in the air causes air vibration of the same frequency as the seismic waves.

5.2 GENERATION AND PROPAGATION OF GROUND VIBRATION FROM OPENCAST BLASTING

When some amount of explosive is detonated at certain depth in single or multiple drill holes, very rapid decomposition of the charge takes place, forming gases at very high temperature and pressure. A true shock wave is formed only when the initial explosive pressure far exceeds the strength of the rock in compression, so much so that any plastic state is by-passed and it can be said to behave hydrodynamically. Such an unstable shock wave rapidly passes through the non-elastic state, due to its instability and decreasing velocity, and settles into a stable quasi-elastic zone in which the oscillatory wave motion propagating at sonic velocity carries insufficient energy to permanently disturb the material in its path. This zone is known as the elastic or semi-elastic wave zone (Ghosh, 1983). The intensity of the shock wave attenuates very rapidly as a large amount of energy is consumed in crushing and producing cracks. Thus, in elastic or semi-elastic zones located away from the source, the intensity drops significantly and thus produces no permanent deformation. The remaining energy goes directly into the surrounding rock as seismic waves. The waves propagate elastically. Two groups of seismic waves are seen to be generated by detonation of explosive charges in drill holes. They are classified as body waves and surface waves. Body waves travel within a medium, while surface waves are restricted to travel along free interfaces, such as the ground surface.

Body waves comprise two discrete components—compression or *P*-wave and shear or *S*-wave. Two types of surface waves are usually produced from normal mine blasting—Rayleigh (*R*) waves and Love (*L*) waves. Other types of surface waves include: Coupled (*C*) waves, Hydrodynamic (*H*), Stoneley waves (*ST*), and Channel waves (Bath, 1979). Some American investigators (Keller, 1955; and others) have reported the presence of Coupled and Hydrodynamic waves in the seismographs of small explosive shots. These surface waves are not usually detected in mine blasts. Figure 5.1 shows the general form of a blast vibration time

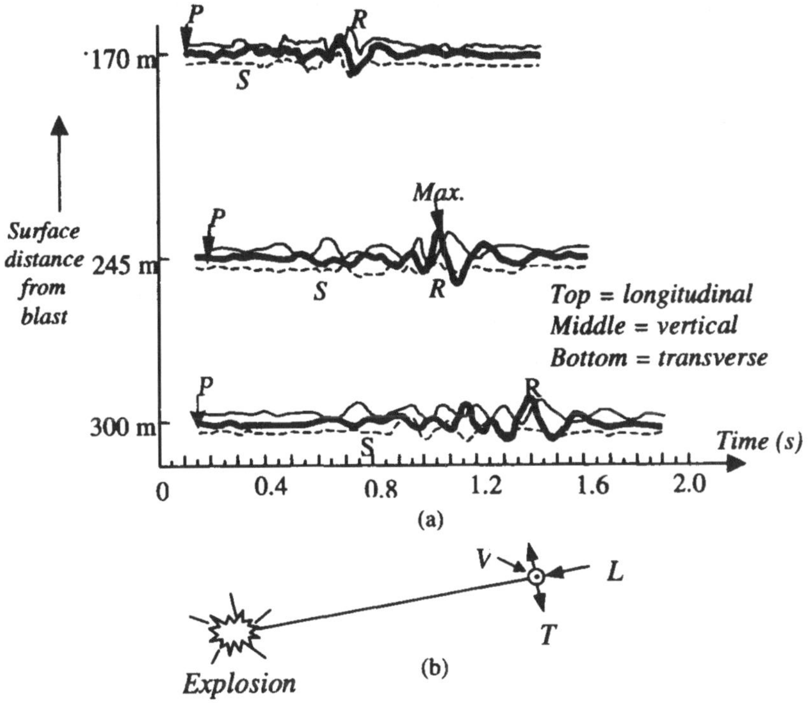

Fig. 5.1 General form of blast vibration time history records: (a) relative magnitudes and arrival times; P, compressive; S, shear and R, Rayleigh wave types (after Dowding, 1985)

history where *P, S* and *R* waves are clearly distinguishable (Dowding, 1985).

After the detonation of an explosive charge in drill holes, the various types of seismic waves propagate at their respective sonic velocities. The *P*-wave is the fastest, followed by the *S*-wave. The velocity of the Love wave would be less than that of the shear-wave velocity of the half-space, but more than the shear-wave velocity of the overlying layer. Rayleigh waves propagate at a velocity slightly less than the *S*-wave velocity. Unfortunately, most explosions are detonated as a series of smaller explosions, which are delayed by milliseconds, and differences in travel paths and delay times result in overlapping arrival of both wave fronts and wave types. This complicated arrival sequence has not been studied enough in blasting seismology.

At high frequencies, surface wave displacements penetrate only to shallow depths, whereas, at lower frequencies, these penetrate to intermediate or a deeper layer. While there is no theoretical limit to the penetration depth of any wavelength, from a practical standpoint, the energy is confined within approximately one wavelength from displaced surface. The amplitude of Rayleigh wave particle motion attains maximum at about one wavelength and decreases exponentially with depth below the surface. Hence, the influence of the material properties on the propagation of the Rayleigh wave is confined essentially to a depth of approximately one wavelength.

5.3 COMMONLY USED BLAST VIBRATION PREDICTORS

A number of investigators have studied ground vibrations from blasting and have developed theoretical analyses to explain the experimental data. If the energy released (E) is taken to be directly proportional to the weight (Q) of the explosive fired, an equation of the simple form $E \propto Q^{1/2}$ is observed.

Early studies on wave propagation phenomena were conducted by Morris (1950). He propounded that the amplitude (A) of particle displacement is proportional to the square root of the weight of the charge (Q) and inversely proportional to the distance (D) from the blast. That is:

$$A = K\frac{Q^{0.5}}{D} \tag{5.1}$$

Where, K is the site constant.

Habberjam and Whetton (1952) suggested a higher power for the charge weight in their formula.

$$A \propto Q^{0.85} \tag{5.2}$$

Subsequent investigators (Duvall and Petkof, 1959; Attewell, 1964; Nicholls et al., 1971; Siskind et al., 1980; and Daemen et al., 1983) have proposed further modifications of the propagation law.

Assuming cylindrical explosive geometry for long cylindrical charges, Daemen et al. (1983); Duvall and Petkof (1959); Duvall and Fogelson (1962); Duvall et al. (1963) concluded that any linear dimension should scale with the square root of the charge weight. Blasts should be scaled to the equivalent distance or scaled distance, which is defined as the actual distance (D) divided by the square root of the charge weight (Q). The corresponding relationship assumes the form.

$$V = K(D/Q^{0.5})^{-B} \tag{5.3}$$

Where B is the slope of the best-fit straight line of the V(peak particle velocity) versus $D/Q^{0.5}$ plot on a log-log scale and K is the intercept on the particle velocity axis when $D/Q^{0.5} = 1$.

Langefors et al. (1958) suggested the following relationship for various charging levels ($Q/D^{3/2}$) to estimate peak particle velocity. The equation is:

$$V = K\left(\sqrt{Q/D^{3/2}}\right)^{B} \tag{5.4}$$

For spherical symmetry, the USBM investigators suggested that any linear dimension should be scaled to the cube root of the charge size. Thus, a scaled distance is the actual distance divided by the cube root of the charge weight. Later investigators, particularly Ambraseys and Hendron (1968) also obtained similar results. An inverse power law was suggested to relate amplitude of seismic waves and scaled distance to obtain the following relationship:

$$V = K(D/Q^{1/3})^{-B} \tag{5.5}$$

The empirical relationship suggested by the Indian Standard (1973) uses the concept in which blast is scaled to the equivalent distance or scaled distance, defined as the charge weight divided by the cube root of the square of actual distance. The relation is expressed as—

$$V = K(Q^{2/3}/D)^{B} \tag{5.6}$$

Davies et al. (1964); Birch and Chaffer (1983); Attewell (1964); Daemen et al. (1983) considered no particular charge symmetry and used a general equation instead:

$$V = KD^{-B}Q^{A} \tag{5.7}$$

Where K, A and B are empirical constants which can be determined by multiple regression analysis of two independent variables (see Chapter-1).

It is to be noted that after the invention of delay detonators (or millisecond relays), the weight of the charge (Q) as given in equations (5.1) to (5.7) became "maximum charge per delay fired in a round of blast".

5.4 GEOMETRICAL SPREADING

Seismic waves propagate away from its source. Thus, as the distance from the source of vibration increases, more and more volume of rock is encountered (Figure 5.2). This results in decreasing energy density with respect to distance. This remains, till there is no other source of energy. Hence there is decay in the amplitude of vibration, which is termed as geometrical spreading or geometrical damping.

In an ideal elastic rock mass, the attenuation of amplitude for different types of waves is given below (Rinehart et al., 1961).

(a) Body waves propagating along the surface, amplitude $\propto R^{-2}$
(b) Body waves propagating through the medium, amplitude $\propto R^{-1}$
(c) Rayleigh waves, amplitude $\propto R^{-0.5}$; R being the distance from the source.

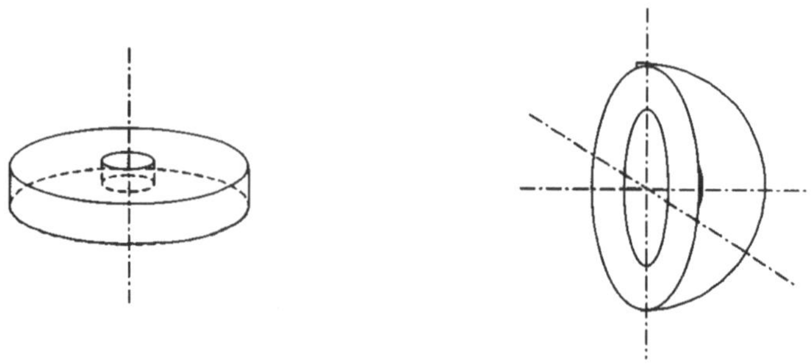

Fig. 5.2 Cylindrical and spherical expansion of volumes

5.5 INELASTIC ATTENUATION OR DAMPING

The actual decay in amplitude of vibration with respect to distance is more than what has been explained due to geometrical spreading. This extra attenuation is due to inelastic nature of the rock and is termed as inelastic attenuation (Bath, 1979).

In a multiple grained rock mass, the available surface area for dissipation of energy is more than that of a single crystal. This causes faster attenuation. Frictional forces are developed due to possible relative motion of grains during wave propagation. This may also cause attenuation, through energy dissipation. Although normally seismic waves do not have sufficient energy to cause relative displacement between the crack surfaces, yet there is some attenuation. This is termed as intrinsic attenuation (Walsh, 1966).

The above mechanism, including dissipation due to relative motion at grain boundaries and across surfaces, is termed as matrix inelasticity. The other inelastic attenuation mechanism includes partial saturation effects such as gas pocket squeezing, enhanced inter crack flow, stress induced diffusion of absorbed volatile, and phase changes in the material. It is also a well-known observation that higher frequency waves undergo greater attenuation than lower frequency waves. The higher frequencies are absorbed mainly because of the inelastic character of the rock mass.

5.6 INELASTIC ATTENUATION FACTOR

Numerous inelastic effects cause energy losses during wave propagation. The inelastic effect leads to a decrease in amplitude in addition to those due to geometrical spreading. There are several causes for inelastic attenuation, some of which are:

(i) Attenuation due to fluid flow, including relaxation, because of shear motions at pore-fluid boundaries;
(ii) Partial saturation effects, such as gas-pockets squeezing;

(iii) Enhanced inter-crack flow;
(iv) Energy absorbed in a system undergoing phase changes;
(v) Large category of geometrical effects, including geological discontinuities, scattering of small pores, large irregularities and selective reflection from thin beds.

Ghosh and Daemen (1983) reformulated the propagation equation of USBM and Ambraseys and Hendron (1968) by incorporating the inelastic attenuation factor $e^{-\alpha D}$. The modified equations are:

$$V = K(D/Q^{0.5})^{-B} e^{-\alpha D} \tag{5.8}$$

and

$$V = K(D/Q^{1/3})^{-B} e^{-\alpha D} \tag{5.9}$$

where K, B and α are empirical constants; α is called the inelastic attenuation factor.

The effect of the inelastic attenuation factor of the Langefors et al. (1958) and Indian Standard (1973) equations has been reported in the work of Pal Roy (1991).

5.7 VALUES OF EMPIRICAL CONSTANTS FOR VARIOUS ROCK MASSES

To standardize the values of site constants for various rock masses and to assess the validity of a particular empirical model, the blasting research group of the Central Mining Research Institute (CMRI), India conducted investigations on different types of exposed rock masses, which included limestone (fissured and highly jointed), granite (hard and fresh), iron ore, coal, dolomite, basalt, sandstone (weathered) and sandstone-alluvium (Pal Roy, 1991). The values of site constants as obtained from equations (5.3) to (5.9) are given in Tables 5.1 to 5.8.

It is important to note that if no assumptions are made about the joint distribution of the concerned random variables, the validity of the prediction and of the estimates of site constants cannot be judged.

Table 5.1 Site constants and index of determination (I_d) of different predictor equations for LIMESTONE

Equation number	B	K	A	α	I_d
(5.3)	1.95481	3352.63	x	x	.898
(5.4)	2.41363	534.564	x	x	.747
(5.5)	1.83095	10520.3	x	x	.909
(5.6)	1.81022	534.562	x	x	.747
(5.7)	1.8965	7890.32	0.738705	x	.914
(5.8)	1.01077	543.730	x	0.003330	.948
(5.9)	1.05301	1268.53	x	0.002785	.931

Table 5.2 Site constants and index of determination of different predictor equations for GRANITE

Equation number	*B*	*K*	*A*	*α*	I_d
(5.3)	1.53311	574.188	x	x	.839
(5.4)	1.80693	255.812	x	x	.859
(5.5)	1.69131	1265.09	x	x	.779
(5.6)	1.35518	255.796	x	x	.859
(5.7)	1.38351	186.754	0.945924	x	.861
(5.8)	1.76428	792.162	x	–0.009301	.889
(5.9)	2.49444	6068.20	x	0.025753	.827

Table 5.3 Site constants and index of determination of different predictor equations for IRON ORE

Equation number	*B*	*K*	*A*	*α*	I_d
(5.3)	1.80138	303.736	x	x	.761
(5.4)	2.50391	30.0096	x	x	.748
(5.5)	1.72116	2471.13	x	x	.769
(5.6)	1.87793	30.0096	x	x	.748
(5.7)	1.58818	495.221	0.091305	x	.773
(5.8)	1.18067	151.538	x	0.000778	.789
(5.9)	1.19988	701.265	x	0.000672	.786

Table 5.4 Site constants and index of determination of different predictor equations for COAL

Equation number	*B*	*K*	*A*	*α*	I_d
(5.3)	1.40761	266.523	x	x	.803
(5.4)	1.75350	90.2828	x	x	.816
(5.5)	1.49335	864.066	x	x	.777
(5.6)	1.31512	90.2830	x	x	.816
(5.7)	1.21122	30.4771	1.03659	x	.820
(5.8)	2.40458	1164.340		–0.010885	.868
(5.9)	3.01538	25525.2	x	–0.015104	.871

Table 5.5 Site constants and index of determination of different predictor equations for DOLOMITE

Equation number	*B*	*K*	*A*	*α*	I_d
(5.3)	1.50848	174.244	x	x	.769
(5.4)	1.99050	112.402	x	x	.762
(5.5)	1.48994	252.811	x	x	.760
(5.6)	1.49287	112.400	x	x	.762
(5.7)	1.50851	170.523	0.767431	x	.769
(5.8)	1.00745	111.203	x	0.243565	.793
(5.9)	1.00955	142.173	x	0.022706	.776

Table 5.6 Site constants and index of determination of different predictor equations for BASALT

Equation number	B	K	A	α	I_d
(5.3)	2.152	895.644	x	x	.903
(5.4)	2.652	176.137	x	x	.921
(5.5)	2.260	4765.470	x	x	.926
(5.6)	1.989	176.137	x	x	.921
(5.7)	2.171	1137.66	1.0342	x	.936
(5.8)	1.739	587.679	x	0.00733	.944
(5.9)	2.456	6745.21	x	0.00318	.926

Table 5.7 Site constants and index of determination of different predictor equations for SANDSTONE

Equation number	B	K	A	α	I_d
(5.3)	.82588	50.4234	x	x	.628
(5.4)	1.02402	20.0686	x	x	.633
(5.5)	.886030	144.404	x	x	.616
(5.6)	.768012	20.0686	x	x	.633
(5.7)	.708569	8.34253	0.603169	x	.635
(5.8)	1.36410	59.0202	x	–0.004618	.705
(5.9)	1.61073	426.746	x	–0.005631	.716

Table 5.8 Site constants and index of determination of different predictor equations for SANDSTONE-ALLUVIUM

Equation number	B	K	A	α	I_d
(5.3)	1.98017	1483.48	x	x	.939
(5.4)	2.52473	310.588	x	x	.893
(5.5)	2.03149	7243.80	x	x	.969
(5.6)	1.89354	310.584	x	x	.893
(5.7)	2.04178	35014.6	0.325227	x	.980
(5.8)	1.23200	697.408	x	0.003048	.960
(5.9)	1.80014	4796.41	x	0.010274	.970

Vibration data may be considered to follow approximately normal distribution, 60% above and below the mean vibration level (Pal Roy and Gupta, 1989). In the case of normal distribution, the square of correlation coefficient will determine the strength of the regression equation. If there are more than one independent variables in the model as in equations (5.7) to (5.9), then in the calculation of correlation coefficient between two variables, there will be an effect of the third variable. This effect can be avoided, if the partial correlation coefficients are considered (see Chapter 1). The partial correlation coefficients measure only the effect of the specified variables while ignoring the influence of other variables. There may be possibility for a positive simple correlation coefficient to be transformed into a negative partial correlation coefficient.

Inelastic attenuation of elastic waves is a characteristic that could be applied for the study of the geo-technical properties of rocks and to predict the change in the shape of a plane stress wave while passing through a rock mass. Numerous inelastic effects cause energy losses during wave propagation in rocks. All such effects cannot be encountered, in detail, in a sample closed-form expression. Thus, for better predictability it is necessary to search for dominant aspects of pulse initiation and propagation.

Consideration of inelastic material damping by means of an explicit exponentially decaying function is a dominant factor in the vibration predictor. The value of α in equations (5.8) and (5.9) should be positive. Therefore, any negative value of α, as obtained in Tables 5.2, 5.4 and 5.7, indicates that the corresponding predictor equation does not fit the data and therefore should not be considered for any calculation purpose (Figure 5.3).

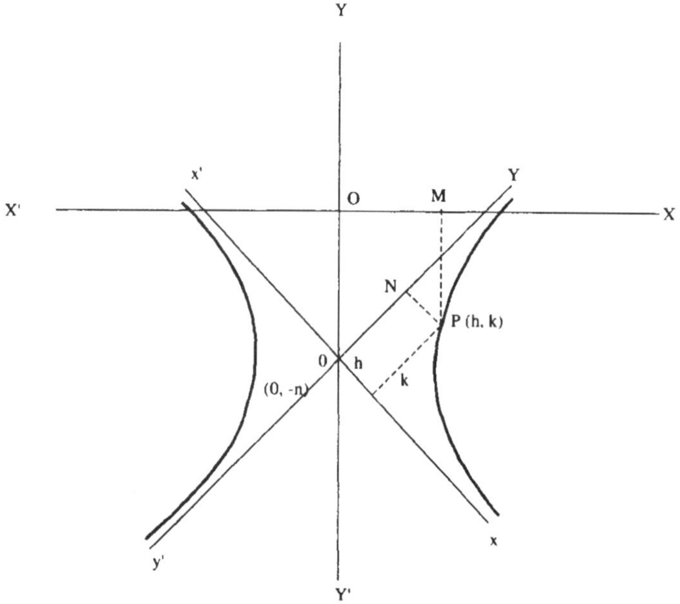

Fig. 5.3 Geometrical representation of CMRI equation

Several causes of inelastic attenuation having different degrees of influence—include matrix inelasticity, attenuation due to fluid flow, dissipation in a fluid saturated rock, shearing flow of fluid layer, partial saturation effects, enhanced intercrack flow, stress-induced diffusion, energy absorption during phase changes and geometrical effects. Many investigators have considered internal friction as a factor causing attenuation and have established that it varies directly with attenuation. Although acoustic velocities and attenuation have been determined extensively for metals and alloys, little systematic work has been done on rocks. In some cases the attenuation coefficient was found to vary linearly, whereas, in others with the second power of frequency.

Ground vibration induced by blasting is a complex mixture of various wave types. It is thus desirable to consider the characteristic of geometrical spreading for various types of waves likely to be associated with blasting. In field measurements, the amplitude decay due to geometrical spreading is usually taken as $1/D^B$, where D is the distance of the measuring transducer from the source and B is a site constant depending on specific site data. For a homogeneous isotropic linearly elastic medium, the amplitude decay due to geometrical spreading for the dominant wave types may be specified by the following three values of B:

(i) $B = 1$ for body waves within a (semi-) infinite medium
(ii) $B = 0.5$ for Rayleigh waves
(iii) $B = 2$ for body waves travelling along a free surface

In addition to geometrical spreading, numerous inelastic effects cause energy losses during wave propagation through a rock mass.

5.8 CMRI PREDICTOR EQUATION

The predictor model developed by the Central Mining Research Institute, India assumes special significance because of its simplified form and the consideration of the zone of disturbance due to blasting (Pal Roy, 1993). The equation is valid only in the zone of disturbance (i.e. when $Q_{max} > 0$, $V > 0$). At the boundary of this zone, and outside this boundary, V is obviously 0 everywhere. When seismic waves propagate through a rock mass, they encounter an increasing volume of rock mass, causing decrease in energy density. In the process of this diminishing energy due to geometrical spreading, they are simultaneously affected by numerous inelastic effects that also cause energy loss during wave propagation and subsequently amplitude is decreased. This equation considers two distinct categories of parameters n and K, where:

n is related to the category of parameters that are influenced by rock properties and geometrical discontinuities, and K is related to design parameters including charge weight, distance from the source, charge diameter, burden, spacing, sub-grade drilling, stemming length and delay interval.

The equation is of the form:

$$V = n + K(D/Q^{0.5})^{-1} \tag{5.10}$$

As n is categorized as a damping parameter that is influenced by rock properties and geometrical discontinuities, in practical situations the value of n is always negative. Putting $n = -n_1$ in equation (5.10), it follows:

$$V = -n_1 + K(D/Q^{0.5})^{-1} \quad (5.11)$$

At the boundary of the zone of disturbance, $V = 0$, and hence from equation (5.11), it follows—

$$D = K(Q^{0.5}/n_1) \quad (5.12)$$

When equation (5.11) is empirically derived at certain confidence level then K and n_1 are known and thus D is known for any amount of maximum charge per delay (Q). It is important to note that equation (5.11) is an empirical equation and thus the zone of disturbance as obtained from equation (5.12) is an empirical estimation that would match with the practical situation at certain confidence level.

5.8.1 Physical and Geometrical Interpretations of CMRI Predictor Equation

The blast vibration model developed by the CMRI simultaneously takes into account the decay of amplitude due to geometrical spreading and inelastic attenuation. The equation always considers a zone of disturbance in terms of scaled distance beyond which vibration will always be zero everywhere. Physical and geometrical interpretations can be presented to substantiate its phenomenological characteristics.

The CMRI predictive model has a special significance because of its simplified form and its consideration of the zone of disturbance due to blasting. When seismic waves propagate through a rock mass, they encounter an increasing volume of rock mass causing a decrease in energy density due to geometrical spreading. As they suffer this diminishing of energy density due to geometrical spreading, they are simultaneously affected by numerous inelastic effects that also cause energy loss during wave propagation, and amplitude is decreased. The decrease in amplitude is obviously non-linear. This may be written as:

$$(V - n) \propto (D/Q^{0.5})^{-1} \quad (5.13)$$

or

$$V = n + K(D/Q^{0.5})^{-1} \quad (5.14)$$

Putting x and y for the following values

$$V = y \text{ and } (D/Q^{0.5}) = x \quad (5.15)$$

Equation (5.14) reduces to

$$x(y - n) = K \quad (5.16)$$

Equation (5.16) is a rectangular hyperbola with its asymptotes intersecting at $(0, n)$ when n is positive, or $(0, -n)$ when n is negative (Figure 5.3). In other words, it can be said to represent a hyperbola when its asymptotes are co-ordinate axes and they intersect at the point (0, n) or (0, $-n$). The value of n actually determines the extent of attenuation or damping for increasing distances due to rock inelasticity and inhomogeneity. In any case study, if n is positive, the equation will not be valid in that case, as principally there should not be any amplification of ground vibration due to rock inelasticity and inhomogeneity.

When $V = 0$ and n is negative (say $n = -n_1$), then from the above equation (5.14), we have:

$$(D/Q^{0.5}) = \frac{K}{n_1} \tag{5.17}$$

For a particular site, when K and n_1 are known, the limiting scale distance can be determined by using a particular value of Q, to determine the zone of disturbance. Outside this zone, the value of V is 0 everywhere. This corroborates with the findings in practice where, whatever be the amount of charge, a finite zone of disturbance is observed beyond which there is no vibration effect.

Equation (5.10) holds for cylindrical explosive geometry. In case of spherical explosive geometry, the equation takes in to the form:

$$V = n + K(D/Q^{1/3})^{-1} \tag{5.18}$$

The charge weights from different predictor equations are compared in figures 5.4 and 5.5. The Ghosh-Daemen model (1983) shows some inconsistency in respect of charge per delay calculations.

Table 5.9 Site constants and Index of determination of CMRI predictor equation (5.10) for different *in situ* rock mass

Rock mass	n	K	I_d
Limestone	–11.0496	489.478	.887
Granite	–.838475	132.063	.860
Iron Ore	–2.09833	98.3577	.911
Coal	–11.2910	277.186	.703
Dolomite	–21.1184	339.377	.945
Basalt	–10.1452	183.815	.923
Sandstone	–18.6286	300.305	.814
Sandstone-Alluvium	–27.0245	456.306	.950

The attenuation relation is not solely a site property. Although it depends on geology, it also heavily depends upon the blast geometry. The general form of predictor equation (equation 5.7) requires regression analysis of two independent variables and thus is not easily applicable in engineering practice. Moreover, it gives lower charges as compared to other equations. The Indian Standard (1973) and Langefors-Kihlstrom (1958) equations are the same, differing only in form.

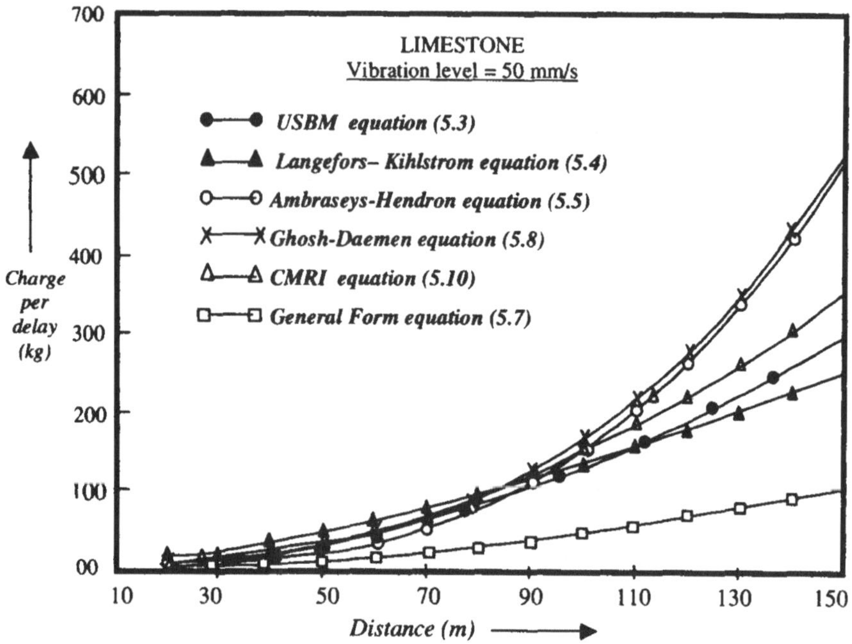

Fig. 5.4 Variation of charge per delay with distance for limestone

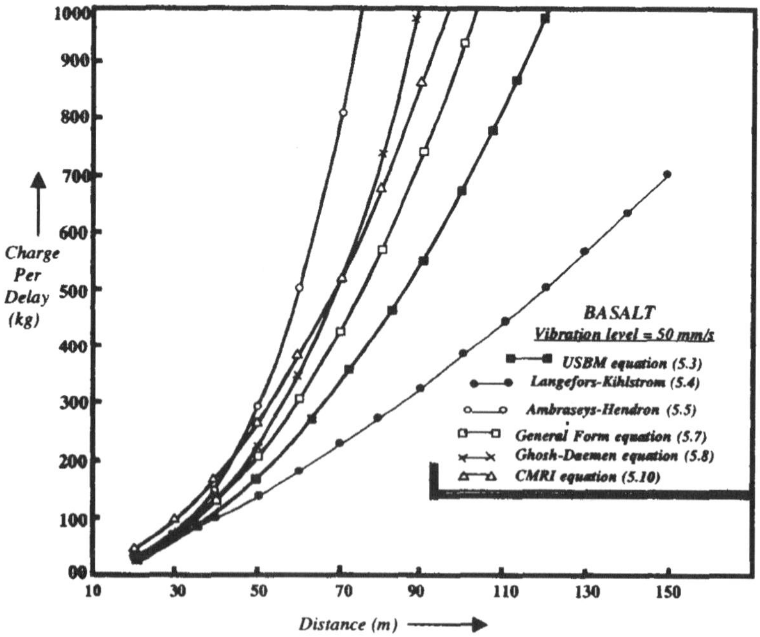

Fig. 5.5 Variation of charge per delay with distance for basalt

The CMRI equation (5.10) involves a very simple charge per delay calculation at any specific distance and vibration level. Figure 5.6 shows the non-linear decay and decrease of particle velocity due to inelastic attenuation whereas figure 5.7 shows the impact of negative value of α in equations (5.8) and (5.9).

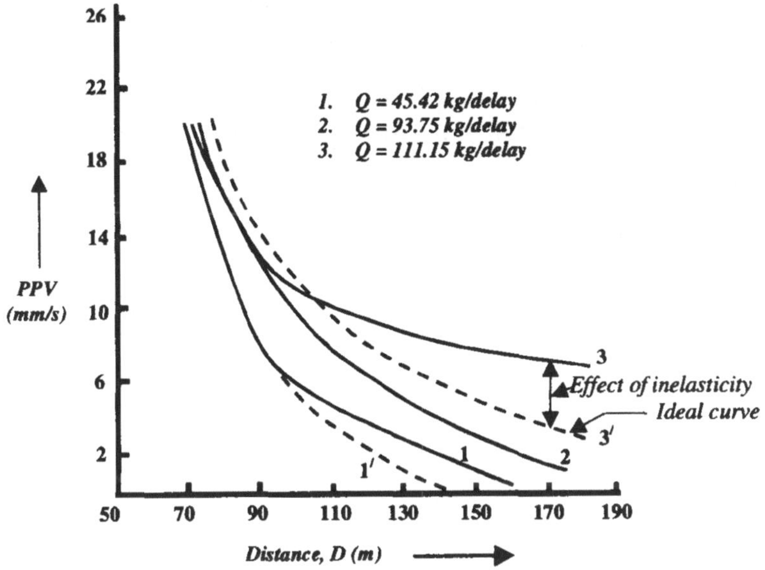

Fig. 5.6 Peak particle velocity as a function of distance at different charges showing the non-linear decay and decrease in inelastic attenuation for increase in charge per delay (data collected from South Jharia Opencast Project, BCCL)

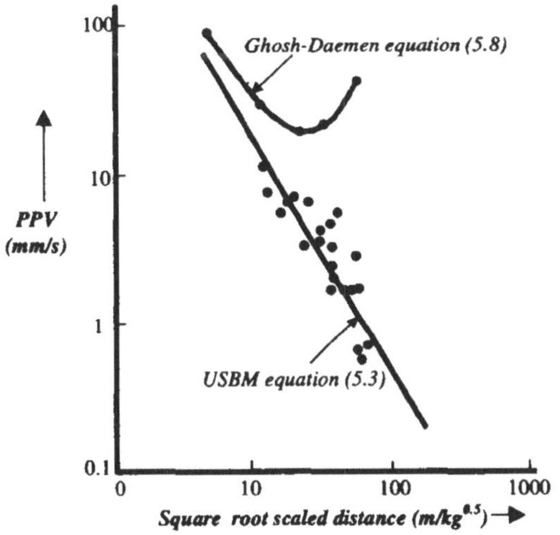

Fig. 5.7 Peak particle velocity as a function of square root scaled distance for Q=75 kg (North-Koel Hydroelectric Project)

5.9 MIXED ANALYTICAL-EMPIRICAL MODEL

Using the concepts of rock breakage and theory of reflection of seismic waves at a free surface, CMRI has developed a mixed analytical-empirical model for the prediction of blast-induced ground vibrations (Pal Roy and Dhar, 1992). This model was developed in two parts. The first part evaluates, analytically, the vertical particle velocity on the surface at any distance from the blast hole. This part along with other parameters incorporates the peak charge pressure on the wall of the cavity. The expression of peak charge pressure was derived empirically using jointly the Langefors-Kihlstrom (1958) and Pearse (1955) equations. The model accounts for the characteristics of the explosives, the rock mass properties and charge loading parameters. Excellent correlation was observed when the results obtained from the model were compared to those in actual field measurements.

The equations are:

$$V_V = \frac{2K_1 P_D r_1^2 l (r_1/r_2)^{-\Gamma\alpha}}{\rho_1 c_1 (l^2 + x^2)^{3/2}} \tag{5.19}$$

$$K_1 = \frac{V_D \rho_2}{\rho_1 c_1} \tag{5.20}$$

$$P_D = \frac{1}{2}\sigma_t \left[\frac{10^3 (Q/K_3)^{1/3}}{K_2 r_2}\right]^2 \tag{5.21}$$

Equations (5.19), (5.20) and (5.21) are dimensionally balanced. When the blasthole contains more than one explosive then V_D and ρ_2 may be taken as the simple average of their respective values.

Where,

V_V = peak particle velocity (mm/s)
K_1 = (characteristic impedance of explosive)/(characteristic impedance of rock)
ρ_1 = density of rock (g/cm^3)
c_1 = P-wave velocity in rock (m/s)
ρ_2 = density of explosive (g/cm^3)
V_D = velocity of detonation of explosive (m/s)
P_D = peak charge pressure (MPa)
r_2 = radius of charge (mm)
l = depth of centre of explosive column from the surface (m)
x = distance of the measuring transducer from the blasthole (m)
Q = weight of explosive in a blasthole (kg)
σ_t = tensile strength of rock (MPa)
r_1 = radius of blasthole (mm)

K_2 = proportionality constant = 1.0 for most rock
K_3 = factor relating to the lifting of one m^3 rock mass by 0.8 kg/m^3 explosive (for most rock)
Γ = charge symmetry parameter (2 for cylindrical charge and 3 for spherical charge)
α = an explosive constant of parametric value lying between 1.2 and 1.5.

In order to validate the proposed model from field measurements, single-hole blasting was carried out in four separate mines of BCCL (Bharat Coking Coal Ltd.), IISCO (Indian Iron and Steel Co. Ltd.), SECL (South-Eastern Coalfields Ltd.) and SAIL (Steel Authority of India Ltd.). Values of peak particle velocities monitored by seismographs were then compared with the results obtained from equation (5.19) with equations (5.20) and (5.21). Parameters relating to rock mass and explosive properties were obtained in the laboratory through samples collected from the field. The results of observations are shown in Table 5.10. The values of α, K_2 and K_3 were mostly standardized to 1.5, 1.0 and 0.8 kg/m^3 respectively. The value of Γ was taken as 3.

5.10 CONFIDENCE ENVELOPES

Any least square regression equation for vibration prediction indicates that there would be 50% probability of actual PPV being higher than the predicted value. In order to improve the confidence level, the regression line is moved up parallel (in log-log scale). If the line is moved up by one standard error, then 84% of the actual values would be below the line. Similarly, moving the line up by 1.65 times the standard error, the confidence level would enhance to 95% (Figure 5.8). These concepts have been used in most of the available software packages like "VIBCON" of PBS, USA; "BLASTWARE" of Instanel Inc., Canada and "VIBAN" of CMRI, India to determine different confidence envelopes. However, in practice, mainly the mean confidence level (i.e. 50% probability) is taken into consideration.

5.11 THRESHOLD VALUES OF GROUND VIBRATIONS FOR STRUCTURES

Different countries have set their own standards on the basis of their extensive field investigations carried out in their mines for several years. A particular standard derived by a particular country may not be universally applicable to houses and other buildings of other countries because of differences in structural conditions and workmanship. As such, it is always desirable to have one's own standards for classifying structural

Table 5.10 Comparison of predicted and measured peak particle velocities from single hole blasting

Name of the Mine	Rock type	ρ_1 (g/cc)	c_1 (m/s)	σ_t (MPa)	ρ_2 (g/cc)	V_D (m/s)	r_1 (mm)	r_2 (mm)	Q (kg)	x (m)	PPV (mm/s) Measured	PPV (mm/s) Calculated
Ramnagar OCP, IISCO	Medium grained sandstone, massive compact and slightly ferruginous	2.30	2900	4.0	1.19	3700	76.5	62.5	31.25	37	3.5	4.0
Belpahar OCP, MCL	Moderately compact, medium grained sandstone	2.21	2500	3.29	1.25	3800	125	62.5	200	147	2.0	1.39
										310	0.2	0.15
Ramkanali OCP, BCCL	Massive medium grained micaceous white sandstone	2.53	3300	5.16	1.11	3900	76.5	62.5	100	25	138.7	149.8
										50	26.02	20.92
										75	7.51	6.33
Kiriburu Iron Ore Mine, SAIL	Iron Ore (Hematite)	3.72	5600	10.7	1.15	3900	76.5	62.5	150	50	22.40	18.78
										100	4.21	2.44
										125	3.21	1.26

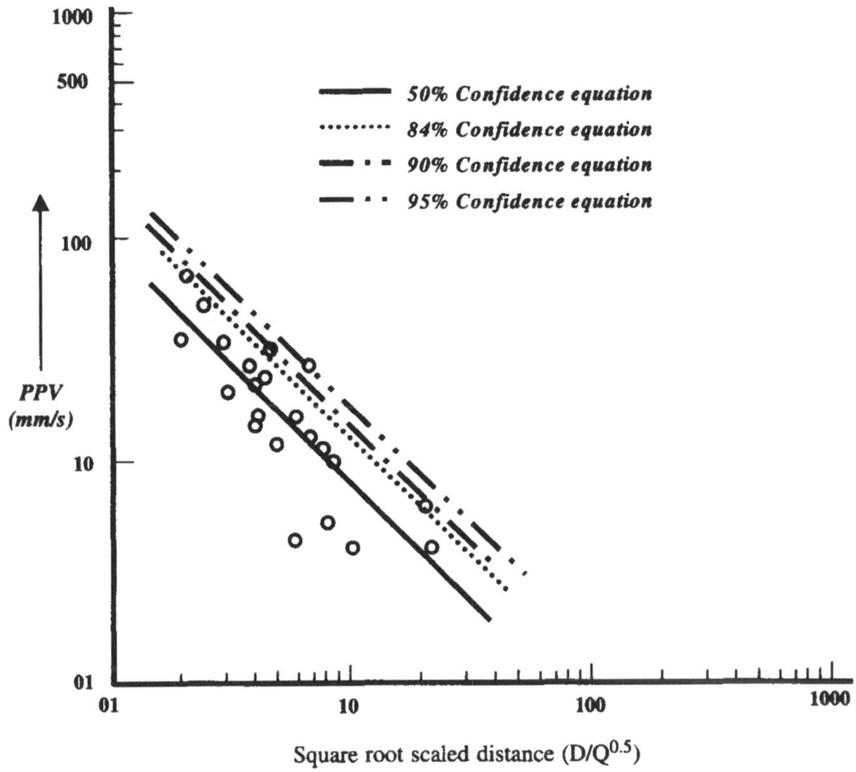

Fig. 5.8 Plot of PPV against frequency at various confidence levels

damage in the vicinity of mining activities to avoid detrimental effects on structures. There are a range of standards available world-over based on various aspects of ground vibrations e.g. amplitude, peak particle velocity, frequency, acceleration, etc. These parameters are used either as a single criterion or in combination; sometimes frequency is combined with amplitude and velocity.

5.11.1 Existing Vibration Standards and Criteria to Prevent Damage

An excellent compilation of different criteria is available in the Ph.D. thesis of Singh (1998). Peak particle velocity has been traditionally used in practice for the measurement of blast damage to structures. In this criterion the shape of the wave form and duration of dynamic loading are not taken into account. However, the energy input to the rock mass is determined not only from the amplitude but also from the time duration of the vibration pulse. The response of the rock material to any loading is strongly affected by the strain rate. In dynamic loading, the stress can exceed the material strength, but it may not damage the material because of the duration of this stress is too short. Blast damage is accounted as a function of both time and applied stress. Single field variable (e.g. stress, strain, particle velocity etc.) at any time cannot be expected to characterize the dynamic fracture process. A damage model based on extensional strain, cumulative crack propagation and continuous damage of the material has been developed to represent the tensile failure of rock. By analysis it was found that the attenuation of the field vibration is dependent not only on frequency but also on amplitude for a given travel time (Yong and Rocque, 1993).

5.11.2 Peak Particle Velocity (PPV, mm/s) as Damage Criteria

A number of researchers suggested the damage criteria based solely on the peak particle velocity (PPV, mm/s). They are listed in Tables 5.11 to 5.19.

Table 5.11 After Langefors et al. (1958)

No damage	< 50 mm/s
Fine cracks	100 mm/s
Cracks	150 mm/s
Serious cracks	225 mm/s

Table 5.12 After Edwards and Northwood (1960)

Safe zone	< 50 mm/s
Damage zone	100 – 150 mm/s

Table 5.13 After Duvall and Fogelson (1962)

Major damage (95%)	50 mm/s

Table 5.14 After Nicholls et al. (1971)

Safe zone (95%)	< 50 mm/s
Damage zone	> 50 mm/s

Table 5.15 Indian Standard Institution (1973)

Soil, weathered or soft rock conditions	70 mm/s
Hard rock conditions	100 mm/s

Table 5.16 Australian Standard (As A-2183) (Just and Chitombo, 1987)

Type of structure	*Ground PPV (mm/s)*
Historical buildings and monuments and buildings of special value	2
Houses and low-rise residential buildings, commercial buildings not included below	10
Commercial buildings and industrial buildings or structure of reinforced concrete or steel construction	25

Table 5.17 Hungarian standard

Type of structure	*Permissible limit (mm/s)*
Construction demanding special protection, military, telephones, airports, dams, bridges which have length of more than 20 m	Extra opinion from expert
Statistically not solid damaged construction, temples, monuments, oil and gas wells and up to 0.17 MPa and below 0.7 MPa pressure in pipes (oil and gas)	2
Panel houses and statistically not fully determined structures	5
Statistically good condition structures, towers, electrical apparatus, water plant	10
RCC and structures concrete, tunnels, canals and other pipelines beneath the soil surface greater than 0.7 m, opening beneath the sub-level	20
Public road, railway and electrical lines, telephone lines, ropeway	50

Table 5.18 Russian standard

Type of structure	*Allowable PPV (mm/s)*	
	Repeated	*One fold*
Hospitals	8	30
Large panel residential buildings and children's institutions	15	30
Residential and public buildings of all types except large panels, office and industrial buildings having deformations, boiler rooms and high brick chimneys	30	60
Office and industrial buildings, high reinforced concrete pipes, railway and water tunnels, traffic flyovers, saturated sandy slopes	60	120
Single storey skeleton type industrial buildings, metal and block reinforced concrete structures, soil slopes which are part of primary structures, primary mine openings (service life up to 10 years), pit bottoms, main entries, drifts	120	240
Secondary mine openings (service life up to 3 years), haulages and drifts	240	480

Table 5.19 After Rosenthal and Morlock (1983)

Distance from the blasting site (m)	*Maximum allowable PPV (mm/s)*
0-91.4	37.75
91.4-1524.0	25.40
1524 and above	19.05

5.11.3 Peak Particle Velocity and Frequency as Damage Criteria

Vibration frequency can be calculated in a variety of ways including Fourier and response spectrum techniques as well as inversion of the time periods read between zero crossings on the trace records. As an advanced module, some researchers included frequency along with the peak particle velocity for the damage criteria. Suggested damage criteria based on peak particle velocity for different frequencies are given in Tables 5.20 to 5.26.

Table 5.20 Australian standard (Ca-23-1967) (Just and Chitombo, 1987)

Type of structure	*Maximum values*
Historical buildings and monuments and buildings of special value	0.2 mm displacement for frequencies less than 15 Hz
Houses and low rise residential buildings, Commercial buildings not included below	19 mm/s resultant PPV for frequencies greater than 15 Hz
Commercial buildings and industrial buildings or structures of reinforced concrete or steel construction	0.2 mm maximum displacement corresponds to 12.5 mm/s PPV at 10 Hz and 6.25 mm/s at 5 Hz

Table 5.21 Swiss standard

Type of structure	*Frequency bandwidth (Hz)*	*Blast induced PPV (mm/s)*	*Traffic/machine induced PPV (mm/s)*
Steel or reinforced concrete structure such as factories, retaining walls, bridges, steel towers, open channels underground tunnels and chambers	10-60	30	—
	60-90	30-40	—
	10-30	—	12
	30-60	—	12-18
Buildings with foundation walls and floors in concrete, walls in concrete or masonry, underground chambers and tunnels with masonry linings	10-60	18	—
	60-90	18-25	—
	10-30	—	8
	30-60	—	8-12
Buildings with masonry walls and wooden ceilings	10-60	12	—
	60-90	12-18	—
	10-30	—	5
	30-60	—	5-8
Objects of historic interest or other sensitive structure	10-60	8	—
	60-90	8-12	—
	10-30	—	3
	30-60	—	3-5

Table 5.22 USBM safe level criteria after Siskind et al., (1980)

Type of structure	*Peak particle velocity (mm/s)*	
	Frequency (<40 Hz)	*Frequency (>40 Hz)*
Modern homes, dry wall interior	18.75	50
Older homes, plaster on wood lath construction	12.5	50

Table 5.23 Swedish standard after Persson et al., (1980)

Type of structure	*Limiting values of parameters*		
	Amplitude (mm)	*Velocity (mm/s)*	*Acceleration (mm/s²)*
Concrete bunker steel reinforced	—	200	—
High rise apartment block-modern or concrete steel frame design	0.4	100	—
Underground rock cavern roof hard rock span 15-18 m	—	70-100	—
Normal block of flats-bricks or equivalent walls	—	70	—
Light concrete building	—	35	—
Swedish National Museums—Building structure	—	25	—
Swedish National Museums—Sensitive exhibits	—	—	5
Computer centre	0.1	—	2.5
Circuit breaker control room	—	—	0.5-2

Table 5.24 German DIN standard 4150 (GIS, 1986)

Type of structure	*Peak particle velocity (mm/s) at foundation*		
	< 10 Hz	*10-50 Hz*	*50-100 Hz*
Offices and industrial premises	20	20-40	40-50
Domestic houses and similar construction	5	5-15	15-20
Buildings that do not come under the above because of their sensitivity to vibration	3	3-8	8-10

Table 5.25 CMRI-proposed standard (Dhar et al., 1993)*

Structure's specification	*Threshold value of PPV (mm/s)*	
	≤ 24 Hz	*> 24 Hz*
Domestic houses, dry wall interior, construction structures with plasters, bridge	5.0	10.0
Industrial buildings, steel or reinforced concrete structures	12.5	25.0
Objects of historical importance, very sensitive structures, more than 50 years old construction and structures in poor state of condition	2.0	5.0

* Mostly intended to minimize perceptions and complaints

Table 5.26 DGMS (India) prescribed permissible limits of ground vibrations

Type of structure	*Dominant excitation frequency, Hz*		
	< 8 Hz	*8-25 Hz*	*> 25 Hz*
(A) Buildings/structures not belonging to owner			
1. Domestic houses/structures (Kuchcha, brick and cement)	5	10	15
2. Industrial buildings	10	20	25
3. Objects of historical importance and sensitive structures	2	5	10
(B) Buildings belonging to owner with limited span of life			
1. Domestic houses/structures	10	15	25
2. Industrial buildings	15	25	50

Japan has strict regulations on blasting in urban areas; residents can virtually veto its use on purely subjective grounds. The permitted vibration amplitude should be 0.5 to 1.0 mm/s in the residential areas (Hagimori et al., 1993).

The New Zealand standard actually states maximum vibration limits depending on frequency, but mine management aims to keep all blasts below 5 mm/s (Felice, 1993).

5.12 CONCLUSIONS

1. Propagation equations proposed by Langefors-Kihlstrom [equation (5.4)] and the Indian Standard [equation (5.6)] are identical. They remain identical even after incorporation of inelastic attenuation (or damping) factor $e^{-\alpha D}$. Explosive charges per delay become identical when calculated from the two equations for various distances and vibration limits. The only difference in the physical appearance of the two equations is in the value of the site constant B. It was found that B (Indian Standard) = .75 × B (Langefors and Kihlstrom).
2. For soft to medium-hard rock masses including limestone, sandstone-alluvium and iron ore, the explosive charge per delay, estimated using equation (5.4), is maximum as compared to other equations up to a certain distance. This limiting distance depends on the vibration limit. At distances (exceeding 150 m) and at vibration level (exceeding approximately 10 mm/s), the explosive charge per delay as obtained from equation (5.4) is the lowest.
3. For moderate to hard rock masses, including dolomite and granite, the charge per delay is the highest when estimated using equation (5.5).
4. Analysis of observations indicates that when explosive charge per delay is estimated using equation (5.3), the blast results are near

optimum as compared to results obtained from other equations. The USBM equation is better applicable for blasting in soft to medium-hard rock masses represented by limestone, sandstone-alluvium, coal and iron ore. The blast results for medium to hard rock masses, represented by basalt, dolomite and granite, are near optimum when explosive charge per delay is estimated using equation (5.3) and equation (5.5).

5. In limestone, sandstone, sandstone-alluvium and iron ore, the permissible charges per delay are low at shorter distances. However, for distances greater than 100 m the charge per delay sharply increases. It is seen that the charge per delay, estimated using equation (5.5) for distances below 100 m, is very low. It would be difficult to conduct any meaningful blast at such a low charge and, therefore, equation (5.5) is unsuitable for estimation of charge when the structure is to be protected within a 100 m distance of the blast site.
6. In sandstone and dolomite, the results indicate that equation (5.7) gives the most acceptable and consistent charge weight per delay.
7. The addition of the inelastic damping factor, as proposed by Ghosh and Daemen [equations (5.8) and (5.9)], represents the exponential decay of velocity. When explosive charges for a given distance and vibration limit are estimated after incorporating the inelastic damping factor, they become less. This shows that the addition of the inelastic damping factor as proposed by Ghosh and Daemen (1983) in the propagation equation is disadvantageous from the optimum blasting point of view.
8. The Ghosh and Daemen model is useful and logical for soft to medium-hard rock masses. For hard rock, the value of the inelastic damping factor (α) becomes negative. When α is negative, the best-fit curve has a tendency to bend slightly upward as the scaled distance increases (Figure 5.7). The physical cause of this type of amplification may be due to the geological characteristics. It may also be due to the fact that in hard strata, within a certain range, amplification of particle velocity occurs in place of attenuation, leading to a negative value of α in equations (5.8) and (5.9).
9. In rock masses where the CMRI equation (5.10) fits well, i.e., in cases where n is negative in equation (5.10), it gives a more consistent and feasible charge per delay than other equations.
10. The CMRI equation (5.10) is the simplest in form for quick evaluation of charge per delay for a particular distance and vibration level. This equation gives a very consistent charge per delay for optimum blasting.

11. Equations (5.19, 5.20 and 5.21) are valid for a single-hole shot. In case of multihole shots with and without delay detonators, there will be interactions between the effects of individual explosions. Each hole will emanate seismic pulse, which will be received by the transducer not as individual but as a complex mixture. The resultant vibration energy may be higher or lower than the sum of individual peaks, which would have been by considering the holes separately. Because of different travel paths, the phase condition would play an important role for the vibratory particle motion. However, this model may be used as a basic tool in the conceptual formulation of a complex blast module. It is obvious from Table 5.10 that the model correlates excellently with the actual field conditions.

REFERENCES

Ambraseys, N. R. and Hendron, A. J. 1968. Dynamic behaviour of rock masses, In: Rock Mechanics in Engineering Practice K.G. Stagg and O.C. Zienkwicz (eds.), John Wiley and Sons, London, pp. 203-207.

Attewell, P. B. 1964. Recording and interpretation of shock effects in rock, Min. Miner. Engg., Vol. 1, USA, pp. 21-28.

Bath, M. 1979. Introduction to seismology, Birkhauser Verlag, Basel, 2nd Rev. Ed., Germany, p. 428

Birch, W. J., and Chaffer, R. 1983. Prediction of ground vibrations from blasting on opencast sites, Inst. Min. Metall., Trans., Sec. A (Mining Industry), UK, pp. A102-A107.

Dhar, B. B., Pal Roy, P. and Singh, R. B. 1993. Optimum blasting for Indian geomining conditions—suggestive standard and guidelines, CMRI Publication, India, p. 40.

Daemen, Jaak J. K. (ed.), 1983. Ground and air vibrations caused by surface blasting, USBM Report prepared by Department of Mining and Geological Engg., University of Arizona, Tucson, USA, AZ 85721.

Davies, B., Farmer, I. W. and Attewell, P. B. 1964. Ground vibrations from shallow sub-surface blasts, The Engineers, London, UK, Vol. 217, pp. 553-559.

DGMS Circular Technical 7 / 1997. Directorate General of Mines Safety, India, pp. 500-508 (Ref. DGMS Circulars by L. C. Kaku, Lovely Prakashan, Dhanbad, India, p. 848).

Dowding, C. H. 1985. Blast vibration monitoring and control, Prentice-Hall, Inc., Englewood Cliffs, NJ, USA, p. 297.

Duvall, W. I. and Petkof, B. 1959. Spherical propagation of explosion generated strain pulses in rock, USA, USBM-RI 5483, p. 21.

Duvall, W. I. and Fogleson, D. E. 1962. Review of criteria for estimating damage to residences from blasting vibration, U.S. Bureau of Mines, R.I. 5968, p.19.

Duvall, W. I., Johnson, C. F. and Meyer, A. V. C. 1963. Vibration from blasting at Iowa limestone quarries, USA, USBM-RI 6270.

Edwards, A. T. and Northwood, T. D. 1960. Experimental study of effects of blasting on structures, The Engineer, Vol. 210, London, p. 40.

Felice, J. J. 1993. Application of modelling to reduce vibrations and airblast levels, Proc. 4th Symposium on Rock Fragmentation by Blasting (ed. H. P. Rossmanith), A. A. Balkema, Rotterdam, Netherlands, pp. 145-151.

Habberjam, B.M. and Whetton, J.T. 1952. On the relation between seismic amplitude and charge of explosive fired in routine blasting operations, Geophysics, vol. 17, No. 1, USA, pp. 116-128.

Hagimori, K., Terada, M., Ouchterlony, F., Furukawa, K. and Nakagawa, K. 1993. Using slot drilling to reduce vibrations and damage from tunnel blasting in urban areas, Proc. 4th Symposium on Rock Fragmentation by Blasting (ed. H. P. Rossmanith), A. A. Balkema, Rotterdam, Netherlands, pp. 97-104.

German Institute of Standards. Vibration of building-effects on structures, DIN 4150, No. 3, May, 1986, pp. 1-5.

Ghosh, Amitava, 1983. A new analytical predictor of ground vibrations induced by blasting, M.S. Thesis, The University of Arizona, USA, p. 322.

Ghosh, Amitava and Daemen, Jaak, J. K. 1983. A simple new blast vibration predictor (based on wave propagation laws), Proc. 24th U.S. Symposium on Rock Mechanics, Texas, USA, pp. 151-161.

Indian Standard, 1973. Criteria for safety and design of structures subjected to underground blast, ISI Bulletin No. IS-6922, India.

Just, G. D. and Chitombo, G. P. 1987. The economic and operational implications of blast vibration limit mining and environment, The Aus. IMM, Australia, pp.117-124.

Keller, G.V. 1955. Dispersion of seismic waves near a small explosion, Trans. Am. Geophys. Union, Vol. 36, No. 6, USA, pp. 1035-1043.

Langefors, U., Kihlstrom, B. and Westerberg, H. 1958. Ground vibrations in blasting, Water Power, Part I-III, pp. 335-338, 390-395, 421-424.

Morris, G. 1950. Vibrations due to blasting and their effects on building structure, The Engineer, London, pp. 394-395.

Nicholls, H. R., Johnson, C. F. and Duvall, W. I. 1971. Blasting vibrations and their effects on structures, U.S. Bureau of Mines, R.I. 656, p. 105.

Pal Roy, P. and Gupta, R. N. 1989. Some statistical methods used in the analysis of blast induced vibration data, J. Inst. Eng. (India)-MN, Vol. 70, pt. MN2, pp. 54-60.

Pal Roy, P. 1991. Prediction and control of ground vibrations due to blasting, Colliery Guardian, 239(7): 215-219.

Pal Roy, P. and Dhar, B. B. 1992. A mixed analytical and empirical model for the prediction of ground vibration due to blasting, Journal of Mining Research, India, July-Sept., 1(2): 15-22.

Pal Roy, P. 1993. Putting ground vibration predictions into practice, Colliery Guardian, UK, 241(2): 63-67

Pearse, G. E. 1955. Rock blasting—some aspects on the theory and practice, Mine and Quarry Engineering, Vol. 21, No. 1, pp. 25-30.

Persson, P. A., Lundborg, N. and Johansson, C. H. 1980. The basic mechanisms in rock blasting, Proc. 2nd Cong. Rock. Mech., ISRM, Belgrade, Yugoslavia, III, pp. 19-33.

Rinehart, J. S., Forttin, J. P. and Burgin, L. 1961. Propagation velocity of longitudinal waves in rocks. Effects of state of stress, stress level of the waves, water content, porosity, temperature, stratification and texture Proc. 4th Symp. Rock Mech., Univ. Park, Penn, USA, pp. 119-135.

Rosenthal, M. F. and Morelock, G. L. 1983. Blasting guidance manual. Office of the Surface Mining Reclamation and Enforcement, U.S. Department of Interior, Washington, DC.

Singh, Pradeep Kumar, 1998. A study on ground vibrations due to rock blasting, Doctor of Engineering thesis, Tech. Univ. of Clausthal, Germany, p.210 and pp. A1-A41.

Siskind, D. E., Staff, M. S., Kopp, J. W. and Dowding, C. H. 1980. Structure response and damage produced by ground vibration from surface mine blasting, U.S. Bureau of Mines, R.I. 8507, p. 74.

Walsh, J. B. 1966. Seismic wave attenuation in rock due to friction, J. of Geophysics Research, Vol. 71, USA, pp. 1591-1599.

Yong, R. L., Rocque, P. Katsabanis, P. and Bawden, W. F. 1993. Blast damage study by measurement of blast vibration and damage in area adjacent to borehole. Proc. 4th Symposium on Rock Fragmentation by Blasting, (ed.) H. P. Rossmanith, A. A. Balkema, Rotterdam, Netherlands, pp. 137-144.

6

Structural Response and Damage Criteria for Safety of Surface Structures

6.1 INTRODUCTION

Over the years, attempts have been made to relate the parameters defining vibratory particle motion with observed structural damage. This has led to the formulation of a number of empirical damage scales relating observed particle motion in terms of peak particle acceleration, velocities, displacements and frequencies. As mentioned in Chapter 5, the damage levels differ from one country to another, depending on construction, type of structure and state of repairing. The threshold limit of vibration for the safety of structures depends on its importance and the degree of safety consciousness prevailing in a country.

There has been limited publication of structural response, resonance characteristics and magnification or attenuation problems of structures affected by blast vibrations. During the course of investigations in the last 20 years, the author has faced many situations in which low frequency vibrations were identified which posed a serious risk to the houses of neighbouring villages and other residential sectors.

When blasting takes place in opencast or underground mines then the incoming blast-induced vibrations, transmitted through the ground, create a dynamic response in structures. The response depends on various factors, such as:

1. The type and characteristics of the vibrations, duration of seismic pulse, frequency of propagating waves, transmitted energy, etc.
2. The type of transmitting medium on which the structure stands, and
3. The type of structure and its vibratory character as well as non-structural entity of the building.

Damage to any structure due to blast vibrations can best be explained in terms of its response to the passing vibrations. One of the critical response factors is the amount of differential movement that occurs between structural members causing strains, which in turn, produce cracking. In 1980, the U.S. Bureau of Mines published its report on its investigations of surface mine blasting (Siskind et al., 1980) which showed that the structural resonance responded to low frequency ground vibration, resulting in increased displacement and strain. Their study reintroduced the dependence of damage on frequency. Prior to this, the safe limit particle velocity was independent of frequency.

Dowding (1985) pointed out that the simplest characteristics of a structure that govern its behaviour under vibratory loading are (1) the masses of the main components (2) the stiffness of the main component, and (3) the amount of energy dissipated through differential movement in cracks, joints, and connections. Experimentally it has been verified (Dowding, 1985; Jimeno et al., 1995) that if the wavelength of the transmitting wave is long enough compared to the size of a structure (low phase angle), the entire structure tends to move generally as one. In this case, it is reasonable to represent the whole structure with a single degree of freedom (SDF) model. On the other hand, if the wavelength is only twice the width of the structure (high phase angle) and the sides of the structure move in opposite direction, use of a SDF model appears to be questionable.

6.2 DAMAGE PREVENTION CRITERIA FOR STRUCTURES

A simple method to predict the structural response of a building to vibrations is the use of Fast Fourier Transform (FFT) analysis. FFT provides information in practice about the frequency band and responsible wavelength which need to be omitted in order to avoid damage and disturbance. It is the simplest scientific way of analyzing resonance and magnification problems for any structure. Figure 6.1 shows that the vibrations in buildings can be magnified due to the response of the structural elements of which they are composed (Jimeno et al., 1995).

In order to distinguish ground vibration and structural response, it is important to place transducers on the compacted ground near the structure and at the same time at different heights on the structure. Owing to the structural response, peak values and trace records are different at each location. The response of any structure to vibration can be calculated if the mass, stiffness and damping are known or can be estimated. Frequency is the number of times in a second the structure oscillates back and forth when excited and its damping is proportional to the rate at which the vibration decays with time during free vibration. The response of a tall structure, such as an apartment building, is typically measured

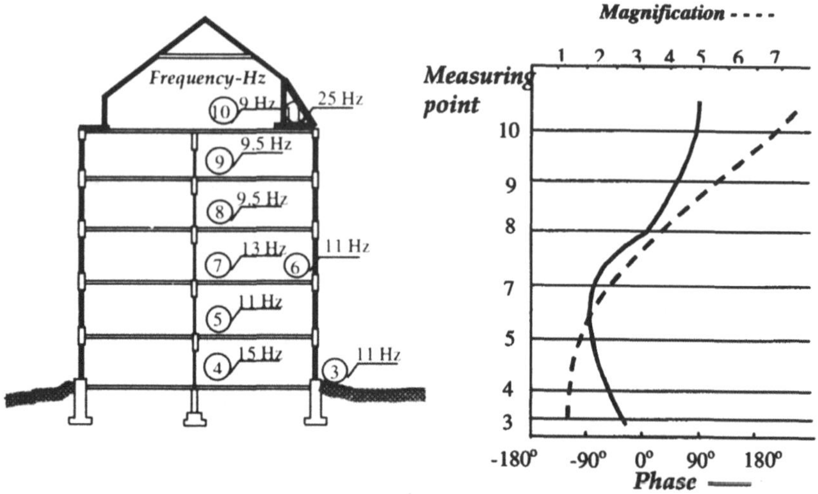

Fig. 6.1 Magnification within a building

by a transducer attached to an upper corner of the structure, and the response of a wall panel is typically measured in the middle of the wall. Siskind et al. (1980) reported that corners represent racking of whole-structure response, and midwalls represent membrane-type flexure with associated secondary noises and rattling. The lighter and more flexible walls have higher natural frequencies and relatively large motions. Typical blast vibration time history records are shown in figures 6.3 and 6.5. Figures 6.4 and 6.6 show the frequency analysis of the trace records of figures 6.3 and 6.5. Figures 6.10 and 6.11 show the plots of data generated during field experiments at two different sites. It is clear from these graphs that the building continued to vibrate freely after the passage of the ground motion.

6.2.1 Response Spectra

The response spectrum calculates how much energy is transferred into a structure and reflects the magnitude, frequency and duration of the ground motion as well as the natural frequency and damping of the structure. The response spectrum is usually presented as a plot of the maximum response (in this case Pseudo Spectral Relative Velocity or PSRV) of a single degree of freedom model of the structure relative to the ground movement against the natural frequency of the structure (Figure 6.2). Considerable work in this field was carried out by Dowding (1985), wherein he showed that the "response spectrum" technique is a more rational approach for predicting damage than the use of a simple particle velocity criterion alone. The response spectrum approach is certainly a model solution with some limitations but it is the only economic approach which allows consideration of each structure and the specific time

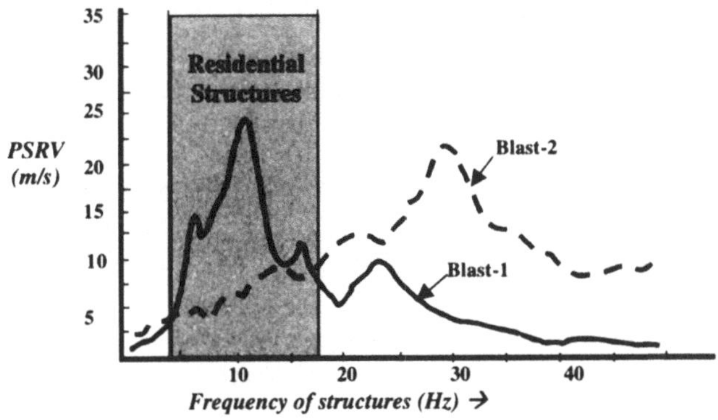

Fig. 6.2 Superimposed PSRV of two separate blasts

history of the recorded vibration effects, particularly its frequency content for each recorded blast (Foster, 1981).

6.2.2 Damage Criteria

Langefors et al. (1958) carried out extensive work establishing the relationship between damage and ground vibrations due to blasting. They concluded that particle velocity was the best criterion for classification of damage. Duvall and Fogelson (1962) also showed that major damage correlated with particle velocity and minor damage with acceleration. Siskind et al. (1980) conducted a series of investigations on ground vibration and air blast damage and response of structures. At 25 mm/s, no damage was noticed but the foundation deformation contributed to the super-structures cracking. A single-storey house with both gypsum wallboard and plaster interior walls was subjected to peak particle velocity of 56.64 mm/s (2.23 in/s). Several cracks propagated over windows and doors. Frequencies were found to be within 7 and 30 Hz.

Bergling et al. (1977) described test blast damage to a concrete and brick residence. Shots were in the range of 1 to 50 m distance and the lowest level at which damage occurred was 110 mm/s. Bogdanoff et al. (1975) described a house of similar construction made on granite-gneiss bedrock. From 38 rounds at a distance of less than 100 m, these authors detected no damage below a vertical peak particle velocity of 90 mm/s. They concluded that at 30 mm/s was safe for this structure (and geology), since many non-damaging shots occurred at this level.

Esteves (1978) described damage to a single concrete and tile residence near a quarry. The first damage observed was plaster cracks at 60 mm/s.

Blast wave signatures and their corresponding FFT (Fast Fourier Transform) analyses from 2 test blasts, one in hilly terrain and another in an operating opencast mine, are shown in figures 6.3 to 6.6. The wave signature recorded even at near-field distance of 13.6 m from the source, as shown in figure 6.3, can be separated after using adequate delay interval (more than 150 ms).

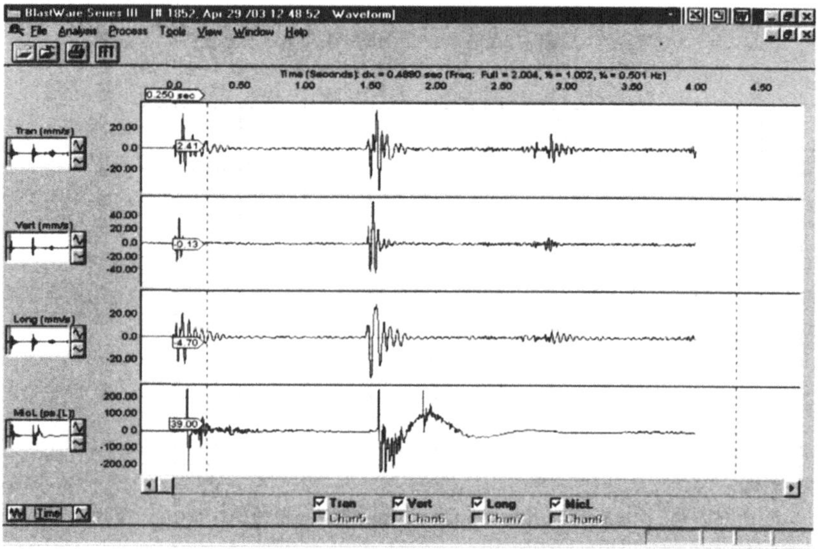

Fig. 6.3 Blast wave signature recorded at a distance of 13.60 m from the blasting source of 10.20 kg explosive used in the formation cutting in a hilly terrain

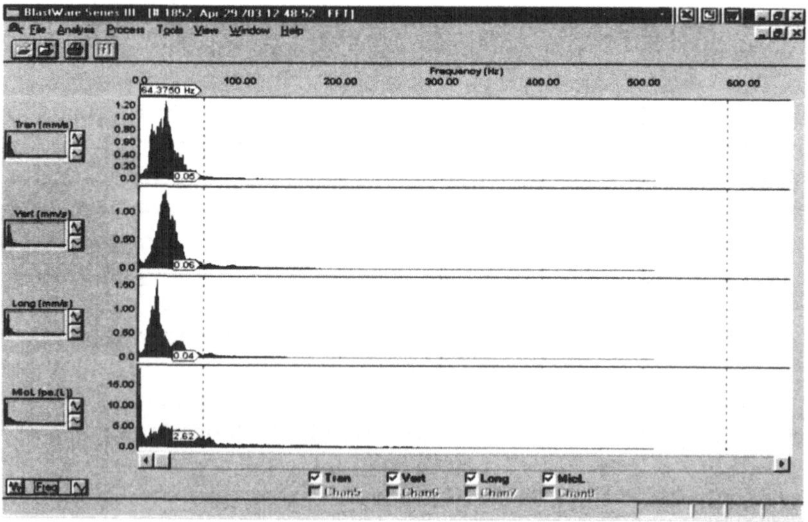

Fig. 6.4 Fast Fourier Transform (FFT) analysis of recorded waveforms of figure 6.3 showing energy concentration at different frequencies

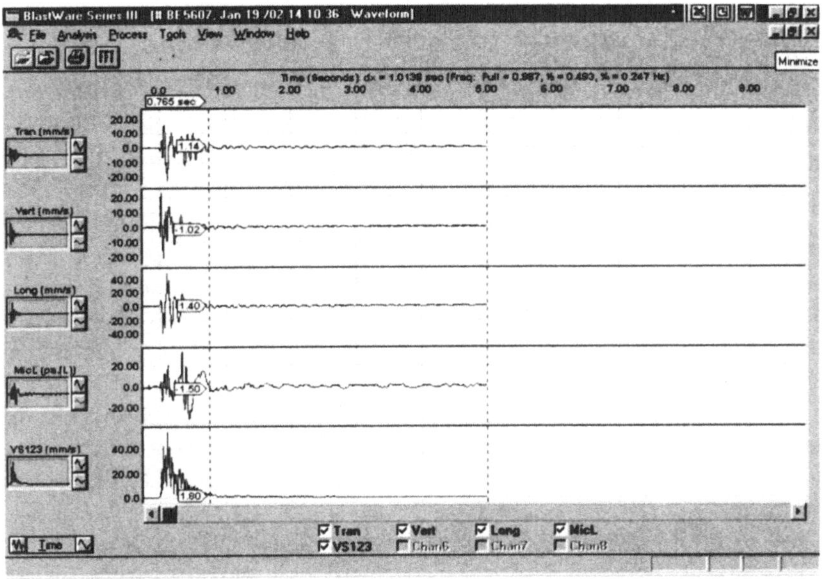

Fig. 6.5 **Blast wave signature recorded at a distance of 150 m from the blasting source of 2500 kg explosive used in top bench in an opencast coal mine. The 5th row represents the vector sum of T, V and L components.**

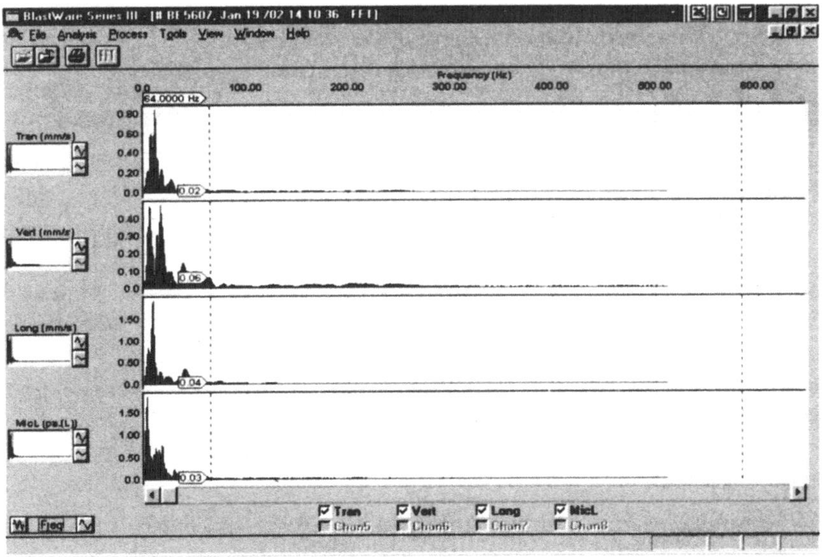

Fig. 6.6 **Fast Fourier Transform (FFT) analysis of recorded waveforms of figure 6.5 showing energy concentration at different frequencies**

6.3 FORMATION OF LOW FREQUENCY VIBRATIONS

It is the view of the author that the generation of low frequency vibrations depends solely on the geological forms or structures of the transmitting medium. It is absolutely independent of the distance of the measuring point. During the course of experiments over a period of nearly 20 years, the author has obtained a dominant frequency band at the level of 5-7 Hz in iron ore mines, irrespective of the distance and blast design. Also, in cases (specially those mines situated in the western part of India) where the entire block was covered by black cotton soil, low frequency vibrations were obtained. However, mixed results, i.e. both low and high frequencies, were obtained in mines situated in the southern and eastern parts of India.

Vazquez and Fernandez (1986) also highlighted the importance of the transmitting medium in the formation of low frequency waves. They characterized thc following situations as possible sources of low frequency vibrations.

- A stratum of hard limestone over softer clay-marls and gypsum
- Interface between the rock and air, or rock and water considering the air and water as elastic materials.
- Semi-decomposed alluvial material over a hard base rock, as in many river basins.
- A phreatic level that divides the rock into two zones, one dry and one water saturated.

6.3.1 Estimation of Natural Frequencies of Structures

When a seismic pulse passes through a structure, it transmits its energy to force the structure to vibrate with the frequency of the transient pulse. This is called forced vibration. After cessation of this forced vibration, the structure vibrates with its own frequency called the natural frequency of the structure. The natural frequency of any structure can be measured directly from the free vibration trace records after the passage of the ground motion. Newmark and Hall (1982) suggested a simple approximation to estimate the fundamental natural frequency (p) of the super structure of any tall building:

$$p = 2\pi \sqrt{\frac{L}{0.05 \times h}} \qquad \ldots(6.1)$$

Where L is width and h is height of the structure.

Three different formulae for buildings with (a) prefabricated or reinforced concrete walls; (b) framework structure of reinforced concrete, and (c) steel structure, are provided by Jimeno et al., (1995).

The dominant frequency of any seismic pulse or dominant frequency band (this is the band which is obtained through FFT calculation) has been the point of concern in controlling potential damage due to vibrations. When resonance takes place, it magnifies the ground vibration which could be even up to eight times at the top of a structure, depending on its height.

Low frequency waves are especially harmful for structures because this is the state in which the structure absorbs most energy, progressively becoming deformed with time, until plastic deformation or rapture occurs. At the same time, these waves have a longer duration. Natural frequency ranges as prescribed in Table 6.7 were obtained in test structures constructed by CMRI at the edge of different operating opencast mines. A typical response spectra analysis of the vertical component of ground vibration is shown in figure 6.7.

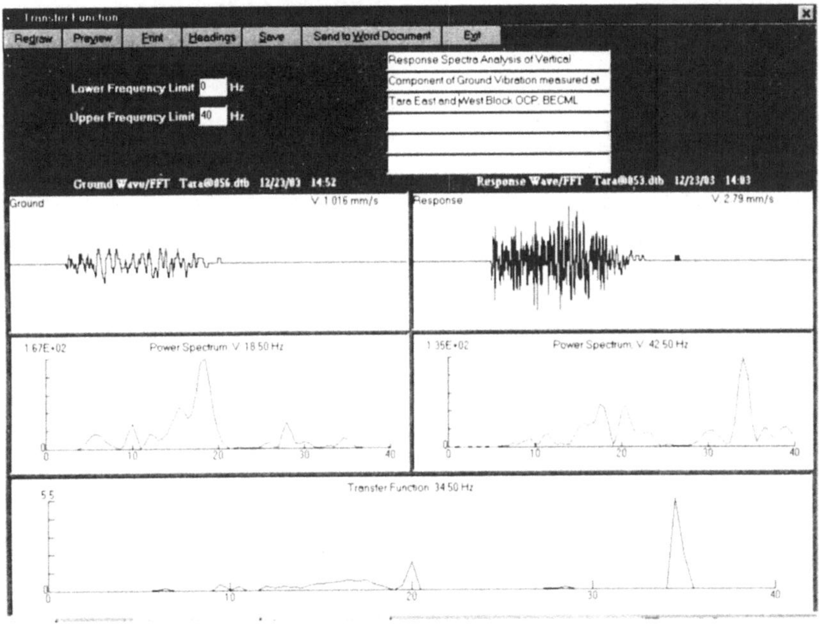

Fig. 6.7 **Response spectra analysis of vertical component of ground vibration measured at Tara East and West Block OCP, BECML**

6.4 DAMAGE STUDIES THROUGH TEST STRUCTURES

For evaluation of damage levels of different types of structures/dwellings in Indian geo-mining conditions, the Central Mining Research Institute (CMRI) undertook a long-term research programme in which blast-induced ground and structure vibrations were monitored regularly on several test structures until they were damaged or severely cracked (CMRI Report of Investigations, 1991).

In this section, various experimental procedures for monitoring vibrations, analysis of the results and conclusions regarding the threshold limit of particle velocity to protect such type of structures/dwellings in close proximity to active opencast mines, are described.

6.4.1 Construction of Test Structures

Test structures were constructed at several locations in close proximity to operating opencast mines. The structures included the low-cost houses of mineworkers normally found in mining areas. The structures at each site were identical, particularly in type and building materials, and comprised of the following three types (Figures 6.8 to 6.10):

(a) Brick walls with RCC roof
(b) Concrete walls without roof
(c) Mud houses with thatched roof or tiles or 'khapra'

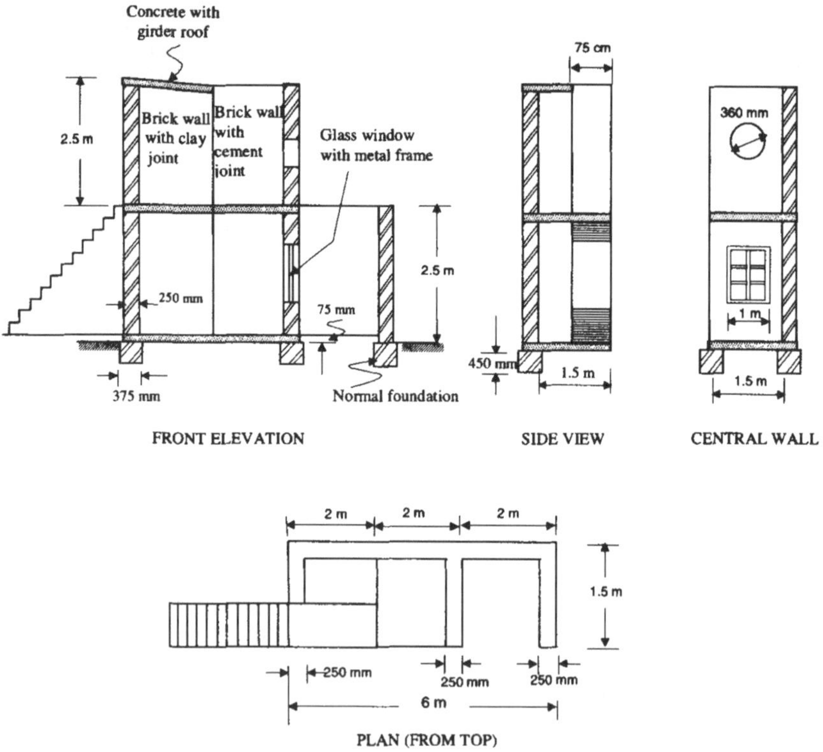

Fig. 6.8 Sketch of double-storey structures constructed by CMRI at the edge of opencast mines

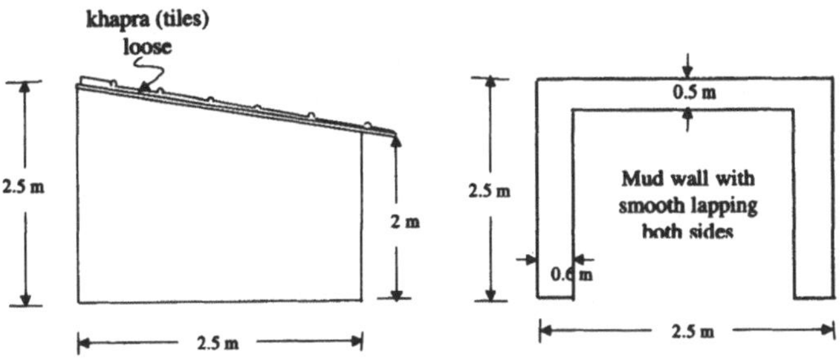

Fig. 6.9 Normal village mud house constructed by CMRI for experimental purposes

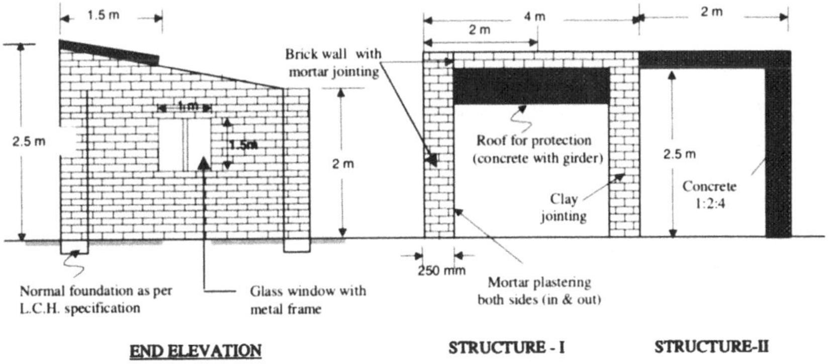

Fig. 6.10 Sketch of single-storey structure constructed by CMRI for experimental purposes

The walls of one of the test structures were made with sand-cement mortar in one-half and clay mortar in the other. Thus, a single house represented two types of construction. This was done to enable a comparative study of the effect of mortars in walls of any structure. In the seven locations, four were double-storey and three were single-storey structures of brick walls. The structural details are given below:

(A) Structures representing low-cost house:

Walls: Made of standard bricks

(a) Half portion with sand-cement mortar and other half with clay mortar

(b) Walls were finished with plaster of sand-cement mortar and finally whitewashed.

Roof: RCC

Dimensions: Length: 4 m
Width: 1.8 m
Height: 2.5 m
Window: rectangular

Thickness of wall: 250 mm

(B) Concrete Structures (without roof):
Concrete Mixture: 1:2:4
Dimensions: Length: 2.0 m; Breadth: 1.8 m; Height: 2.5 m
Thickness of wall: 250 mm
Concrete mixtures were poured at the sites to prepare the walls. Finally, the walls were plastered with a sand-cement mixture and whitewashed.

(C) *Mud house:* Made of clay with a thatched roof of tiles or 'Khapra'.
Dimension: Length: 2.5 m; Breadth: 1.8 m; Height: 2.5 m
Thickness of wall: 66 mm

6.4.2 Blast Monitoring and Analyses

The instruments used during investigation were: (a) Seismographs SINCO-3, SINCO-6, VS-1100 and VS 1600 (all USA-make); (b) Vibration indicators SMU-31 (German make). These instruments were calibrated in the laboratory on a vibrating disc of IMV (Japanese-make), before as well as after field use, to verify their accuracy and to make necessary corrections when required.

The seismometers/transducers of the instruments were mounted on the ground close to the structure, cemented ground floor at plinth level, windows, open walls, and on the first floor and roof of the structure. The purpose of these different placements was to study the structure's response to blast-generated ground motions. Pre-crack, if any, developed before and after the blast was thoroughly inspected.

The blasting faces at each site of investigation approached gradually towards the test structures. Vibration was measured in terms of peak particle velocity. The frequency of the ground vibration was estimated from the wave forms recorded by seismographs and from the frequency summary table given by SINCO-6 (Slope Indicator Company). This frequency summary table actually revealed which frequencies had the most vibration energy during the event. Vibration monitoring continued at each site until the test structures became damaged or severally cracked. With a view to studying the generation of low-frequency waves at distances far from the source, vibrations were also recorded at nearby houses other than the test structures. During investigations, at one of the sites, the blasting face was as close as 2.7 m to a test structure, causing its complete damage. The quantity of explosive used in blasting rounds ranged from 36 to 4000 kg. However, the maximum charge/delay varied from 10.8 to 625 kg in a blasting round.

Vibration measured within or on a structure is normally termed as structural response. With a view to determining structural response, ground and structure vibrations were simultaneously measured. One seismometer was placed on the ground adjacent to the structure and another on top of the structure.

Typical records of ground and structure vibrations are shown in figures 6.11 (a to d). Gain setting and recording speed of the seismographs were identical so that ground and structure vibrations could be compared. Ground vibrations were seen to be amplified by the structure and duration of the structure vibration to that of the ground vibration, was computed. The results are given in Table 6.1.

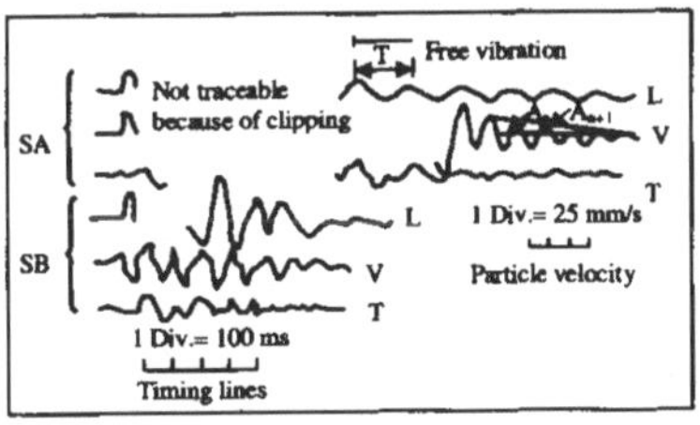

SA: Structural response; SB Ground vibration Max. charge/delay: 225 kg; Distance: 50 m (a: Site-A)

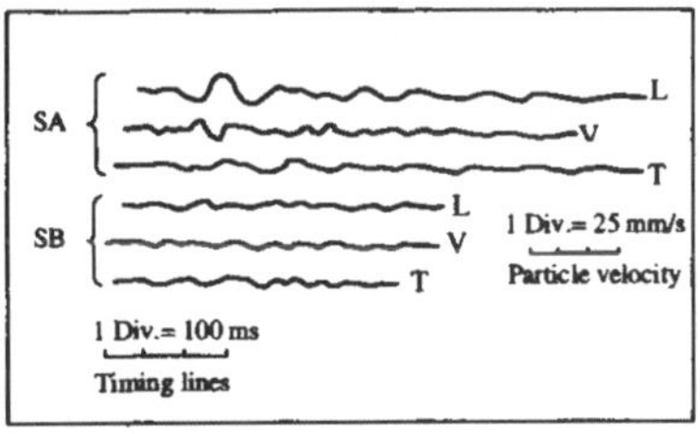

SA: Structural response; SB: Ground vibration Max. charge/delay: 187.5 kg; Distance: 130 m (b: Site-B)

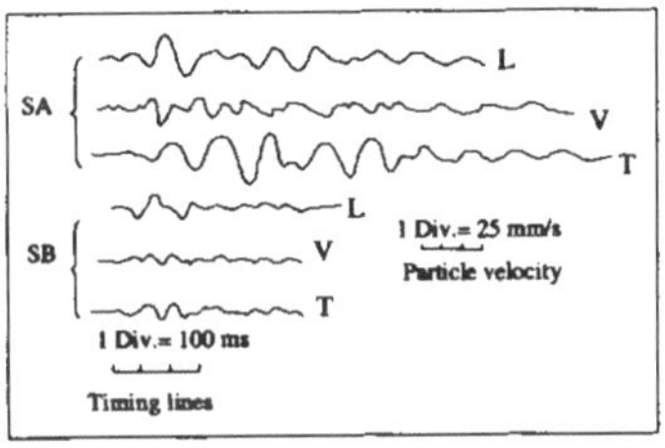

SA: Structural response; SB: Ground vibration Max. charge/delay: 237.5 kg; Distance: 150 m (c: Site-A)

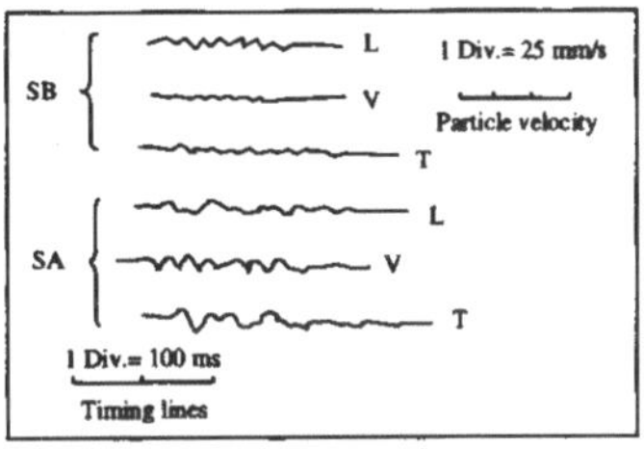

SA: Structural response; SB: Ground vibration Max. charge/delay: 31.12 kg; Distance: 87 m (d: Site-B)

Fig. 6.11 Comparison of ground and structural vibrations

In test structures, the magnitude of structure vibration was observed to be 1.0 to 2.8 times higher than the incoming ground vibration. The duration of structure vibration was also 1.0 and 2.2 times longer compared to ground motion. The two factors which influence dynamic response of a structure to ground vibration are its natural frequency and damping characteristics of the structure. The natural frequency is the frequency at which any structure freely vibrates after the cessation of ground motion.

When the ground and structure vibration are compared, the free vibration portion may be easily distinguished. This case was observed in typical mine blasting (Figure 6.11). Natural frequency f_n is determined by $f_n = 1/T$, where T is the time period. The natural frequencies of the test structures varied between 8 and 14 Hz. These values are within the range as reported by other researchers (Siskind et al., 1980; Mederis, 1978).

Table 6.1 Structural response to ground vibration due to blasting (measured on coal measure rocks)

Distance (m)	*Charge per delay (kg)*	*Ground vibration* *PPV (mm/s)*	*Frequency (Hz)*	*Structure vibration PPV (mm/s)*	*Amplification factor*
130	187.5	15.0	12-17	40.1	2.67
118	237.5	21.9	16	55.9	2.55
115	250.0	16.0	—	40.0	2.50
50	450.0	92.0	—	240.0	2.60
45	362.5	90.0	—	240.0	2.66
76	300.0	35.0	—	37.0	1.06
46	375.0	119.5	—	140.0	1.17
61	77.7	35.6	25-30	35.6	1.00
87	31.12	11.3	20-25	20.6	1.82
92	30.58	7.3	20-25	20.6	2.82
60	38.8	38.6	20-30	39.8	1.03
30	91.5	63.4	25-30	75.7	1.19

The amplification factor is maximum when the predominant frequency of the ground motion reaches the natural frequency of the structure. The structure at this frequency absorbs most of the energy and oscillates for a longer period, which results in structural damage.

Damage criteria based on PPV helped many opencast mine operators to extract minerals safely and with minimum complaints from the residents nearby. However, human annoyance and complaints of structural damage were lodged by villagers located at distance greater than one kilometre, although PPV measured on the ground was within safe limits. In-depth model and field studies indicated inadequacies in damage criteria based solely on PPV (Mederis, 1978; Siskind et al., 1980; Dowding, 1985). The latest trend is based on a combination of PPV and frequency of ground vibration (Chapter 5).

Ground vibrations were, therefore, also measured at distances greater than one kilometer. A record of such measurements is shown in figure-6.12. Each component of ground motion, i.e. longitudinal (L), vertical (V), and transverse (T), is comprised of high-frequency waves corresponding to body waves and low frequency to surface waves. Waveforms consisted of high frequency initially and low frequency thereafter. These waves were analysed for PPV and dominant frequency. Results of analysis are shown in Table 6.2.

High-frequency waves varied between 22 and 80 Hz and low-frequency waves between 6 and 12 Hz. Peak particle velocities (PPVs) associated with low-frequency waves were 1.09 to 3.39 times higher than those of high-frequency waves. For comparable distances, frequencies were higher when vibrations were measured on hard rock versus those measured on

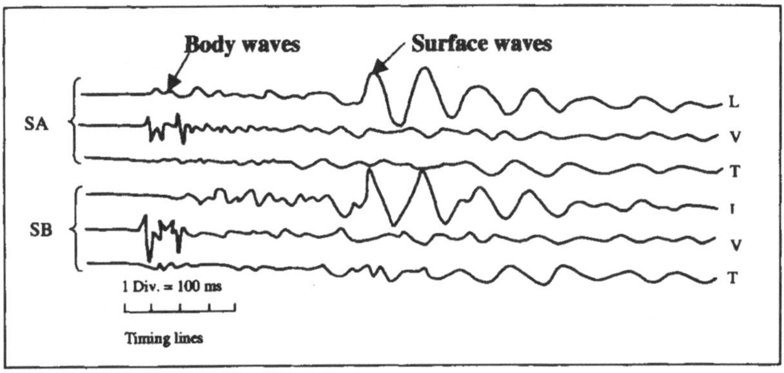

Fig. 6.12 Time histories of ground vibration from overburden bench blast at distances greater than 1 km. SA and SB denote vibrations recorded at two different locations

Table 6.2 Analysis of waves for PPV and frequency recorded at distances greater than 1 km (coal measure rock)

High-frequency waves		*Low-frequency waves*		*Ratio of PPVs*
Frequency (Hz)	*PPV (mm/s)*	*Frequency (Hz)*	*PPV (mm/s)*	
28-36	0.76	6-9	1.23	1.56
24-38	1.05	6-9	1.30	1.23
40	0.83	8-12	1.84	2.21
26-45	1.60	9-10	2.74	1.71
30-40	4.24	8-10	4.90	1.15
26-32	4.76	6-12	6.61	1.38
40-75	1.60	7-12	1.75	1.09
75-80	2.07	6-10	3.25	1.57
22-44	0.96	6-10	1.74	1.87
24-35	1.78	7-9	2.61	1.46
—	2.06	6-8	7.00	3.39
25-32	3.74	6-8	5.09	1.36

soil or soft rock. Frequency analysis by the Fast Fourier Transform (done by the instrument SINCO-6) revealed that low-frequency in the range of 5-10 Hz accounted for 65 to 97% of the energy (Table 6.3). Such low-frequency waves were responsible for higher structural response vis-à-vis damage to buildings.

6.4.3 Classification of Damage

All structures/houses develop cracks from natural causes, including ground settlement and periodic change of humidity, temperature and wind velocity. Soil moisture changes are notorious for causing foundation cracks. The width of old cracks changes seasonally and often daily.

Table 6.3 Analysis of blast-induced ground vibration in an opencast coal mine

Rock	*Distance from the blast (m)*	*Maximum charge per delay (kg)*	*Peak particle velocity (mm/s)*	*Dominating frequency range (Hz)*	*Normalized energy (%)*
Shale	1350	400	2.97	5-10	84
Coal	1300	425	1.10	5-7	89
Shale	1350	400	2.83	40-53	49
Shale	1350	750	3.40	5-7	89
Coal	1340	784	1.66	5-7	70
Coal	1360	784	2.89	5-7	86
Shale	1340	1163	4.50	5-7	65
Shale	1360	1163	6.99	5-10	97
Coal	900	354	2.31	5-7	73
Coal	870	870	1.94	19-39	81

However, the number of cracks continues to increase with age. This is independent of blasting. For correct analysis of damage from blast vibrations, it is thus essential to have the structures constructed near enough to blasts with careful pre- and post-blast inspections. The degrees of damage (threshold, minor and major) were studied by Dvorak (1962); Edwards and Northwood (1960); Langefors et al. (1958). Their classifications of damage are given in Table 6.4.

Table 6.4 Classification of structural damage due to blasting

Classification	*Description of damage*
1. Threshold	Loosening of joints, small plaster cracks at joints between construction elements, lengthening of old cracks
2. Minor	Loosening and falling of plaster, cracks in masonry around openings near partitions, fall of loose mortar
3. Major	Cracks of several millimetres in walls, ruptures opening vaults, structure's cracking, fall of masonry, e.g. load support stability affected.

The cracking location and the wall material have an influence on the particle-velocity level at which usual cracking begins. Exact information concerning either wall covering/plaster or location of crack inspection has to be recorded correctly. If the entire structure is not inspected thoroughly, threshold values of cracking can be biased by the location and manner of the most intense inspection. The location of the transducer is as important as the measured velocity itself.

6.4.4 Damage Observed in Test Structures

Test structures, described above, were constructed within 400 m from operating opencast mines. Vibrations were monitored for blasting faces advancing towards the structures. Initially, when blasting faces were away, the level of vibration observed was non-damaging. As the faces were advanced, the level of vibration increased. The structures were thoroughly inspected for cracks before and after each blast. Vibration levels at which damage to the structures in the form of new cracks, extension/widening of old cracks, etc. were recorded, are listed in Table 6.5. For the same blast, the vibrations measured at the ground, first floor and roof of the structure differed. The damaging values of vibration monitored at different locations in the structure are given in Table 6.8. The damage and cracks observed in the structures are discussed below (Plates 6.1 to 6.4).

6.4.4.1 Mud House with Tiles

Mud houses constructed at all the sites were affected very much by the weather. Numerous cracks developed over the smoothened surfaces even before the blasting. The cracks further widened and extended with the passage of time (Plate 6.1). It was difficult to distinguish whether widening/extension of cracks was due to blast vibrations or due to change in the weather. Although new cracks developed at a vibration level of 107.4 mm/s (PPV), the cracks became prominent at PPV of 121.0 mm/s.

6.4.4.2 Concrete Structure

Concrete walls were found to vibrate for longer duration than brick and mud walls. Since the concrete walls were free at the top, i.e. without a roof, they showed no cracks at vibration levels 182 mm/s while the other two types, mud and brick, were severely damaged. Cracks developed in the concrete walls only at a vibration level of 400 mm/s (PPV) while the other two types collapsed (Plates 6.2 and 6.3).

6.4.4.3 Brick Structure

Cracks in brick structures were first observed at junctions of walls and roof and at window corners. With an increment in level of vibrations, the earlier cracks widened and extended—leading to severe damage. Brick walls with clay mortar and sand-cement mortar reacted almost the same. There was no distinct change observed in development of cracks in the structure.

The lowest level of vibration which caused development of hairline cracks was 50 mm/s (PPV) observed on the ground floor of the structure. The old cracks widened/extended with an increase in vibration level to 85 mm/s and above. Window glass panes cracked at PPV of 86.6 mm/s. Vibration levels around 100 mm/s and above caused more and more cracks to develop in the structure (Plate 6.4). The brick and mud houses were severely damaged at 182.4 mm/s and virtually collapsed at PPV of 400 mm/s.

Table 6.5 Summary of blast studies on test structures

Mine	*Type of test Structures*	*Period of monitoring*	*No. of shots*	*Total explosives in the shot (kg)*	*Max. charge in a delay (kg)*	*Distance from shot (m)*	*Frequency of ground vibration (Hz)*
WM OCP							
Phase I	S	21.02.86 – 26.05.86	10	780 – 3845.8	216.6 – 625	200 – 7.0	11 – 40
Phase II	D	05.11.86 – 07.05.87	36	100 – 4000	25 – 400	128 – 20.0	11 – 50
SJ OCP	S	01.08.86 – 14.03.87	17	75 – 297.43	11.62 – 50	179 – 26.0	6.25 – 35
GN OCP	S	04.11.86 – 30.06.87	14		10.8 – 2125	155 – 9.0	12.5 – 25
WB QUARRY-A	D	12.02.88 – 14.08.88	12	145 – 1568	100 – 200	60 – 2.7	
WB QUARRY-B	D	12.02.88 – 14.04.88	12	919 – 1774	18.3 – 162.5	110 – 7.0	17 – 32
BJN OCP	D	26.04.88 – 31.05.88	17	36 – 152.35	18 – 41.55	110 – 23.0	7 – 28

Note : 1. S—Single storey structure (brick + cement, brick + clay as mortar), concrete structure, mud house shaded with tiles and straws
2. D—Double storey

WB- A—West Bokaro Quarry A   WM-II—West Mudidih Phase II   BJN—Badjna
WB- B—West Bokaro Quarry B   SJ—South Jharia
WM- I—West Mudidih Phase I   GN—Gopinathpur

Table 6.6 Recorded frequencies of ground and structural vibrations at different sites in Hz

Site	*Ground*	*Brick Structure*	*Concrete Structure*	*Mud House*
WB OCP:— S	11-40	6-25	8–33	
- —D	11-50	6-25	8-33	16-20
WB-QA	17-32	7-15	7-17	NA
SJ OCP	6-35	10-45	8-50	NA
GN OCP	12-25	25-33	NA	NA
BJN OCP	7-28	Ground floor		
		7-40	NA	
		Roof		
		5-28		NA

Table 6.7 Natural frequencies of test structures recorded during the study

Structure	Natural frequency in Hz
Single storey brick structure	12-14
Double storey brick structure	8-10
Concrete structure	9-16

Table 6.8 Minimum values of blast-induced vibration at which damage occurred in the structures

	PPV in mm/s	Place of measurement	Type of structure
Appearance of first crack	50.0	Ground floor	Brick structure
Appearance of first crack	76.0	First floor	Brick structure
Appearance of first crack	400.0	Side walls	Concrete structure
Widening of existing cracks	90.0	Ground floor	Brick structure
Deep cracks	240.0	Ground floor	Brick structure
Falling of plaster	280.0	Ground floor	Brick structure
Severe cracks	192.0	Sidewalls and corners	Mud house

6.4.4.4 Findings of Investigation

The investigations on test structures revealed the following:

1. The magnitude of vibration on test structures was 1.0-2.82 times greater than on the ground. Duration of structure vibration was also 1.0-2.2 times longer than that of ground vibration.
2. Natural frequency of test structures varied between 8 and 16 Hz.
3. Within the range of natural frequency of the structure, the amplitude of particle velocity on the structures amplified approximately six times versus ground vibration.
4. High-frequency waves varied between 22 and 80 Hz and low-frequency wave ranged between 6 and 22 Hz. PPVs associated with low-frequency waves were 1.09 to 3.39 times higher than for high-frequency waves.

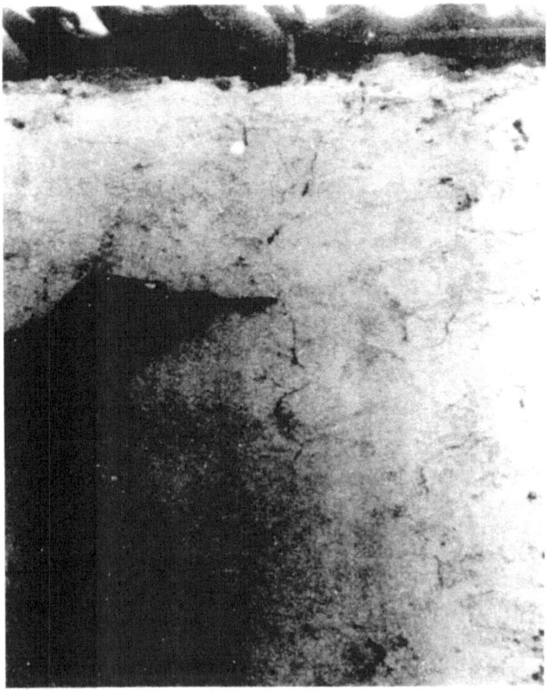

Plate 6.1 Superficial cracks in smoothened surface of mud house at an experimental site

Plate 6.2 Frontal view of double-storey house at an experimental site

Plate 6.3 Severe cracks near roof of single-storey concrete structure

Plate 6.4 Severe cracks near window of brick structure

6.4.4.5 Damaging Levels of Vibration in Test Structures

1. The lowest level of ground vibration at which new cracks developed in the brick structure at one site was 50 mm/s and the frequency was greater than 20 Hz.
2. At PPV of 85 mm/s and above, almost all existing cracks were either extended and/or widened.
3. Window glass panes cracked at PPV of 86.6 mm/s.
4. Vibration levels around 100 mm/s and above caused more and more cracks and falling of loose mortar.

Based on the investigations carried out by USBM researchers, Dowding (1985) showed that the typical natural frequency values of structures lay between 5 and 15 Hz, which decreased as the number of floors increased.

The ceiling and walls vibrate independently from the superstructure and usually have natural frequencies between 12 and 20 Hz. Dowding et al. (1980) reported that, like natural frequency, absorption was also another important parameter. They mentioned that the common values of these coefficients in residential type structures varied by around 5%.

6.5 FIELD STUDIES ON RESIDENTIAL AND INDUSTRIAL STRUCTURES

Besides the above mentioned long-term R&D research programme carried out on various test structures, the author also alongside generated vast database on different kinds of existing structures (made by their owners) at the edge of operating opencast mines in different parts of India. In such studies, basically structural responses along with their resonance characteristics as well as magnification and attenuation problems were considered. Experimental data from 20 different mines in India were analysed to develop one simple general equation for use in field conditions. Interpretation of low frequency waves and their active role in enhancing damage probability of structures was also taken into account. Particular situations were examined where the blasting was expected to generate low and high dominant band frequencies. Table 6.9 shows the abbreviated form of different structures considered during field investigations. Table 6.10 shows the details of field results on various kinds of structures as abbreviated in Table 6.9. Severity of blast damage can be viewed in the Plates 6.5 and 6.6.

Table 6.9 Abbreviated form of different structures where vibrations were monitored

Abbreviated form	*Type of construction*
BC	Brick and Cement construction with concrete roof
BMC	Brick + Mud and Cement plastered, with wooden ceiling
CN	Framework structure with reinforced concrete
M	Mud structure with tiled roof
BL	Brick + Lime and lime-plastered, with wooden ceiling

6.5.1 Results and Analysis

The vibrations in buildings can be magnified due to the response of the structural elements of which they are composed. It is therefore, very important to control the delay intervals of the detonators. By lowering charge weights per hole using a number of in-hole delay detonators and thereby increasing blasting times, harmful modulated vibration frequencies may be generated if they are close to those of resonance. The data shown in Table 6.10 were separately analysed, taking into consideration the blast geometry, blast size, location, monitoring stations and other relevant aspects. The results, so obtained, are given in Table 6.11.

Table 6.10 Details of vibration measurements at the ground level (compacted ground surface) and on structure adjacent to it due to surface blasting

Experimental site	*Total charge* (kg)	*Maximum charge per delay* (kg)	*Distance* (m)	*PPV (mm/s) and frequency* Compacted ground	Structure	*Height difference* (m)	*Amplification factor Structure type**
1	2	3	4	5	6	7	8
AMLAI OCP SECL	14350.00	1800.00	940	4.03 (5 Hz)	5.37 (5 Hz)	0.92	1.33 BC
DHANPURI OCP, SECL	56875.00	5600.00	3000	1.596 (5 Hz)	1.640 (15 Hz)	0.76	1.03 BMC
	19528.00	5200.00	1450	3.73 (5 Hz)	4.03 (5 Hz)	0.92	1.08 BMC
SHARDA OCP, SECL	632.75	382.00	325	2.41 (9 Hz)	2.48 (8 Hz)	0.61	1.03 CN
	2137.50	737.50	300	13.72 (9 Hz)	14.73 (9 Hz)	0.61	1.07 CN
UMRER OCP WCL	1350.00	187.50	130	15.00 (15 Hz)	40.10 (12 Hz)	4.0	2.67 BC
			118	21.9 (16 Hz)	55.9 (12 Hz)	3.05	2.55 BC
DVC BERMO Mine	731.50	300.00	312	2.00 (15 Hz)	8.90 (15 Hz)	2.44	4.45 BC
	775.00	225.00	450	0.67 (25 Hz)	1.312 (17 Hz)	1.2	1.96 BMC
RAMPURA AGUCHA Mine, HZL	2150.00	445.00	1380	0.445 (5 Hz)	0.925 (9 Hz)	2.44	2.07 M
	4520.84	645.00	1190	0.880 (7 Hz)	1.978 (8 Hz)	3.65	2.25 BC
KIRIBURU Iron Ore Mine, SAIL	2029.30	764.40	120	11.78 (6 Hz)	14.90 (7 Hz)	0.76	1.26 BC
	1825.00	650.00	200	9.99 (7 Hz)	11.33 (8 Hz)	0.76	1.13 BC
JAMBAD OCP, ECL	260.13	140.70	145	6.71 (12 Hz)	14.32 (15 Hz)	3.20	2.13 BC
	230.25	118.75	128	7.75 (10 Hz)	12.82 (18 Hz)	2.74	1.65 M
RAJASHREE CEMENT MINE, IRIL	1925.00	175.00	135	10.36 (28 Hz)	27.29 (25 Hz)	3.76	2.63 CN
DEMPO Iron Ore Mine, DMC	390.00	65.00	350	0.60 (7 Hz)	1.19 (5 Hz)	1.22	1.98 BMC
CHASNALLA OCP, IISCO	1237.50	225.00	400	4.03 (7 Hz)	12.53 (5 Hz)	3.96	3.11 BC
	1475.00	337.50	250	3.88 (5 Hz)	10.59 (7 Hz)	1.52	2.73 BC

(*Contd.*)

(*Contd.*)

Experimental site	*Total charge* (kg)	*Maximum charge per delay* (kg)	*Distance* (m)	*PPV (mm/s) and frequency* *Compacted ground*	*Structure*	*Height difference* (m)	*Amplification factor Structure type**
1	2	3	4	5	6	7	8
IPI-SP Limestone Mine	125.00	41.70	275	0.65 (32 Hz)	1.64 (27 Hz)	1.83	2.52 BC
RAMNAGAR Colliery, IISCO	712.50	50.00	115	5.22 (10 Hz)	7.01 (7 Hz)	3.66	1.34 BMC
	2300.00	50.00	230	0.89 (5 Hz)	3.53 (7 Hz)	3.05	3.97 BC
	2300.00	50.00	240	1.49 (8 Hz)	9.84 (10 Hz)	7.0	6.60 BC
Diamond Mining Project, PANNA	2268.48	200.16	850	0.25 (22 Hz)	0.60 (16 Hz)	3.66	2.40 BC
NICHITPUR Colliery BCCL	2637.50	250.00	100	7.61 (7 Hz)	31.62 (7Hz)	4.57	4.15 BC
	2443.75	125.00	120	8.35 (8 Hz)	21.62 (8 Hz)	1.78	2.59 BC
	2443.75	125.00	110	7.01 (17 Hz)	9.54 (20 Hz)	1.40	1.36 BC
LINGARAJ Opencast Mine, MCL	530.00	80.00	200	4.18 (35 Hz)	7.61 (17 Hz)	3.05	1.82 BC
	798.00	114.00	225	9.91 (20 Hz)	16.26 (15 Hz)	3.50	1.64 BMC

*See Table 6.9 for abbreviated form of the type of structures

6.5.2 Development of Predictor Equations

In the preceding sections, it has been discussed how and why a seismic pulse, propagating through the ground, changes its behaviour while passing through structures. The intensity, duration as well as the characteristics of the propagating waves also change. Table 6.11 confirms that the frequency of the propagating wave also changes when it is modulated in the structure. It was also observed that in the case of surface blasting and surface effects, the vibration is amplified with respect to the height of the structure, whereas in the case of underground blasting and surface effects the reverse is the case. In this case, the greater the height of the structure, the lower the value of vibration. The reason could be due to the modulation of the high frequency seismic pulse with the low natural frequency of the structure.

Plate 6.5 The ceiling of this house collapsed due to heavy quarry blasting as claimed by the owner. Although no noticeable sharp crack appeared in roof and walls of the rooms

Plate 6.6 Severe damage occurred at a villager's house. Its distance at the time of taking this photograph was 210 m and the quarry was supposed to extend another 60 m. Persons residing in this house were shifted to the company's quarters

The data in Table 6.10 were thoroughly analysed to obtain the best-fit equation (6.2) with a moderately high index of determination value. The main feature of this equation is that one can predict the anticipated vibration level of any structure at any specific height by knowing the value of the ground vibration near to it. This could help the mine management in determining the level of risk associated with any structure for a particular size of blast.

Table 6.11 Results of surface blasting and its effects on surface structures

Type of structure (See Table 6.9)	*Rate of amplification measured on the structure with respect to the ground vibration*
BC	Maximum: 6.60
	Minimum: 1.13
BMC	Maximum: 3.05
	Minimum: 1.03
CN	Maximum: 2.63
	Minimum: 1.03
M	Maximum: 2.07
	Minimum: 1.65

The data in Table 6.10 were regressed to obtain the following best-fit equation (6.2)—

$$V_S = 1.312 \times V_g \times \exp(0.0215 \times d_1 + 17.63 \times d_2) \quad (6.2)$$

I_d (index of determination) = 0.833

Where,

V_s = Peak particle velocity on the structure (mm/s)

V_g = Peak particle velocity on compacted ground at the foundation level of the structure (mm/s)

d_1 = (Total charge fired in the round/Maximum charge per delay); both in kg

d_2 = (Sensor height above ground level/Sensor distance from the blasting source); both in m

Duration of vibration: Vibration persisted 1.5 to 2.5 times of the total hook-up time of surface delays or total-firing time of holes.

Rate of amplification: Taller structures showed higher amplification.

Frequency of vibration: Dependent on the geological formation of the rock mass. The frequency was generally high at ground level. In most cases (> 24 Hz) and low in some cases (≤ 24 Hz).

Modulated frequency: Low frequency (< 24 Hz) on the structure.

6.6 EFFECT OF REPEATED BLASTING

It is inevitable that repeated loading will be raised as an issue in blast cracking claims where the measured particle velocities are below normal criteria. Such degradation through repeated loading is called fatigue in material testing. A structure gets weakened when it is repeatedly shaken by blasting vibrations even at a low intensity. And when a structure gets weakened, it becomes susceptible to damage even at low intensity of vibration or acceleration.

Vibration generated due to opencast blasting is a short-term transient phenomenon, lasting usually a second or two, frequently less. The question of repeated blasting effects arises regularly as a point of concern, such as—it didn't damage my house this time, but what will happen if it is repeated? The effects from pile driving and recurring industrial operations are of similar concern. Generally, these vibration levels are relatively low with low damage potential. However, repeated blasting effects from opencast mines may be grave, especially in the low-frequency range. Wall panels, ceilings and roofs have different natural-frequency levels and therefore their response would be different when subjected to repeated blasting impacts (Konya and Walter, 1990).

Walter (1980) investigated the effect of impact vibration on a structure. Vibration continued for 13 months, 24 h a day. The test structure was an ordinary room approximately 8 by 8 by 8 ft of dry wall construction. The vibrator was mounted on the ceiling. The wall panels had a natural frequency of 12.5 Hz and the ceiling panel was 60 Hz. Vibration frequencies measured in the wall panels ranged from 10 to 18 Hz with particle velocity ranging from 1.25 to 4.00 mm/s.

Inspite of the long time involved, approximately thirty million seconds and the many impacts imparted to the structure, no damage occurred. Inspection of the structure revealed no observable effects. It was concluded that low-level vibration even in the natural frequency response range of the structure has little potential for causing damage.

An extensive fatigue test was conducted for the U.S. Bureau of Mines by the U.S. Army Corp. of Engineers, Civil Engineering Research Laboratory, CERL. An 8 by 8 by 8 ft (1 ft equals 0.3048 m) test room was mounted on a biaxial shake table which was programmed with one horizontal component and the vertical component of a quarry blast from USBM bulletin 656. The predominant blast frequencies were 26 and 30 Hz respectively.

Vibration tests were run at 2.50, 12.50, 25, 50, 100, 200, 400 and 800 mm/s. At each vibration level the test was run a series of times starting with 1 run, then 5 runs, then 10, 50, 100 and 500 runs with inspection after each series. No damage occurred until the fifth run at 100 mm/s was done.

The investigations carried out by the U.S. Bureau of Mines (RI-8896) in a wood frame house by subjecting it to 587 production blasts after which it was shaken mechanically to determine the threshold of fatigue cracking. The first crack appeared after 56,000 cycles, which was the equivalent of 28 years of blasting twice a day at a vibration level of 12.5 mm/s. The damage was a cracked wallboard tape joint. These studies showed that the fatigue effects such as cracking may occur at vibration levels that are relatively high. In fact, there should be concern about the damage potential of a single vibration event of this level.

6.7 VIBRATION RESPONSE OF RESTRAINED PIPELINES

Several authors studied near-field blast vibration response of restrained pipelines either buried or exposed. Dowding (1985) reported that the stretching strains due to shear wave propagation parallel to the structure became so small that they can be neglected. Compressive and shear waves arrive almost simultaneously when blasting is usually close. In such cases, bending (ε_b) and stretching (ε_c) strains are additive. Early researchers (US Bureau of Mines and National Research Council of Canada) reported that restrained concrete walls cannot be cracked by vibrations with PPV less than 254 mm/s and strains less than 100 μ mm/mm when measured on the concrete. As such, there is no point of damaging restrained buried pipelines either separately or inside reinforced concrete casing unless the strain exceeds 100 μ mm/mm. Pal Roy and Sawmliana (2000) conducted near-field blast vibration response of restrained pipelines during blasting at sensitive locations for Stage-II (4x500 MW) commissioning work in a super thermal powerplant of the National Thermal Power Corporation Ltd. (NTPC) of India. The highest value of strain obtained was 67.61 μ mm/mm, corresponding to 141 mm/s peak particle velocity (PPV) monitored at a distance of 9.0 m from the blasting site due to firing of 33.18 kg of total explosive charge using short delay detonators.

6.8 DISCUSSIONS

In the test structures, while experimenting, continuous inspections were carried out with sophisticated instruments such as SINCO-3, SINCO-6, VS-1600, SMU-31 vibration monitors and a Bruel and Kjaer Analyser to monitor the appearance of cracks at different vibration levels. The minimum values of PPV (mm/s) at which different cracks appeared or widened are shown in Table 6.8.

It is worth mentioning that the values shown in Table 6.10 were monitored in residential houses, school buildings, poorly constructed mud houses and other places of interest. Many of these houses were in a poor state of repair and constructed without any proper supervision or quality materials, whereas the test structures were constructed under the supervision of CMRI scientists with proper design (Figures 6.7, 6.8 and 6.9). It can also be seen in Table 6.10 that the values of peak particle velocity measured on the ground near any structure were amplified 4.45, 3.97 and 6.6 times on respective structure at heights 2.44, 3.05 and 7.0 m above ground level. Siskind (2000) also reported the similar type of amplification. This rate of amplification may be more depending on the height and the type of structure. In a recent study conducted by CMRI in a dragline face, the amplification with respect to the ground vibration was as high as 8 times on a multi-storey structure. Thus it can be said that a blast producing PPV level of 5.0 mm/s at the foundation level having low frequency, could even amplify to produce 40.0 mm/s vibration on a high structure. These facts reasonably justify the threshold values of ground vibrations posed by CMRI and DGMS (Tables 5.25 and 5.26 of Chapter 5) in conjunction with the results obtained in Table 6.8.

6.9 CONCLUSIONS

From the above study, it may be concluded that the structural response of different types of structures should be determined thoroughly before establishing any damage threshold values of vibration, otherwise, it may be misleading and noneffective. It was also revealed that due to surface blasting, both low- and high-frequency vibrations can be generated, depending solely on the geological forms or textures of the transmitting medium. Ground vibration was significantly amplified when measured on the structure and its rate of amplification would depend on the height of the structure. Predictor equation (6.2) can be used to determine the vibration on any structure when the blast geometry and the level of ground vibration are known. This may be helpful to the mine management to determine the size of blast that will not pose any problem for any residential sector.

REFERENCES

Bergling, J. O., Eklund, K. and Sjoberg, C. 1977. A villa built of light concrete and concrete exposed to vibrations from blasting, Swedish Council for building Research, Building Research summaries, Sweden, R. 32.

Bogdanoff, I., Larson, T. and Nilson, R. 1975. Light weight concrete building exposed to vibrations from blasting operations, Swedish Council for Building Research, Building Research Summaries, Sweden, R. 42.

CMRI S&T Project Report-GAP/VI/2, 1991. Investigations into the influence of blasting pattern and geo-technical properties of the surrounding rock mass on the ground vibration, fragmentation, flyrock etc., May, Central Mining Research Institute, Dhanbad, India, p. 75.

Dowding, C. H. 1985. Blast vibration monitoring and control, Prentice-Hall, Englewood Cliffs, NJ, USA, p. 287.

Dowding, C. H., Murray, P. D. and Atmatzidis, D. K. 1980. Blast vibration implications of cyclic shear behaviour of model plaster panels, ASTM Geo-technical Testing Journal, 3(2): 81-90.

Dvorak, A. 1962. Seismic effects of blasting on brick houses, Prace Geofysikeniha Ustanle Ceskoslovenski Akademine Ved., Czech Republic, No. 169.

Duvall, W. I. and Fogleson, D. E. 1962. Review of criteria for estimating damage to residences from blasting vibration, U.S. Bureau of Mines, R.I. 5968, p.19.

Edwards, A. T. and Northwood, T. D. 1960. Experimental study of effects of blasting on structures, The Engineer, Vol. 210, UK, p. 40.

Esteves, J. M. 1978. Control of vibrations caused by blasting, Laboratorio Naional De Engenharia Civil, Lisbon, Memoria, Portugal, p. 498.

Foster, Alan, G. 1981. Structural response and human response to blasting vibration effects—Is there a connection? Proc. 7th Conference on Explosives and Blasting Technique, ISEE Annual Meeting, Phoenix, Arizona, USA, pp. 10-18.

Jimeno, L. C., Jimeno, L. E. and Carcedo, A. J. F. 1995. Drilling and blasting of rocks, A.A. Balkema, Rotterdam, Netherlands, p. 391.

Konya, Calvin J. and Walter, Edward J. 1990. Surface Blast Design, Prentice-Hall Inc., USA, p. 303.

Langefors, U and Kihlstrom, B. 1963. The modern technique of rock blasting, John Wiley and Sons Inc., New York, USA, p. 405.

Mederis, K. 1978. Blasting damage criteria for low-rise structures, Proc. 4th Annual Conference on Explosives and Blasting Technique, ISEE, Miami, Florida, USA, pp. 280-290.

Newmark, N. M. and Hall, W. J. 1982. Earthquake spectra and design, Earthquake Engineering Research Institute, Berkeley, CA, USA, p. 103.

Pal Roy, P. and Sawmliana, C. 2000. Near-field blast vibration response of sensitive pipelines, J. of Explosives Engineering, November-December Issue, USA, pp. 38-44.

Siskind, D. E., Staff, M. S., Kopp, J. W. and Dowding, C. H. 1980. Structure response and damage produced by ground vibrations from surface mine blasting, U.S. Bureau of Mines, R.I. 8507, p. 74.

Siskind David, 2000: Vibrations from blasting, Inter. Soc. of Explosives Engineers, USA, p. 120.

Vazquez, J. R. and Fernandez, E.G. 1986. Progress in studying low frequency waves caused by blasting, Proc. Mini Symposium on Explosive and Blasting Research, Society of Explosives Engineers, Mini-Symposium, Feb. 13-14, Atlanta, GA, USA, pp. 72-79.

Walter, E. J. 1980. Natural variation of vibration levels associated with blasting, Proc. 6th Conference on Explosives and Blasting Technique, USA, pp. 331-340.

7

Influence of Blasting on Surface Structures and Underground Workings

7.1 SURFACE BLASTING—UNDERGROUND EFFECTS

7.1.1 Introduction

To enhance production and productivity, there is a trend to extract top coal seams, whether developed or virgin, by opencasting wherever it is economical. In coal and non-coal sectors, around the world, at several mines, underground workings and open-pit operations take place simultaneously. In India, for example, this combination of methods exists in several subsidiaries of Coal India Limited (CIL), a major portion of which falls in Eastern Coalfields Ltd. (ECL) and Mahanadi Coalfields Limited (MCL). In Chile, this combination of methods is used at Andina, Mantos Blancos, and the El Soldado mines to extract porphyry and strata-bound copper ore and an underground operation beneath the open-pit at the giant Chuquicamata copper mine.

This system enables extraction of a major portion of the reserve, which might not be possible with underground mining methods. But the heavy blasting practised in opencasting poses environmental problems. The blasting operations, if uncontrolled, may have a detrimental effect on surface or underground structures. The methodology of investigating the stability of underground roof and pillars under the impact of surface blasting and the results of these studies, conducted globally, are discussed in this chapter.

When an underground mine is developed beneath an open-pit, it is possible for blast vibrations from the open-pit to damage the underground excavations. Aspects of blast vibrations, blast damage mechanisms and the influence of rock quality on the severity of such damage should be given priority while studying damage characteristics. The rock in which an excavation is situated has an influence on the level of ground vibration to which it can be subjected without sustaining damage.

The mode (e.g. cracking, spalling or loosening of rock blocks) and the severity of damage are influenced principally by the magnitude of ground vibrations in the vicinity of the underground void and the quality of the rock in which it has been excavated (Hoek and Brown, 1980). Broadly speaking, higher levels of ground vibration would be expected to cause more damage than lower levels because of the increased shaking imparted to the rocks surrounding the excavation and the associated increase in dynamic stress (Dowding, 1979).

7.1.2 Blast Vibrations

The detonation of an explosive in a blasthole excites the rock and sets up a seismic wave, which radiates from the blast site. The passage of this seismic wave through the rock mass represents a transfer of blast-vibration energy as rock particles in the path of the wave vibrates in about their equilibrium positions. When blasting is conducted in an open-pit over an underground excavation, this transfer of energy is accomplished predominantly by primary waves, in which the rock particles move back and forth along the line of advance of the wave (Figure 7.1). Primary waves travel faster than other elastic waves, such as secondary waves and Rayleigh waves; they are, therefore, the first to arrive at monitoring stations (Tunstall, 1997).

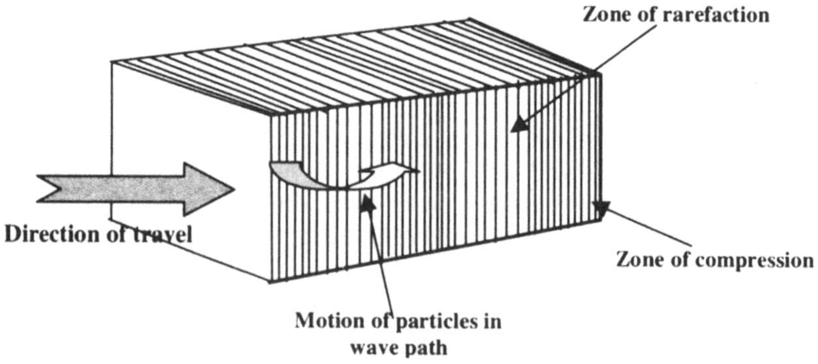

Fig. 7.1 Energy transfer in primary wave

The use of a single index of vibration, such as peak particle velocity or peak particle acceleration, is common in earthquake engineering and in the development of residential damage criteria (Siskind et al., 1980). Figure 7.2 depicts typical blast vibration time history record for surface

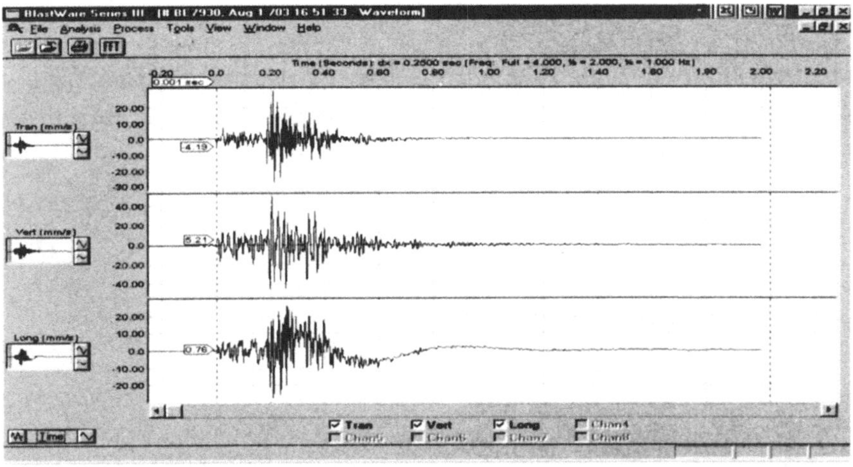

Fig. 7.2 Typical blast vibration time history record for surface blasting and underground measurement (depth of cover-112 m, total charge fired-3000 kg; Nonel Initiation)

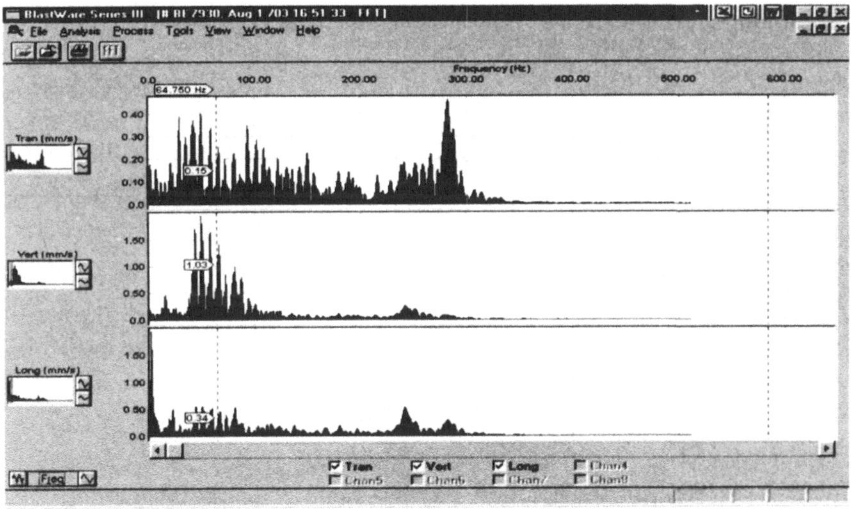

Fig. 7.3 Frequency analysis of the trace record shown in figure 7.2

blasting and underground measurement in the roof of developed gallery while Figure 7.3 shows its frequency analysis.

7.1.3 EFFECT OF ROCK QUALITY

Tunstall (1997) has elaborated quite categorically that the quality of rock in which an excavation has been created can have a significant influence on the amount of damage caused to it by adjacent opencast blasting. Of the various rock mass classification systems, the numerical method of classification developed by Barton, et al., (1974), popularly known as NGI classification, is the most reliable and effective tool available for the assessment of rock mass. The overall rock mass quality Q and rock mass rating (RMR) are given by the following equations—

$$Q = (RQD/J_n) \times (J_r/J_a) \times (J_w/SRF) \quad (7.1)$$

$$RMR = 19 \ln Q + 26 \quad (7.2)$$

where,

RQD = rock quality designation = $115 - 3.3J_v$
RMR = rock mass rating (Bieniawski, 1973)
J_n = joint set number
J_r = joint roughness number
J_a = joint alteration number
J_w = joint water reduction number
J_v = total number of joints/m^3 of rock
SRF = stress reduction factor

Both Q and RMR contain parameters that reflect the joint intensity and joint strength, and thereby determine the resistance of pre-existing rock blocks in the excavation perimeter to shear displacement from seismic loading.

7.1.3.1 Influence of the Joint Intensity

Joint intensity is crudely related to the *in situ* block size. Joint intensity is accounted for in the *RMR* system by the rock quality designation (*RQD*) and joint spacing parameters.

7.1.3.2 Influence of Joint Strength

Joint strength has a significant effect on the stability of blocks in the perimeter of an excavation. Joint condition is accounted for in the *RMR* system by the joint condition and groundwater parameters.

7.1.3.3 Influence of Intact Rock Strength

Intact rock strength determines whether fresh cracking or spalling will occur at high ground vibration levels. Intact rock strength is accounted for in the *RMR* system by the rock material strength parameter.

7.1.3.4 Orientation of Joint Sets

The orientation of the joint sets with respect to each other (and to the tunnel axis) influences the resistance of an excavation to blast damage. This factor is taken into account by an adjustment for orientation in the original *RMR* system developed by Bieniawski (1973). It is to be noted that equation (7.2) is the indirect method of obtaining *RMR* when Q is known or determined.

7.1.4 Damage Classification

Dowding and Rozen (1978) classified observed damage into the following two major categories:

(a) Minor damage: characterized by new cracks and minor rock fall with estimated peak particle velocities in the range 200-400 mm/s;

(b) Major damage: characterized by severe cracking, major rock falls and tunnel collapse at estimated peak particle velocities in excess of 400 mm/s.

In a separate study of the effects of surface blasts on unlined underground military silos, Labrenche (1983) identified the following three types of damage characteristics at increasing distances from the blasts:

(a) the fracturing of intact rock and tensile spalling close to the blast,

(b) the acceleration of the existing blocks into the tunnel further away, and

(c) gravity induced rock falls at greater distances from the blasts.

In coal mines, Fourie and Green (1993) categorized the degree of damage according to the following criteria:

Class I—Minor damage: Damage manifested as falling of loose materials from the pillar sides and roof

Class II—Intermediate damage: New cracks are formed in intact rock material. Gradual spalling develops on the sides of pillars accompanied by periodic falls to reveal fresh surfaces

Class III—Major damage: The formation of new cracks accompanied by immediate falling of freshly loosened materials. Gradual spalling develops on the sides of pillar accompanied by periodic falls to reveal fresh surfaces.

Kidybinski (1986) reported that damage to underground coal mine openings in the form of small roof falls or floor heave may occur when the PPV lies in the range of 50-100 mm/s and large roof falls at PPV of 100-200 mm/s (Table 7.1). Tunstall (1997) suggested that peak particle velocity of 175 mm/s did not contribute any damage to the underground opening where very good quality rock (RMR = 85) was encountered. On the other hand, poor quality rock (RMR = 49), which had been loosened by previous open-pit blast vibrations, sustained minor visible damage at a peak particle velocity of 46 mm/s and major damage at a peak particle velocity of 379 mm/s. Singh et al. (1999) observed that cracks in the coal roof occurred at peak particle velocity of 296.69 mm/s but spalling of coal chips from pillars and roof started at a vibration level of 125.3 mm/s. Lewandowski et al. (1999) set a conservative criteria of targeted maximum PPV of 50 mm/s for the safety of coal in underground heading. They further clarified that this conservative value of PPV was decided after investigations indicated a possible limit of 250 mm/s.

Table 7.1 Destructive effect of seismic events (after Kidybinski, 1986)

Peak particle velocity (mm/s)	*Roadway deterioration anticipated*
50-100	Slight deterioration, small roof falls and / or floor heave
100-200	Large roof falls and / or floor heave
200-500	Significant damage
More than 500	Complete destruction

7.1.5 Blast Damage Index (BDI)

Blast Damage Index (BDI), which is denoted by 'D_{ib}', is defined by the ratio of the induced stress due to blasting to the dynamic tensile strength of the rock mass being blasted (Yu and Vongpaisal, 1996). It is a dimensionless indicator analogous to the reciprocal of safety factor and may vary between 0–3.5. It can be used as an index to quantify the damage of roof and pillars due to underground blasting for the safety of man and machinery. The Blast Damage Index (D_{ib}) is defined as follows:

$$D_{ib} = \frac{3.6 \times 10^{-3} \cdot V_R \cdot \rho_r \cdot C_P}{K \cdot U_{CS}} \tag{7.3}$$

where,

V_R = Vector sum of peak particle velocities in three orthogonal directions (mm / s)

ρ_r = Density of rock mass (g / cc)

C_P = *P*-wave velocity (m / s)

K = Site quality constant=*RMR* (20-100)

U_{CS} = Uniaxial compressive strength of rock mass (MPa).

Table-7.2 shows different range of values of Blast Damage Index (D_{ib}) and the corresponding damage established by Yu and Vongpaisal (1996).

Table 7.2 Blast Damage Index (D_{ib}) and corresponding type of damage (after Yu and Vongpaisal, 1996)

D_{ib}	*Type of Damage*
≤ 0.125	*No damage to underground excavation.* Maximum allowable value for key permanent workings, e.g. crusher chambers, shafts, permanent shops, ore bins pump house etc.
0.25	*No noticeable damage.* Maximum tolerable value for long-term workings, e.g. shaft accesses, rescue stations, lunch room etc.
0.50	*Minor and discrete scabbing effects.* Maximum tolerable value for intermediate term workings, e.g. main drifts, main haulage ways, etc.
0.75	*Moderate and discontinuous scabbing damage.* Maximum tolerable value for temporary workings, e.g. cross cuts, drill drifts, stope accesses, etc.
1.00	*Major and continuous scabbing failure, requiring intensified rehabilitation work.*
1.50	*Severe damage to an entire opening, causing rehabilitation work difficult or impossible.*
≥ 2.00	*Major caving, normally resulting in abandoned accesses.*

7.1.6 Case Studies

7.1.6.1 S&T Study by CMRI

The Central Mining Research Institute (CMRI) has established *RMR*-based (of roof rock) damage threshold values of underground workings due to nearby surface blasting impacts, after extensive field experiments at seven different sites (Singh, 2003). The sites were—

1. West Mudidih colliery, seam IV due to surface blasting in West Mudidih opencast project of Bharat Coking Coal Ltd.
2. West Mudidih colliery, seam III due to surface blasting in West Mudidih opencast project of Bharat Coking Coal Ltd.
3. Kusunda colliery, combined seam V, VI, VII, VIII due to surface blasting in Kusunda opencast project of Bharat Coking Coal Ltd.
4. Bansra colliery, Purandip seam (R-VIII bottom) due to surface blasting in Bansra opencast project of Eastern Coalfields Ltd.
5. Kottadih colliery, seam R-III/II due to surface blasting in Kottadih opencast project of Eastern Coalfields Ltd.
6. Orient Mine No. 2, HR seam IV due to surface blasting in Lajkura opencast project of Mahanadi Coalfields Ltd.
7. Hingir Rampur colliery, HR seam IV due to surface blasting in Samleshwari opencast project of Mahanadi Coalfields Ltd.

In all, 202 blasts were experimented and 622 vibration data were recorded at those seven sites. The plan distance from the blasting site in opencast mine to the monitoring points in underground workings varied from zero (vertically below point) to 630 m away and the depth of cover varied from 11 m (blasting source and monitoring station were in the same seam IV of West Mudidih colliery) to 159 m in Hingir Rampur colliery. The explosive charge weight per delay fired in a round differed widely (9 kg to 3,892 kg) depending on the size of a blast. Similarly the total charge weight in a round of blast also varied from 47 kg to 24,360 kg. The propagating medium was mostly medium-hard sandstone and the roof rock up to 2 m was coal at 4 sites and coal-shale, coal-shale-sandstone bands at 3 remaining sites.

The following blast damage criteria were established (Figure 7.4). The study also indicated that a value of BDI less than 1.0 would create no damage whereas more than 2.0 would create major damage. Minor damage would be in between 1.0 and 2.0 of BDI values.

7.1.6.2 Blasting in Ground Excavation and mines (Singh and Pal Roy, 1993)

Singh and Pal Roy (1993) extensively discussed three case studies carried out by the Central Mining Research Institute during the early nineties. A well-compiled version of such studies is also available in Siskind's book (2000). These works are presented below:

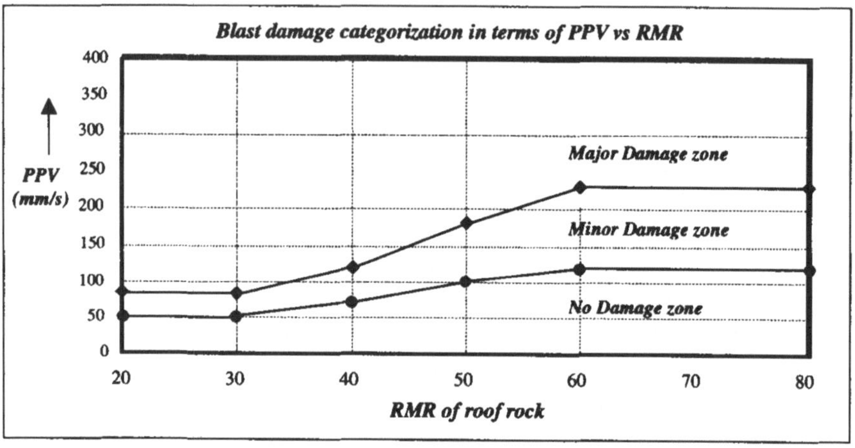

Fig. 7.4 Different damage levels based on PPV and RMR

7.1.6.2a Site-A

Opencast mine # 1 on the rise side of underground mine # 2 under New Majri Colliery

The underground mine had a 3.0 to 4.5 m parting between two developed seams in a total coal seam thickness of 15 to 18 m. Coal pillars were 20 m centre to centre. Extensometers and convergence indicators were installed in both sections for possible bed separations and convergence. Vibrations were measured in both coal sections on the roofs, floors and coal pillars. Ten blasts were measured, 300 to 2250 kg fired instantaneously, and at distances (depths of cover) of 150 to 286 m.

The minimum and maximum vibrations observed in the roof were 0.73 and 12.73 mm/s whereas in the floor they were 0.76 and 9.87 mm/s. The extensometers and convergence indicator showed no changes.

Vibration analysis: On statistical analysis of field data, the following best-fit empirical relationships involving the maximum charge weight per delay (Q), distance (D) and the vibration (V) were established.

(a) $V = 1909.72\ (D/Q^{1/3})^{-1.961}$ (7.4)

$r = -0.85$

(b) $V = 2454.7\ (D/Q^{1/3})^{-2.199}$ (7.5)

$r = -0.96$

where r is the correlation coefficient.

Equation (7.4) was related to the vibration measured in the parting, i.e., floor of top section and roof of bottom section. Equation (7.5) was obtained after analyzing the vibration data measured in the roof of the top section.

7.1.6.2b Site-B

Underground water dams on the rise side of Kargali Colliery near quarry # 3

Five dams were located in two sections of the 21 m thick Kargali coal seam. The top section along the middle bed about 11 m from the roof and the bottom section along the seam floor. The nine surface mine blasts were 550 to 1440 kg total and 62.5 to 187.7 kg/delay, and slant distances from the monitoring locations of 99 to 201 m.

It was noted that underground vibrations were lower than those on the surface by a factor of 1.2 to 3.12x, a result globally reported by other researchers. The highest underground velocities were 23.35 mm/s and no damage to the dams was found. There was also absence of spalling of coal from roof and pillar areas.

Vibration analysis: On statistical analysis of the vibration data, the following empirical relationships were established:

(a) $$V = 324.79\ (D/Q^{1/2})^{-1.23} \quad (7.6)$$

$$r = -0.91$$

(b) $$V = 685.36\ (D/Q^{1/3})^{-1.372} \quad (7.7)$$

$$r = -0.69$$

Equation (7.6) was obtained from the surface measurements and equation (7.7) was established from the data measured on underground coal pillars.

7.1.6.2c Site-C

Ramagundem Area-II of M/s Singareni Collieries, GDK-7 Incline

Two coal seams were surface mined over an underground mine roadway. The depth of cover was 115 to 130 m and about 8 m less from the bottom of the blast holes. Seven rounds were monitored, 1600 to 8388 kg total and 400 to 1500 kg/delay. Vibrations were measured in the roof and coal pillars.

Maximum vibrations were 48 mm/s in the roof and 31.61 mm/s in the coal pillars. There was no spalling either from the roof or pillars except for a few small pieces of coal and dust, which were probably already loose there and pillars. Inspection of the roadways before and after the blast revealed that they remained unchanged.

Vibration analysis: The best fit relationship between the peak particle velocity (V), monitoring distance (D) and the total charge weight (Q_T) was obtained. Instead of maximum charge per delay, the total charge was found responsible for varying amplitudes of PPV. The reason might be due to total strata movement instead of localized disturbance. The relationship obtained was:

$$V = 96.091 \ (D/Q_T^{\ 1/2})^{-1.8016} \quad (7.8)$$
$$r = -0.70$$

The frequency bands in which the major percentage of energy was spent varied between 5-53 Hz. These are listed in Table 7.3.

Table 7.3 Frequency content of vibration waves measured in GDK-7 incline of Ramagundem Area, SCCL

Vibration energy (%)	*Range of frequency (Hz)*
72	11-39
18	8-39
90	5-18
53	19-53
70	5-10
68	8-18
72	15-27
49	19-53
83	5-27
86	5-18
77	5-10
57	5-10
81	5-18

7.1.6.3 Seismic Effects of Quarry Blasting (Thoenen and Windes, 1942)

Thoenen and Windes (1942) measured mine and pillar vibrations from quarry blasting at three locations in an underground limestone mine to determine if the mine roof had a characteristic resonant (or natural) frequency. They concluded that "as the roof did not react as a unit, there was little, if any, possibility of its vibrating in synchronization with an imposed vibrating force, as such as might be caused by mine blasting, or of failure of a roof due to synchronous vibrations. This does not mean that loose pieces of the roof will not react to vibrations and fall. However, the tests indicate that the roof as a whole does not react dangerously to imposed vibrations."

7.1.6.4 Effect of Depth (Siskind and Stagg, 1985; Siskind, 2000)

Siskind (2000) reported that at distances beyond the immediate blast zone of about 15 to 75 m, the strongest vibration waves were surface waves. These were produced by the interaction of *P* and *S* body waves with geologic structural surfaces, the ground-air interface being the most important one. A characteristic of surface waves was their rapid decrease of amplitudes with distance away from that surface. For surface mine blasting, this implies that vibrations rapidly decrease with depth as well as with horizontal distance, which may even be true in as small a depth as the basement floor of a home (Siskind and Stagg, 1985). As a result, vibrations at large depths will generally be dominated by the relatively low-amplitude body waves. Siskind (2000) also concluded that while specifying 'velocities' the more appropriate parameter was 'acoustic impedance', which was the product of propagation velocity (i.e. sonic wave velocity and not particle velocity) and density. Densities of geologic materials do not vary as much as propagation velocities therefore, most comparisons are done using the velocities in the various media.

7.1.6.5 Rock Blasting (Langefors and Kihlstrom, 1963; Hendron, 1977)

Langefors and Kihlstrom (1963) and Hendron (1977) reported that in hard rock "falls of stones in galleries and tunnels" at 300 mm/s and "cracks in rock" at about 600 mm/s. Hendron (1977) also concluded after conducting his tests near nuclear blasts that visible damage in unlined tunnels could rarely occur below a free-field particle velocity of 300 to 600 mm/s.

7.1.6.6 Summary of Other Important Works

7.1.6.6a Sakurai and Kitamura (1977)

The authors studied the impacts of surface mining blasts (not so big sizes) on two unlined tunnels in granite and two concrete-lined tunnels, one in granite and one in liparite. They were at 32 and 61 m depths and separated by a minimum vertical distance of 25 m. The authors measured tensile strains and vibrations on the tunnel roofs and sides. Based on a tensile strength of 1.96 MPa, they concluded that damage would occur at a particle velocity of 350 mm/s. This prediction is comparable to a case of observed crack initiation in concrete lining at 338 mm/s (Siskind, 2000).

7.1.6.6b Rupert and Clark (1977)

The authors scientifically monitored underground responses due to blasting in a nearby surface mine in the same seam. They monitored vibrations and at the same time used roof bolt pads, horizontal roof strain indicators, convergence pins, and thermometers for observing any change in the mine condition due to surface blasting impacts.

They reported "only minor damage in the form of localized thin spall and possible collapse of previously fractured coal ribs from vibrations above 50 mm/s". They also noted an absence of major damage or changes in the mine condition (roof bolts and convergence).

7.1.6.6c Jensen et al. (1979)

In this USBM-sponsored project, the authors, in addition to measuring mine roof, floor and surface vibrations, also monitored convergence, inspected for rock falls, and borescoped roof strata at three locations. They conducted 26 blasts of 125 to 5851 kg/delay and up to 23,1334 kg of total explosive charge detonated directly over the mined-out coal seam at depths of nearly 45 m.

The authors reported no roof failures even at roof vibrations of 445 mm/s and only a few loose stones at 127 mm/s.

7.1.6.6d Fourie and Green (1993)

Over a year, the authors monitored for responses to 32 surface mine blasts of varying sizes and found that the integrity of the underground mine remained unaffected. For correct judgment, the coal pillars were cleaned and whitewashed to identify any new spalling from blast impacts.

7.1.6.7 Summary of Vibration Risks in Underground Works

Blast vibrations may induce substantial dynamic loading on mine openings. There is a straightforward procedure to calculate the imposed dynamic loads due to a given vibration. It is difficult to assess and to generalize the effect on the overall stability of workings. The best way, possibly, is to do a physical assessment of the underground workings in order to develop a risk strategy and to understand the mechanism which would control the deterioration of the mine. To understand the long-term effects of vibrations in the stability of old workings, a continuous programme of vibration monitoring and statistical damage analyses may be the correct and feasible procedure for determining any damage criteria. It is well known that horizontal strain and tilt are potentially most damaging to surface structures. However, in mining areas, people mostly get confused whether the alleged damage of surface structures over old underground workings are due to subsidence or nearby surface mine blasts. Figure 7.5 and Plates 7.1 to 7.3 depict a few such typical circumstances, which generally create confusion about the reality of the alleged damage.

After reviewing technical research works carried out in the world, Siskind (2000) has made a general recommendation on risks in underground works due to blasting. He concluded that "although, there is much variation between the structure and geologic conditions represented by several authors worldwide, the general observation is that major failure such as roof collapse and pillar failure would require vibrations greater than 300 mm/s". He also commented that in some cases, loose pieces were dislodged even at lower vibration levels of about 12 to 30 mm/s. However, low-level vibrations, below 25 mm/s, were found to be totally harmless to underground workings, even active ones where rockfalls are a personnel hazard.

7.2 UNDERGROUND BLASTING—SURFACE EFFECTS

7.2.1 Introduction

The operation of drilling and blasting for mining activities beneath a shallow depth of cover has led to many complaints from property owners.

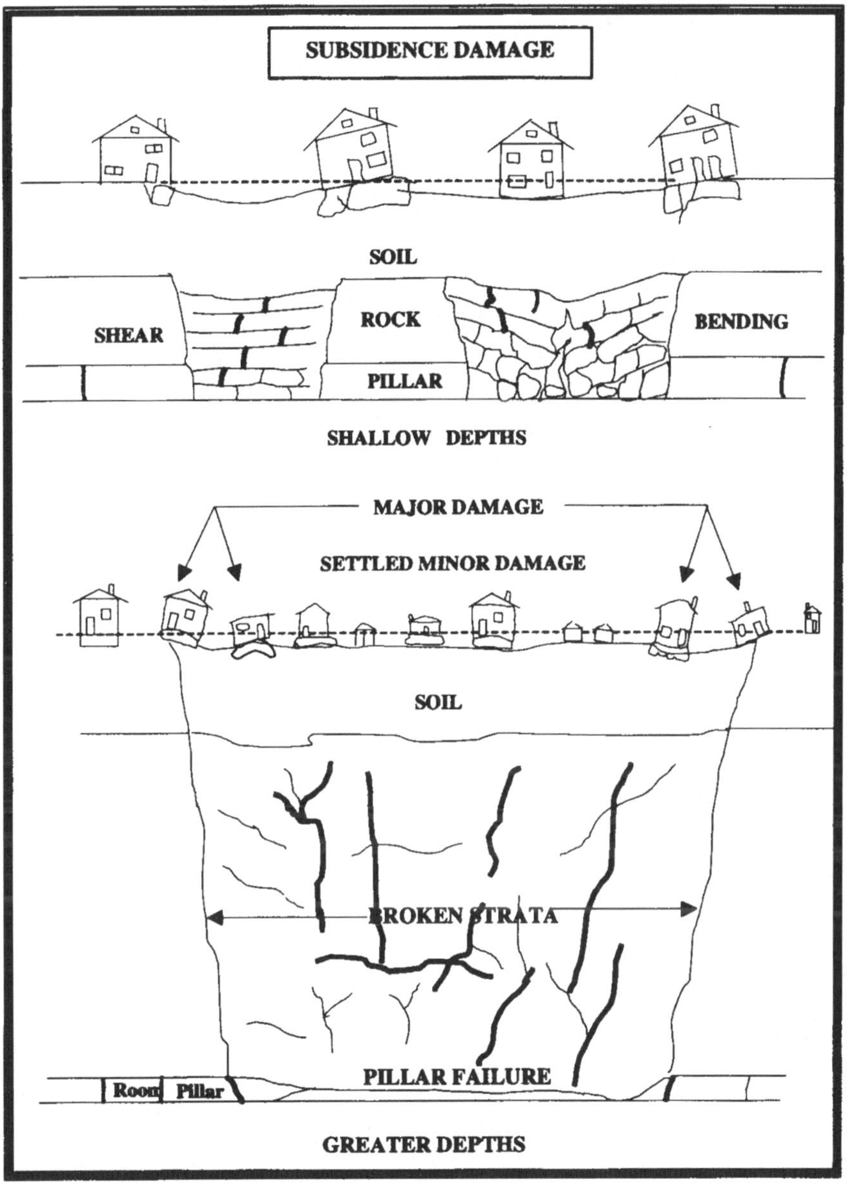

Fig. 7.5 Scientific reasoning of building damage due to subsidence (after Clark and Cavin, 1981)

The management of mines is often threatened with cessation of blasting unless a solution to contain blast-induced vibration within tolerable limits can be found. The answer to this dilemma is scientific investigation of

Plate 7.1 Pictorial view of tensile crack appeared due to land subsidence, a fact generally confused with that of underground blasting impacts

Plate 7.2 View of typical dwellings in a village above underground workings

Plate 7.3 Extensive cracks appeared on the surface due to land subsidence

whether blasting at the site of complaint genuinely poses a problem or is a fabrication arising from vested interests. The Central Mining Research Institute has conducted a number of such investigations covering mines in BCCL (Bharat Coking Coal Limited), ECL (Eastern Coalfields Limited), WCL (Western Coalfields Limited), and SECL (Southeastern Coalfields Limited). These studies established a reliable technological basis for planning and conducting blasting operations so that the number of complaints regarding damage and the nuisance to mine management would be minimized and at the same time enabled them to conduct comprehensive analyses of the vibrations produced due to underground blasting. Guidelines for a safe-blast design, methods to reduce vibrations without altering the efficiency of a blasting round and empirical means to predict maximum vibration amplitudes were established. A comparative study on the effect of machine-cut blasting and blasting off the solid (BOS) was also conducted and is discussed below (Singh and Pal Roy, 1993).

7.2.2 Problems Related to Underground Blasting

The private owners/residents of dwelling houses in close proximity to underground blasting operations, though familiar with the importance of extracting minerals, look upon mining activities as an intrusion in their lives and wastage of their properties. Sometimes they are exhorted by local leaders with a vested interest to raise a hue and cry against mining activities. Thus, a number of socioeconomic problems beleaguer the mining industry and very often claims for damage are levelled to extract money from the organization concerned. In the absence of an instrumented study, the mine owners hesitate to take bold steps in the execution programme. Such a scientific document may serve as the owners' best defence against unwanted damage suit. Central Mining Research Institute (CMRI) under the aegis of the Council of Scientific Industrial Research (CSIR), Government of India investigates such problems and offer plausible advice for a safe method of blasting.

7.2.3 Factors Affecting Ground Vibration

The generation and propagation of vibration to the surface depend on several factors, which include the following:

- Depth of cover from underground blasting source to the monitoring point
- Explosive charge weight fired instantaneously or maximum charge fired in a delay
- Blasting pattern—whether machine cut or blasting off the solid (BOS)
- Type of explosive used
- Type and structure of the overlying rock mass
- Geological discontinuities in the superincumbent strata
- Direction of shot-holes to the plane of overlying strata
- Type and method of monitoring system
- Location of measuring transducer—whether on the ground or on the structure itself

7.2.3.1 Case Studies

Not many published research reports are available on vibration from underground blasting. Snodgrass and Siskind (1974) studied vibrations from underground blasting to estimate expected vibration levels over a wide range of geologic conditions and mining configuration. Olson et al. (1973) studied ground vibrations from tunnel blasting in granite and found that cube root scaling grouped the vibration data more effectively than the square root scaling, established by earlier studies in quarries and underground mines by the U.S. Bureau of Mines (USBM). They concluded that more data was needed to determine which factor should be used to design underground blasts with safe vibration levels.

The blasting research group of the Central Mining Research Institute, India has conducted case studies in substantial number of mines, covering various coalfields of ECL, BCCL, WCL, SECL of Coal India Limited subsidiaries and the Singareni Collieries Company Limited (SCCL). The studies included machine cut blasting as well as BOS and depth of cover from the source of blasting to the surface ranged between 17 and 166 m. The explosive charge/delay or charge fired instantaneously in a round of blast varied from 0.2 to 8.0 kg. In the case of tunnel blasting, the maximum charge/delay was 26.5 kg. The vibrations measured were within 26.2 mm/s in all the coal mines. The maximum vibration of 18 mm/s was observed in a phosphate mine (see Table 7.4).

The results of vibration under varying geologic and mining conditions at different sites of investigation are listed in Table 7.4. The vibration results at each site were separately analysed to derive a best-fit propagation equation. A comparative study of the empirical constants of such propagation equations is presented in Table 7.5.

7.2.4 Structural Response

A simple method to predict the structural response of a building to vibrations is the use of Fast Fourier Transform (FFT) analysis. FFT provides information in practice about the frequency band and responsible wavelength that need to be omitted in order to avoid damage and disturbance. It is the simplest scientific way of analyzing resonance and magnification problems for any structure (Pal Roy and Chakraborty, 2000).

It has been found in many cases that while the villagers complain and demand compensation but at the same time they show resistance to the management or any scientific body for monitoring of blast impacts on their structures. This act of non-cooperation from the villagers' side always keeps the management in ambiguity. On the other hand, the management itself requires the knowledge of anticipated structural response beforehand to assess the possible damage, if any, while conducting a blast on demand. As a solution to this practical problem, equation (7.9) can be used to predict the anticipated vibration on any structure at a desired height generated due to underground blasting. Figure 7.6 shows a typical blast vibration time history record and its consequent frequency analysis for underground blasting and surface measurement. The measured values of structural response and consequent summary of observations are given in Tables 7.6 and 7.7.

In the case of underground blasting and its effect on the surface structures, the equation obtained by analyzing the data given in Table 7.6 is:

$$V_{SS} = 0.646 V_{gg} \cdot \exp(0.279 d_{11} + 3.31 d_{22}) \qquad (7.9)$$

I_d, index of determination = 0.874

where,

V_{SS} = peak particle velocity on the surface structure (mm/s)

Table 7.4 Vibration results under varying geomining conditions at different sites of investigation

Site of investigation	*Geology of overlying formation*	*Depth of u/g blasting (m)*	*Explosive charge weight fired/delay or instantaneously (kg)*	*Vibration (PPV) (mm/s)*	*Method of blasting*
(1)	(2)	(3)	(4)	(5)	(6)
Orient Colliery, Ib valley Area, MCL	Immediate roof: carbonaceous shale, overlain by shale, gray sandstone, while sand-stone and subsoil at the top	67-72	0.9-2.7	0.17-.52	*U/C*
Hindgir Rampur Colliery, SECL	Sandy shale and coal shale	41-76	2.7-3.6	1.05-8.66	*BOS*
Bijuri Colliery, SECL	Immediate roof: massive sandstone, about 2 m thick alluvium at the top	19-32	1.8-2.4	3.60-8.60	*BOS*
Maldeota Phosphate Mine, PPCL	—	60-166	1.6-8.0	0.94-18	*BOS*
Banaso Irrigation Tunnel (Bihar)	Jointed hard granite gneiss with mica schist patches. Top 3 m reddish soil and below that 3 m sandy soil.	23-38	18.5-26.5	2.5-9.0	*BOS Tunnel Blasting*
Burhar Mine No. 3 (SECL)	Hard gray sandstone and yellow clay of about 19 m with band of SST	48-52	0.6-4.2	2.5-12.5 f = 40-50 Hz	*BOS*
Chachai Colliery (SECL)	Fine to course grained sandstone, shaly coal subsoil of about 10 m thickness	72-100	3.6-4.2	0.4-3.0	*BOS*
Tirat/Quardih Colliery (ECL)	Immediate roof: hard shale, floor: hard sand-stone, top 3 m alluvium	74-77	0.2-0.8	0.026-0.15	*U/C*
Shivpuri colliery (WCL)	Immediate roof: shaly coal overlain by sandstone layers	23-37	0.9-2.4	3.15-8.3	*U/C*
Mine 1 and 2, Orient Area, MCL	Immediate roof: shaly coal overlain by hard sandstone layers	17-61.67	1.48-2.78	0.46-26.20	*BOS*

BOS—Blasting off the solid; U/C—Undercut blasting; f—frequency

Table 7.5 Comparison of empirical equations derived at various sites of study [form of equation $V = K(D/Q^b)^n$]

Sl. No.	*Site of investigation*	*Site constant (K)*	*Attenuation constant (n)*	*Scaling factor (b)*	*Correlation coefficient (r)*
1.	Tirat/Quardih (ECL)	19416.0	–2.67	0.50	–0.61
2.	Shivpuri Colliery (WCL)	284.84	–1.26	0.50	–0.69
3.	Bijuri Colliery, Hansdeo Area (SECL)	306.12	–1.14	0.33	–0.67
4.	Maldeota Phosphate Mine (PPCL)	4306.0	–1.85	0.50	–0.83
5.	Konar Irrigation Tunnel, Banaso (Bihar)	233.14	–2.17	0.50	–0.94
6.	Hingir Rampur Colliery, Ib Valley (MCL)	173.3	–0.99	0.33	–0.33
7.	Burhar Mine No. 3, Sohagpur Area, (SECL)	1226.35	–1.46	0.50	–0.88
8.	Orient Colliery, Mine No. 3, Ib Valley (MCL)	93.327	–1.37	0.33	–0.69
9.	Bundia Colliery, Ib Valley (MCL)	86559.0	–2.58	0.33	–0.76
10.	Chachai Colliery, Sohagpur Area (SECL)	30835.3	–2.98	0.66	–0.56
11.	Mine 1 and 2, Orient Area, MCL	370.54	–1.361	0.33	–0.88

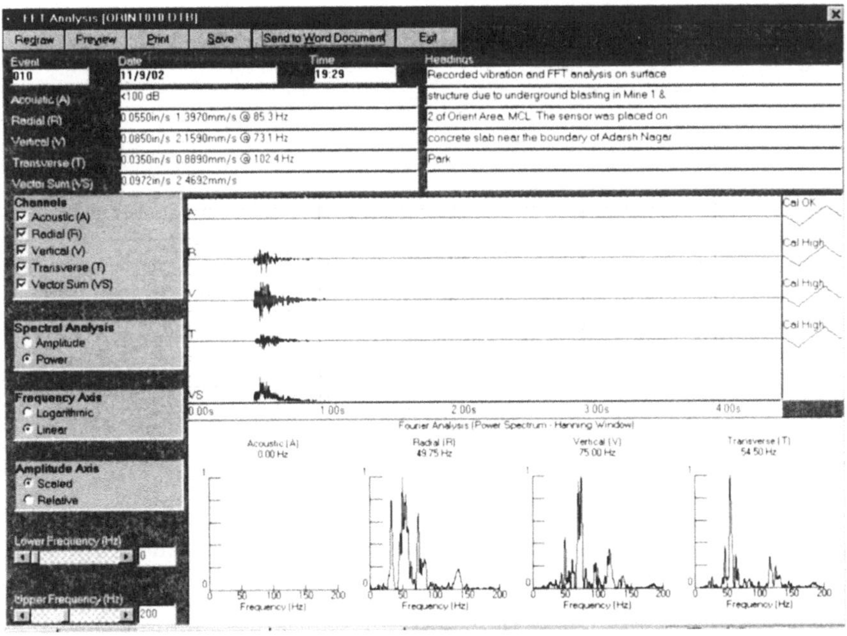

Fig. 7.6 **Blast vibration time history record and its frequency analysis for underground blasting and surface measurement (depth of cover = 46 m and radial distance of the sensor from the source = 75 m)**

Table 7.6 Details of vibration measurements at the ground level on surface (compacted ground surface) and on structure adjacent to it due to underground blasting

Experimental Site	*Charge per round (kg)*	*Max. charge per delay (kg)*	*Radial distance from the source (m)*	*PPV (mm/s) and frequency (Hz)* Ground surface	*PPV (mm/s) and frequency (Hz)* On Structure	*Height difference (m)*	*Attenuation factor/ Structure type**
GIRMINT Colliery, ECL	5.40	1.80	142	0.149 (76-107 Hz)	0.119 (76-107 Hz) - D.F.B.	1.83	1.25 Window ceiling
	1.80	1.80	140	0.224 (76-107 Hz)	0.209 (76-107 Hz) -D.F.B.	1.83	1.07 Window ceiling
JAMURIA Colliery, ECL	3.60	1.80	77.76	2.610 (54-75 Hz)	1.148 (40-53 Hz)	15.0	2.27 Water tank
	3.60	1.80	110.00	0.567 3 m high (40-53 Hz)	0.477 15 m high (28-39 Hz)	12.00	1.19 Water tank
BHOWRA(N) Colliery, BCCL	3.60	1.20	45.20	3.88 (70 Hz)	2.00 (68 Hz)	1.80	1.94 BC
DEULBERA Colliery. MCL	5.55	1.48	95.00	0.98 (85 Hz)	0.30 0.50 (54-75 Hz)	13.5 1.16	3.27 1.96 BL
	5.55	1.48	95.00	1.402 (54-75 Hz)	0.567 (54-75 Hz)	7.5	2.47 BL

*D.F.B. Dominant Frequency Band; For Structures' specifications: See Table 6.9 of Chapter 6

Table 7.7 Results of underground blasting and its effects on surface structures

Type of structure	*Rate of attenuation measured on the structure with respect to the ground vibration*
For all structures identified in Table 6.9 of Chapter 6	Maximum → 3.27 times Minimum → 1.07 times
Duration of vibration	: Maximum up to 1 second
Rate of attenuation	: Taller structures showed faster attenuation
Size of structure	: Larger structures showed smaller response
Frequency of vibration at ground level	: (40-107 Hz)
Modulated frequency on the structure	: (28-107 Hz)

V_{gg} = peak particle velocity on compacted ground at the foundation level of the surface structure (mm/s)

$d_{11} = (T_{mm}/T_{CC})$

$d_{22} = (H_{SS}/D_{SS})$

T_{mm} = maximum charge per delay (kg)

T_{CC} = total charge fired in the round (kg)

H_{SS} = height of the sensor on the structure from the foundation level (m)

D_{SS} = radial distance of the sensor from the blasting source (m)

7.2.5 Findings of the Investigation

Propagation of shock-wave vibration depends on the type and structure of rocks forming the overlying strata. Massive and dense rocks carry the waves to far-off places with less attenuation. The soft and different layers of rock intervening between the surface and the blasting site in underground dampen the vibration to a great extent by reflection and refraction. In one colliery where the immediate rock above the underground workings was massive sandstone, the level of observed vibration was much higher compared to that of overlying strata formed of layers of shale and sandstone.

The location of sensors also affects the level of vibration observed. A pick-up mounted on loose soil resulted in reduced amplitudes (around 1.5 times less) compared to a sensor placed on hard compact ground for the same blasting round (observed during the study of one mine). Vibrations measured with sensors mounted on the surface just vertically above and at points 5-10 m off the vertical were not the same for the same round of blasting. In the case of the former, i.e., at the point, just vertically above, the measured peak particle velocity was 2-2.5 times higher than that of the latter.

In identical conditions, i.e., same charge/delay and same depth of cover in a coal mine, the vibration produced on the surface by solid blasting was about 2.5-3 times higher than that with undercut blasting. The reason could be constriction in the case of solid blasting compared to undercut blasting.

The top row of holes in the case of undercut blasting produced 1.3-2.75 times more vibration than the bottom row of holes. The possible reason might be that the row was very near to the roof rock.

A stronger explosive of the P_3 group (density: 1.55 g/cc, VOD: 2400 m/s, RWS or relative weight strength 54% B.G.) produced 1.5 times more vibration than the P_5 group of explosives (density: 1.3 g/cc, VOD: 2000 m/s, RWS 41% B.G.) in identical conditions.

The results of vibrations and propagation equations derived at various sites (Tables 7.4 and 7.5) indicated that each site needed separate investigation for the estimation of optimal charge in a blasting round. As such, the influence of geo-technical properties on the velocity and frequency content in the wave motions should be given specific attention for a generalized conclusion.

7.3 UNDERGROUND BLASTING —UNDERGROUND EFFECTS

7.3.1 Introduction

A lot of progress has been made in recent years on the structural stability of surface due to opencast blasting operation. In this context, several authors, worldwide, have tried to establish different norms and levels of basic dominating parameters to contain the effects of surface blasts within tolerable limits. However, no systematic work has been done on the effects of underground workings due to underground blasting. In view of the demanding situation, CMRI has investigated some mines with the following objectives (Singh and Pal Roy, 1993):

(a) To study the nature and intensity of blast-induced vibration produced in the mine roof and coal pillars;
(b) To determine the effect of blasting in the lower section upon the higher section and vice versa;
(c) To measure the nature and extent of fracture developed in the surrounding rock mass;
(d) To develop predictive models for determining the level of vibration in order to design a safe blasting pattern;
(e) To study the effect of vibration on roof support and ventilation stoppage in its vicinity with a view to determining the zone of influence of underground blasting;
(f) To make a comparative study on the effects of blasting with machine cut, auger holes and blasting off solid (BOS);

7.3.2 Effect of Vibration on Parting between Two Contiguous Seam Workings

The effect of underground blasting (machine cut, auger holes and BOS) on the parting between two contiguous seam workings as well as on the mine roof and coal pillars was studied by the Blasting Research Group of the Central Mining Research Institute (CMRI) in joint collaboration with the R&D cell of the Directorate General of Mines Safety (DGMS).

The vibration generated due to blasting in a developed face of bottom seam working was monitored by placing instruments in the roof and floor at various distances from the face. Sensors were mounted in both the top and bottom seam workings. The top seam was developed earlier than the bottom seam and the parting between the two contiguous seam workings was about 2 m thick, consisting of sandstone. Vibrations were monitored with microcomputer-based seismographs in the roof as well as on the floor of the gallery. In addition to vibration measurement, investigations were also carried out to monitor the deformation of strata with convergence pegs, strain bar, borehole extensometers and load cells at different points in both sections. The instruments showed no deformation for 12 m development of the face. The results of vibration monitored are given below:

Vibration in roof: (i) in the same seam working under development: 1.6-100 mm/s; (ii) in the top seam (developed): 0.9-40 mm/s.

Vibration in floor: (i) in the same seam working: 2.4-18.2 mm/s; (ii) in the top seam (developed), i.e. parting between the two seams: 1.6-395 mm/s.

Empirical equations: The following empirical equations were established to predict the level of vibrations for:

(i) *Mine roof of seam (under development)*

$$V = 574.6\ (D/Q^{1/2})^{-1.55} \quad (7.10)$$

$$r = -0.98 \quad (7.10a)$$

(ii) *Roof of top seam (developed)*

$$V = 47.3\ (D/Q^{1/2})^{-0.99} \quad (7.11)$$

$$r = -0.95 \quad (7.11a)$$

(iii) *Floor of seam (under development)*

$$V = 588.6\ (D/Q^{1/3})^{-1.39} \quad (7.12)$$

$$r = -0.96 \quad (7.12a)$$

(iv) *Floor of top seam (developed), i.e. parting between the two seams*

$$V = 436.1\ (D/Q^{1/2})^{-1.36} \quad (7.13)$$

$$r = -0.96 \quad (7.13a)$$

7.3.3 Comparative Study of Effect of Blasting with Machine Cut, Auger Holes and Blasting off the Solid

A comparative study was jointly carried out by DGMS and CMRI scientists to assess the effect of blasting with particular reference to blast-induced vibration in the development of working faces with machine cut, auger holes and BOS. Unfortunately all the three methods were not available in the same mine. Hence the effect of auger holes blasting and BOS was investigated in the Churcha colliery of SECL, and the effect of machine cut blasting and BOS was studied in the Sonawani mine of SECL. Empirical equations were derived for each method of blasting at Churcha and Sonawani mines. The drilling and charging patterns are shown in figures 7.7-7.10.

Churcha colliery

(a) Auger holes blasting

(i) Pillar vibration:

$$V = 805.03\ (D/Q^{1/2})^{-1.92} \quad (7.14)$$

$$r = -0.9 \quad (7.14a)$$

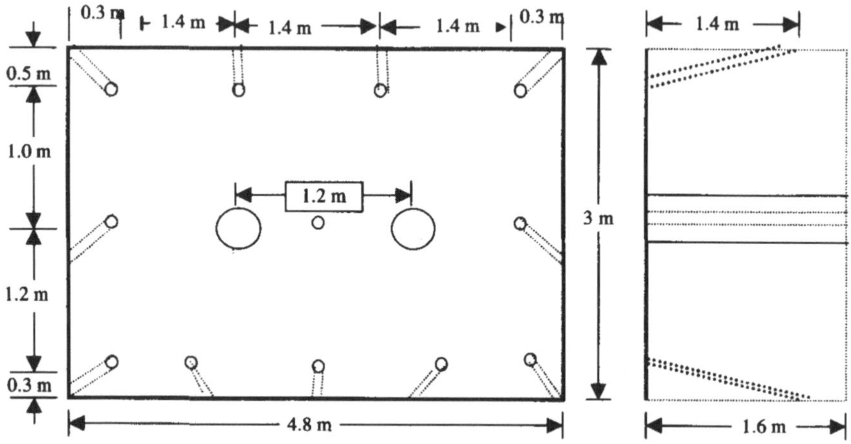

No. of holes	*: 12*	*Avg. coal obtained*	*: 18-20 tonnes*
No. of auger holes	*: 2*	*P/F tonnes/kg*	*: 2.7-3.0*
Charge/hole		*D/R tonnes/no.*	*: 1.38-1.55*
Top row	*: 450 gm*	*Av. pull obtained*	*: 1.1-1.15 m*
Middle row	*: 600 gm*		
Bottom row	*: 600 gm*		
Total charge	*: 6.6 kg*		
Firing sequence	*:*		
Middle and bottom row – Ist round			
Top row - IInd round			
Explosive type	*: P_1*		
Detonators	*: Instantaneous copper electric*		
Maxm. Charge/delay	*: 4.8 kg*		

Fig. 7.7 Drilling and charging pattern for augur blasting, Churcha Colliery, SECL

(ii) Roof vibration:

$$V = 605.08 \ (D/Q^{1/3})^{-1.2} \tag{7.15}$$

$$r = -0.75 \tag{7.15a}$$

Propagation equation for floor vibration could not be established because of too few observations.

(b) Blasting off the solid

(i) Pillar vibration:

$$V = 748.97 \ (D/Q^{½})^{-1.89} \tag{7.16}$$

$$r = -0.79 \tag{7.16a}$$

Due to very few observations in the roof and floor, statistically no predictor equation was determined.

Sonawani colliery

(a) Machine cut blasting

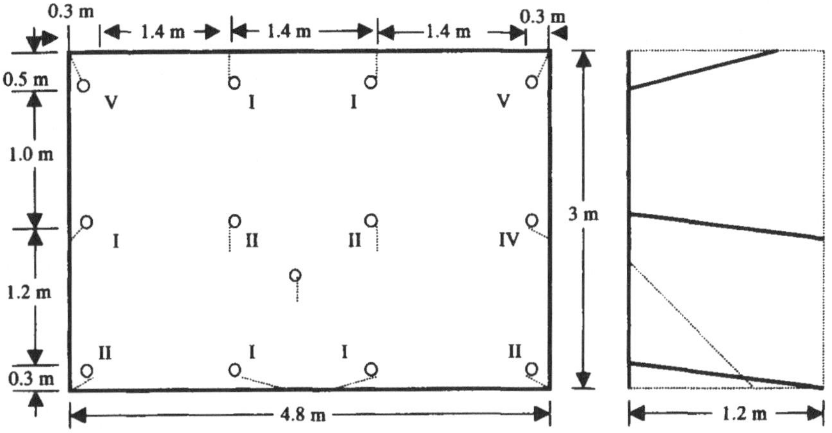

No. of holes: 12
Stab hole: 1
Charge/hole: Delay No. Z, I, III (400 g); II, IV, V (600 g)
Total charge: 6.8 kg
Explosive used: Pentadyne (P_5)
Detonators used: Non-incendive copper delays
Maximum charge/delay: 2.4 kg

Avg. coal obtained: 11-18 tonnes
P/F tonnes/kg: 2.35-2.6
D/R tonnes/no.: 1.23-1.38
Pull: 1-1.1 m

Fig. 7.8 Drilling and charging pattern for BOS, Churcha Colliery, SECL

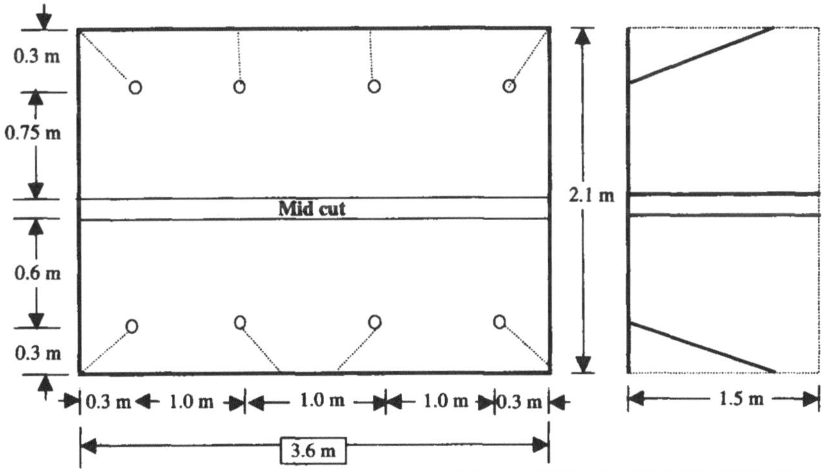

No. of holes: 8
Depth of holes: 1.1 m
Depth of cut: 1.5-1.7 m
Total charge: 2.64 kg
Type of explosive used: P_1
Type of detonator: Instantaneous copper electric
Maximum charge/delay: 2.64 kg

Avg. coal obtained: 8-10 tonnes
P/F tonnes/kg: 3.03-3.78
D/R tonnes/no.: 1.0-1.25
Pull: 0.8-1.0 m

Fig. 7.9 Drilling and charging pattern for mid-cut, Sonawani Incline, SECL

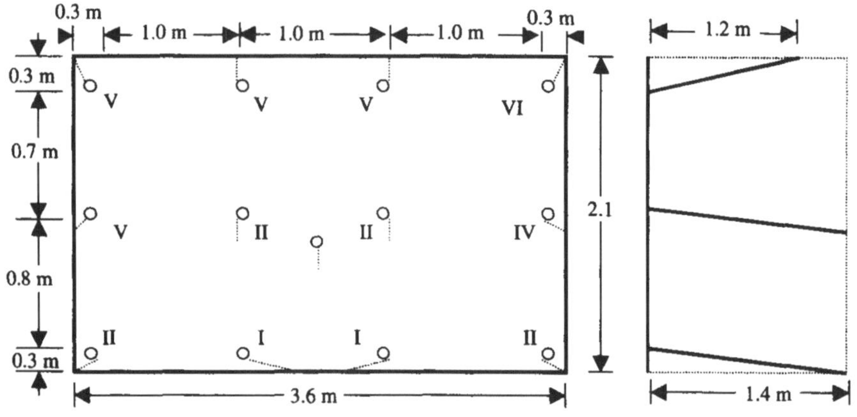

No. of holes: 12	*Avg. coal obtained: 11-12 tonnes*
Stab hole: 1	*P/F tonnes/kg: 1.9*
Charge/hole: Delay No. Z, IV, V, VI (400 g)	*D/R tonnes/no.: 0.92*
I, II, III (600 g)	*Pull: 1.2 m*
Total charge: 6.4 kg	
Explosive used: Pentadyne (P_5)	
Detonators used: Non-incendive copper delays	
Maximum charge/delay: 1.2 kg	

Fig. 7.10 Drilling and charging pattern for BOS, Sonawani Incline, SECL

(i) Roof vibration:

$$V = 744.37 \ (D/Q^{½})^{-1.15} \quad (7.17)$$

$$r = -0.89 \quad (7.17a)$$

(ii) Pillar vibration:

$$V = 1232.4 \ (Q^{2/3}/D)^{+1.36} \quad (7.18)$$

$$r = 0.67 \quad (7.18a)$$

(b) Blasting off the solid

(i) Roof vibration:

$$V = 2483.63 \ (D/Q^{1/3})^{-1.72} \quad (7.19)$$

$$r = -0.87 \quad (7.19a)$$

Vibrations induced by the auger method of blasting were observed to be generally on the higher side than those induced by solid blasting (BOS). The possible reason could be the instantaneous firing of auger holes versus millisecond delayed blasting in the case of BOS. Further, in the former method of blasting, the total charge fired instantaneously was more than that of the maximum charge per delay fired in the latter.

A comparison of machine cut (CCM) versus solid blasting (BOS) revealed that sometimes the former generated more vibration for the same quantity of charge, while sometimes just the opposite. As such, no definite conclusion could be drawn from the available data generated by CMRI and DGMS.

7.3.4 Air Overpressures in the Underground Workings

Air overpressures were measured at 25-30 m distances from blasting faces. At Churcha colliery, the sound pressure level (SPL) ranged between 143.4 and 155.4 dB-L. In auger holes blasting with an explosive charge of 4.8 kg, fired in a round, the maximum SPL (Sound Pressure Level) was 155.4 dB-L at 25 m distance in front of the blasting face, whereas in the case of BOS, the maximum SPL was 150.2 dB-L at the same 25 m distance from the face when the charge fired in a round was 6.8 kg.

At Sonawani colliery, the measured values of SPL during machine cut blasting ranged between 155.2 and 158.5 dB-L for 2.3 to 2.4 kg of charge fired in a round. The distance from the blasting face varied from 9.0 to 15.0 m. In the case of BOS with 2.64 kg of charge, the measured values of SPL were between 145.4 and 152.0 dB-L at a distance of 10 m from the blasting source.

7.3.5 Zone of Disturbance Caused by Blasting

Excessively high ground vibrations from underground production blasts alone or coupled with the increased stress due to openings created by extraction/winning of coal may cause damage to the roof rock and may affect the stability of roof bolts and mine supports and cause ventilation stoppage. The area of underground workings affected by such vibration needs to be assessed and proper supports ensured.

Based on the studies conducted by CMRI in conventional methods of blasting in mining and construction projects, a peak particle velocity of 100 mm/s was found safe for the stability of coal mine roof and pillars from underground blasting. Thus the area of underground workings experiencing a vibration of 100 mm/s or more should be treated as a zone of disturbance. The disturbed zones of underground workings in the Test Panel of GDK-10 incline of Ramagundem Area were estimated as 80 m for 100 kg charge and 60 m for 80 kg charge. A graphical plot of the same is available in Singh and Pal Roy's (1993) book. Table 7.8 shows the zone of disturbance as calculated from the empirical equations derived for conventional methods of blasting at Churcha and Sonawani Collieries (SECL).

It can be seen from the Table 7.8 that for an identical charge, the zone of disturbance in the Sonawani colliery was almost double to that in the Churcha colliery. There were two reasons for this phenomenon. First, the coal of Sonawani colliery was comparatively harder than that of Churcha colliery; secondly they differed in face geometry.

Table 7.8 Zone of disturbance (m) for Churcha and Sonawani collieries

Charge/delay (kg)	*Zone of disturbances (in m)*			
	Churcha colliery		*Sonawani colliery*	
	Auger	*BOS*	*CCM*	*BOS*
2.0	4.18	4.10	8.10	9.15
2.5	4.68	4.58	9.05	10.22
3.0	5.12	5.02	9.92	11.20
3.5	5.53	5.42	10.71	12.10
4.0	5.92	5.80	11.45	12.94
4.5	6.23	6.15	12.15	13.72
5.0	6.61	6.48	12.81	14.46
5.5	6.94	6.80	13.43	15.17
6.0	7.25	7.10	14.03	15.84
6.5	7.54	7.39	14.60	16.49
7.0	7.83	7.67	15.15	17.11

As a general guideline, the following damage quantifying indices based on BDI [equation (7.3)] and S.D. (scaled distance) were derived by Pal Roy et al. (2003) for the safety of underground roof and pillars due to underground blasting. These are shown in Table 7.9.

Table 7.9 Damage quantifying indices based on BDI and SD (Pal Roy et al., 2003)

Classification of zones	*Range of BDI values*	*Range of S.D.*
Safe zone	<1.0	>2.5
Alarming zone	1.0–1.75	>2.0–2.5
Minor damage zone	>1.75–2.50	1.50–2.0
Major damage zone	>2.50–3.50	<1.5

7.4 CONCLUSIONS

Blast vibrations may induce substantial dynamic loading on mine openings, which is difficult to assess and generalize in respect of the effect on the overall stability of workings. Physical assessment of the underground workings, for developing a risk strategy and to understand the mechanism which would control the deterioration of the mine, was found to be a rational approach to understand the long-term effects of vibrations in the stability of old workings. General observation of Siskind (2000), as described in section 7.1.6.7, can be an effective guidance to prevent roof collapse and pillar failure. Influence of geo-technical properties on the velocity and frequency content in the wave motions are needed with specific attention for determining the damage threshold values of underground workings.

REFERENCES

Barton, N., Lien, R. and Lunde, J. 1974. Engineering classification of rock masses for design of tunnel support, Norwegian Geotechnical Institute, Publication No. 106, Oslo, Norway, p. 115.

Bieniawski, Z. T. 1973. Engineering classification of jointed rock masses, The Civil Engineer in South Africa, December, pp. 335-344.

Clark, D. A. and Cavin, R. E. 1981. Wave propagation in a subsurface environment due to blasting operations, Proc. 7th Conference on Explosives and Blasting Technique, Int. Soc. of Explosives Engineers, Annual Meeting, Pheonix, Arizona, USA, pp. 35-51.

Dowding, C. H. and Rozen, A. 1978. Damage to rock tunnels from earthquake loading. The Geotech. Engg. Div., American Society of Civil Engineers, 104 (GT2), pp. 175-191.

Dowding C. H. 1979. Damage to rock tunnels from earthquake shaking, Tunnels and Tunnelling, June, UK, pp. 15-20.

Fourie, A. B. and Green, R. W. 1993. Damage to underground coal mines caused by surface blasting. Inter. J. of Surface Mining and Reclamation, Vol. 7, No.1, Netherlands, pp. 11-16.

Hendron, Jr., A. J. 1977. Engineering rock blasting in civil projects, Structure and Geotechnical Mechanics: A volume honouring Nathan M. Newmark, W. J. Hall (ed.), Prentice-Hall, Inc., Englewood Cliffs, NJ, USA, pp. 242-277.

Hoek, E. and Brown, E. T. 1980. Underground excavations in rock, Institution of Mining and Metallurgy, London, p. 527.

Jensen, D. E., Munson, R. D., Oriard, L. L., Reitman, J. D. and Wright, R. S. 1979. Underground vibration from surface blasting at Jenny mine, K. Y. Woodward-clyde consultants, Orange CA, Final Contract RPT J0275030 for the US Bureau of Mines, p. 99.

Kidybinski, A. 1986. Design criteria for roadway supports to resist dynamic loads, Int. Jr. Min. Enggs., Vol. 4, pp. 91-109.

Labrenche, D. A. 1983. Damage mechanisms in tunnels subjected to explosive loads, In Seismic design of embankments and caverns, (ed.) T. R. Howard, American Society of Civil Engineers, New York, pp. 128-141.

Langefors, U. and Kihlstrom, B. 1963. The modern technique of rock blasting, John Wiley and Sons. Inc., New York, USA, pp. 265-295.

Lewandowski, T., Keith, G., Croucher, M. and Richards, A., 1999: The impact of surface blasting on underground opening—Geotechnical Measurement, Proc. 6th International Symposium on Rock Fragmentation by Blasting, Fragblast-6, South African Institute of Mining and Metallurgy, 8-12 August, p. 131.

Olson, J. J., Willard, R. J., Fogelson, D. E. and Hjelmstad, K. E. 1973. Rock damage from small charge blasting in granite, USBM-RI 7751, p. 44.

Pal Roy, P. and Chakraborty, A. K., 2000: Critical evaluation of ground vibrations and consequent structural damage under varying blast configurations, Proc. Explosives and Blasting Technique, Holmberg (ed.), A. A. Balkema, Rotterdam, Netherlands, pp. 71-78.

Pal Roy, P., Sawmliana, C., Bhagat, N. K. and Madhu, M. 2003. Induced caving by blasting: innovative experiments in blasting gallery panels of underground coal mines of India, Mining Technology (Trans. Inst. Min. Metall., A), April, Vol. 112, pp. A57-A63.

Rupert, G. B. and Clark, G. B. 1977. Criteria for the proximity of surface blasting to underground coal mines, Proc. Rock Mechanics Symposium, Colorado School of Mines, USA, pp. $3C_31$- $3C_310$.

Sakurai, S. and Kitamura, Y. 1977. Vibration of tunnel due to adjacent blasting operation, International Symposium on Field Measurements in Rock Mechanics, Zurich, Switzerland, April 4-6, pp. 61-74.

Singh, B. and Pal Roy, P. 1993. Blasting in ground excavation and mines, A. A. Balkema, Rotterdam, Netherlands p. 177.

Singh, P. K., Singh, R. B., Singh, T. N. and Singh, D. P. 1999. Impact of surface blasting on the stability of underground workings, Proc. 6th International Symposium on Rock Fragmentation by Blasting, Fragblast-6, South African Institute of Mining and metallurgy, 8-12 August, pp. 139-144.

Singh, P. K. 2003. Blast vibration damage to underground coal mines from adjacent open-pit blasting, International Journal of Rock Mechanics and Mining Sciences, Vol. 39, No. 8, pp. 959-973.

Siskind, D. E., Staff, M. S., Kopp, J. W. and Dowding, C. H. 1980. Structure response and damage produced by ground vibrations from surface mine blasting, U.S. Bureau of Mines, R.I. 8507, p. 74.

Siskind, D. E. and Stagg, M. S. 1985. Blast vibration measurement near and on structure foundations, U.S. Bureau of Mines, R.I. 8969, p. 20.

Siskind, David, 2000: Vibrations from blasting, International Society of Explosives Engineers, Cleveland, OH, USA, p. 120.

Snodgrass, J. J. and Siskind, D. E. 1974. Vibrations from underground blasting, USBM-RI 7937, p. 30.

Thoenen, J. R. and Windes, S. L. 1942. Seismic effects of quarry blasting, U.S. Bureau of Mines Bulletin 442, p. 83.

Tunstall, A. M. 1997. Damage to underground excavations from open-pit blasting, Trans. Instn. Min. Metall. (Section A: Mining Industry), Vol.106, January-April: A19-A24.

Yu, T. R. and Vongpaisal, S. 1996. New blast damage criteria for underground blasting, The Canadian Institute of Mining Bulletin (CIM), March, pp. 139-145.

8

Blast Monitoring and Characterization

8.1 INTRODUCTION

Detonation of larger volumes of commercial explosives in the mining and quarrying industry, along with unplanned growth of urbanization, necessitates a well-organized and scientific approach to rock blasting problems.

Blasting operations have a direct impact on environment and socio-economic conditions in and around mining and rock excavation activities, and the standards meant for industrialized countries may not be applicable to developing countries as their problems of development are different. However, there should be a uniform approach in blast monitoring procedures and characterization. In order to maintain a balance between safety and production, a blast design has to take into account the hazards of blasting including vibration, flyrock, noise and public annoyance. The annoying circumstances as a result of these impacts at many times lead to unfavourable mining operations, confrontations and sometimes even litigations. This chapter highlights some useful observations that are globally accepted and broadly recommended for conducting any sort of vibration study related to mining, quarrying and tunnelling.

8.2 SENSOR SETTING

The sensors being the first element of the measuring system, should be planted well in contact with the ground so that they vibrate as part of the earth, emitting the signal which represents the true ground motion. This contact can be achieved by many practical ways namely—(i) Placing the sensors on the ground and fixing them with the help of plaster of paris for firm attachment; (ii) Screwing them to blocks of aluminium or other non-ferric material (Plate 8.1); (iii) Setting the sensors inside a box and burying it in the earth specially when the ground is not consolidated.

Plate 8.1 Non-ferric sensor attachments fabricated at CMRI for vibration measurements in underground roof and pillars

For the determination of structural response, the sensors should be placed: one, on the ground near the structure to be protected (at a distance of 2 to 3 m away from the foundation of the structure); and the other, on the structure itself, bearing in mind that the latter would reflect the response of the construction (i.e. structural response) and not record the ground movement (Pal Roy, 2003).

8.3 SENSOR TYPES

For the recording of ground vibrations, vibration seismographs and the acceleration seismographs are commonly used. Vibration seismographs have electromagnetic type transducers which emit an electric signal that is proportional to the velocity of the vibrating particle (Plate 8.2). The electric signal is generated by a moving coil within the field of a stationary magnet. The range of application is limited by the actual resonance frequency of the vibration seismograph, which is usually between 5 and 15 Hz and up to a maximum of 200 to 300 Hz. They are not advisable when low frequencies exist.

The acceleration seismographs are based on the difference of potential to the mass of the crystal by the acceleration of vibratory movement.

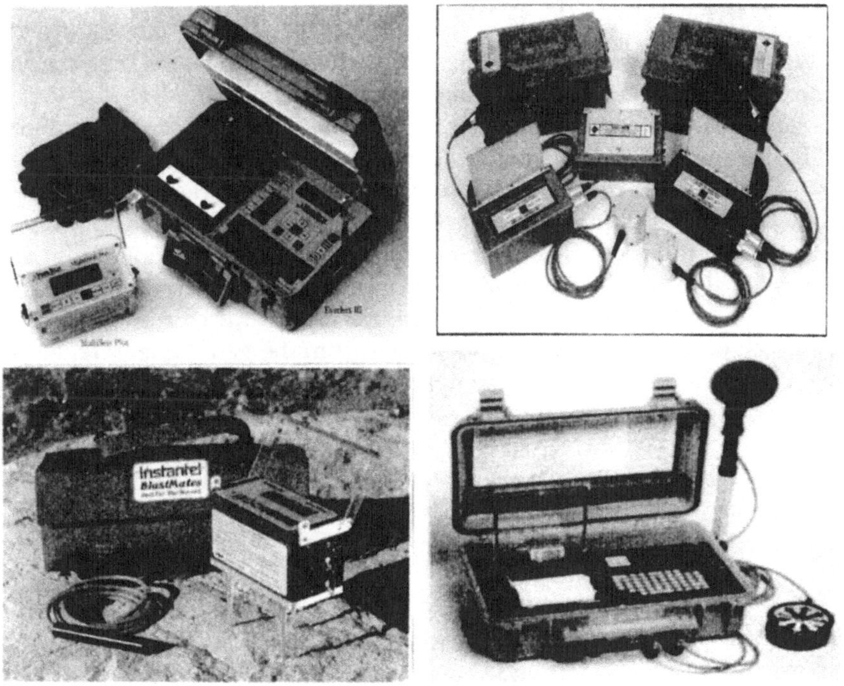

Plate 8.2 View of a few modern days seismographs (Courtesy: Instantel Inc.; GeoSonics; White Industrial Seis. Inc. and Nomis Seismographs)

8.3.1 Recording System

The recording system comprises of several components for carrying out the following functions (Jimeno et al., 1995):

- Pulse detection by sensors,
- The transmission of the electric signals emitted by the sensors through the conductor wires, and
- Recording of the signals with a seismograph for its scientific study and analysis.

The most frequently referred ranges of typical blast ground vibration parameters are given in Table 8.1 (Cording et al., 1975).

Table 8.1 Ranges of different characteristic parameters of blast vibrations

Parameter	*Range*
Displacement (mm)	$10^{-4} - 10$
Particle velocity (mm/s)	$10^{-4} - 10^{3}$
Particle acceleration (mm/s^2)	$10 - 10^{5}$
Length of pulse (s)	0.5 – 2.0
Wave length (m)	30 – 1500
Frequency (Hz)	0.5 – 200
Strain (μ mm/mm)	3.0 - 5000

8.4 WHAT KIND OF SENSOR SHOULD BE USED?

A typical accelerometer is capable of recording much higher frequencies than a typical and even a high frequency geophone. The geophone has the deficiency in providing the high frequency information of the vibration. At the same time, it is mathematically impossible for these high frequencies to be carried in the differentiation process. As a result, trace differentiated from a geophone signal always lacks these high frequencies and, thus is of very low frequency. The peak amplitudes of an acceleration signal are generally associated with the higher frequencies of the signal. This is why the high frequencies with which peak amplitudes are associated in the acceleration signal are simply absent from the geophone data. Now, while moving further away from the blast, the high frequencies of the vibrations are naturally filtered out, giving the difference between the results.

However, a distance would be reached where the accelerometer will not be able to pick up the low frequencies of the vibrations, whereas the geophone would be well suited to these. This again would introduce differences between the results obtained with the two types of gauges.

In a nutshell, accelerometers are good at recording high frequency vibrations, whereas geophones are well suited to low frequency vibrations (between 5 and 300 Hz). Therefore, the selection depends upon what one tries to measure, the distance between the measurements and the blasts, etc. (Jimeno et al., 1995).

There would often be large differences between PPV readings taken with accelerometers and geophones. These variations are due to differences in frequency response characteristics, especially at low frequencies (<10 Hz). Two types of frequency response need to be considered, namely amplitude and phase response. The degree of variation depends, both, upon the type of transducer used and the frequency content of the vibration being monitored. Certainly, there could be an error even up to 40%.

8.5 RICHTER SCALE VERSUS PPV

There has been lots of confusion on measurements of peak particle velocity and Richter scale number during earthquake. A number on the Richter scale represents the total magnitude of an earth-motion event while the readings on the seismograph indicate the intensity of ground motion at a given location. The magnitude of a particular event would remain the same, but the intensity of vibration would vary with the distance from the epicentre the readings were taken. In other words, there is no practical way that one could relate seismograph readings (taken at one location) to the Richter scale.

The blast vibration intensities are recorded in inches or millimetres per second of particle velocity, while earthquake intensities are assigned a number on the Modified Mercalli Scale (MM Scale), a scale that relates the intensity to the apparent effects from the motion. In greatly simplified form, it ranges from MM I, which is only detectable by very sensitive instruments, to MM XII which represents total destruction of structures. One can readily assign particle velocities to the 12 different MM levels but it would not be technically correct to do so.

There is such a great difference of characteristics between blast vibration and earthquake motion that it is not possible to relate the two in any logical manner. The frequency of blast vibration could range from 5 hertz (at greater distances) to 1 kHz (in close proximity). Earthquake frequencies are too low and they are not normally measured in hertz but rather in the number of seconds for a full cycle or period. The greater the difference in frequencies between the two vibrations being compared, the greater will be the differences in the effects they generate. The acceleration is especially sensitive to frequency, that is why this is inappropriate as a general tool for rating damage potential of vibrations with different frequencies (Oriard, 2002). One might see rather high acceleration levels from blasts, but the displacements would be so low that they would not cause damage. On the other hand, an earthquake may cause damage at rather low acceleration levels because the displacements could be quite high (Plates 8.3 and 8.4). In earthquake-prone areas, bridges and other highway structures are designed to survive specific earthquake acceleration levels. These could be quite low, of the order of 2 to 4 g or so. Unless one understands that the displacements are only going to be of the order of a few ten thousandths of an inch and would not damage the structure, it is unlikely that blasting will be allowed.

If one tries to relate the size of the entire blast with Richter Scale, some of the differences between blasts and earthquakes, such as energy release of the explosive, spatial separation of charges, frequency, time duration of energy release, depth of the source of energy etc must be disregarded.

Doug Anderson (2002) while presenting a paper in one of the ISEE conferences indicated that the Richter magnitude would only relate to the total blast if all explosives were detonated simultaneously. He also noted that it would not strictly relate to the charge weight per delay (unless there was an unusually large delay between charges, making each an individual blast). The actual magnitude number would then fall somewhere between those calculated for the largest charge weight per delay.

Plate 8.3 Complete collapse and severe structural damage of single and multi-storey buildings in old Bhuj township (Gujarat Earthquake in India, January, 26th, 2001)

Plate 8.4 Complete collapse and extensive damage of tall structures in Air Force Colony in New Bhuj City (Gujarat Earthquake in India, January 26th, 2001)

8.6 APPROACH OF TRACKING BLAST VIBRATIONS

In order to ensure that cracking and other damage do not result from blast vibrations, one always needs to follow a provocative environmental approach which would reduce complaints and damage-claims. Recommended, international guidelines include the following (Siskind, 2000):

Plate 8.5 Underground monitoring of strata convergence by rheostat type convergence indicator

1. Design for a safe vibration level, either an industry standard or a regulatory limit;
2. Prepare a site map;
3. Conduct an information and site relation programme for neighbours, prior to blast;
4. Perform pre-blast inspections and document the condition of nearby properties;
5. Monitor with seismographs or use an alternative, such as scaled distance;
6. Develop a site programme plot and update regularly;
7. Using the propagation measurements, track trends in PPVs and frequencies;
8. Conduct post-blast inspections of structures to document any changes.

When the chosen "safe limit" is an appropriate one and the other steps are carried out properly, the above mentioned guidelines should be sufficient to ensure protection of homes and important establishments near mining operations.

8.7 PEAK PARTICLE VELOCITY VERSUS STRAIN

To calculate failure vibration level, for monolithic materials, Siskind (2000) readjusted the usual $S = \rho c V$ to the following form:

$$\text{PPV (in/s)} = S/(9.25 \times 10^{-5} \times \rho \times C_p) \qquad \text{...(8.1)}$$

$$strain = (\Delta x / y) \sin\phi \cos\phi$$

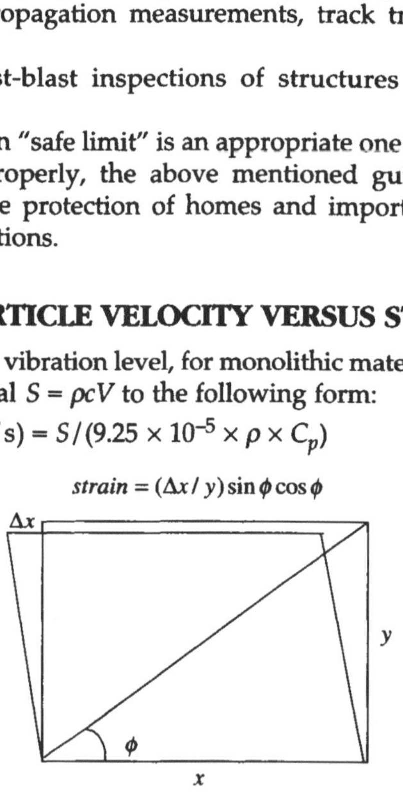

Fig. 8.1 In-plane shear strain computation from relative horizontal displacement (Δx) occurring over a height (y) in a wall panel x wide and y high (Siskind, 2000)

Where V is the peak particle velocity in inch per second (1 in/s = 25.4 mm/s), S is the tensile strength in pounds per square inch (1 lb/in^2 = 0.0689 kg/cm^2), ρ is the specific gravity, and C_p is the P-wave propagation velocity in inch per second (1 in/s = 25.4x10^{-3} m/s). Using conversion formula of Siskind (2000) [equation (8.1)], it can be calculated that a weak concrete with tensile strength of 200 lbs/in^2 (13.78 kg/cm^2), a specific gravity of 2.0 and a propagation velocity of 10,000 ft/s (3048 m/s) would crack at a PPV of about 9 in/s (228.6 mm/s) which is roughly consistent with Dowding's (1985) concrete cracking criteria.

8.8 LOW FREQUENCY RESPONSE

Crum and Siskind (1993); Crum and Pierce (1995) and the other USBM studies led to the conclusion that there was little additional risk to residences from low frequencies. At exact resonance frequencies, a few of the highest amplifications exceeded the highest in RI 8507 (4.3x), which could justify some caution. Indian researchers obtained almost similar results (Pal Roy, 1998). Below the natural or resonant frequencies, structure responses decreased rapidly, particularly when based on different displacements. It is noted that the differential displacements are the vector differences between high and low responses which in turn determine strains.

However, this does not imply that low frequencies are always harmless. For example, the September 19, 1985, earthquake waves that caused extensive damage in Mexico City were of amplitude 0.2 g at 0.5 Hz (Siskind, 2000). At high frequencies, 0.2 g would only be marginally risky (e.g. 15 mm/s PPV at 20 Hz.). However, because the dominant wave frequency was 0.5 Hz, the vibration had a peak velocity of 624 mm/s, and even worse, a peak displacement of 20 cm.

8.9 SUMMARY ANALYSIS OF VIBRATION RISKS IN U/G WORKS

After reviewing an extensive analysis of the technical research carried out all over the world, Siskind (2000) has made a general recommendation on risks in underground works due to blasting. He recommended 300 mm/s or higher vibration level for roof collapse and pillar failure. However, a vibration level below 25 mm/s was recommended as totally harmless to underground workings, even in hazardous situations.

REFERENCES

Anderson, D. 2000. Quarterly Newsletter of Instantel, Vol. 2: Q2, Canada, p. 4

Cording, E. J., Hendron, A. J., Hansmire, W. H., MacPherson, H., Jones, R. A. and O'Rourke, T. D. 1975. Method for geotechnical observations and instrumentation in tunnelling, Vol. 2, The National Science Foundation, Grant 6Z 33644X.

Crum, S. V. and Siskind, D. E. 1993. Response of structures to low-frequency ground vibrations, A preliminary study, Proc. 9th Annual Symposium on Explosives and Blasting Research, Inter. Soc. of Explosives Engineer (ISEE), San Diego, CA, pp. 149-161.

Crum, S. V., Pierce, W. E. 1995. House responses to low frequency ground vibrations from coal mine overburden blasting: A technical update, Proc. 11th Annual Symposium on Explosives and Blasting Research, International Society of Explosives Engineers (ISEE), Nashville, TN, USA, pp. 76-88.

Dowding, C. H. 1985. Blast vibration monitoring and control. Prentice-Hall, Englewood Cliffs, NJ, p. 287.

Jimeno Lopez, C., Jimeno Lopez, E. and Carcedo F. J. A. 1995. Drilling and blasting of rocks, A. A. Balkema, Rotterdam, Netherlands, p. 391.

Oriard, Lewis L. 2002. Explosives engineering, construction vibrations and geotechnology, Intern. Soc. of Explosives Engineers (ISEE), USA, p. 680.

Pal Roy, P. 2003. Some useful observations on the prediction of blasting vibrations, Proc. National Seminar on Status of Environmental Management in Mining Industry, January, 17 and 18, IT BHU, Varanasi, U.P., India, B. K. Shrivastava, R. Kumar and A. Jamal (eds.), pp. 437-440.

Pal Roy, P. 1998. Characteristics of Ground vibrations and structural response to surface and underground blasting, Geotechnical and Geological Engineering, U.K., Vol. 16, pp. 151-166.

Siskind, David 2000. Vibrations from blasting, International Society of Explosives Engineers (ISEE), USA, p. 120.

9

Ringhole Blasting in Coal in Blasting Gallery Panels

9.1 INTRODUCTION

The Blasting Gallery (BG) method of winning coal was originally developed by French scientists and in India, it was proposed in practical field conditions by Charbonnage de France (CdF). This method was primarily introduced in India for higher production and low cost of excavation to substitute conventional hand section and stowing methods for thick coal seam. The Singareni Collieries Company Ltd. (SCCL) adopted this method in the year 1989 and since then have been successfully operating GDK-10, GDK-8 and VK-7 Incline mines in Andra Pradesh, a southern state of India. It is also successful at Chora-10 Pit Colliery of Eastern Coalfields Ltd. (ECL) in West Bengal whereas at East Katras Colliery of Bharat Coking Coal Ltd. (BCCL) in Jharkhand State, the method has been partially successful due to some geological constraints of the area. In this method of working the average percentage of coal extraction can be more than 85%.

An important aspect of BG method of extraction is the fragmentation of blasted coal which should be at optimum level so that they can easily be lifted totally by remote controlled LHDs, thereby mitigating the possibility of occurrence of spontaneous heating in BG districts before exhaustion of the panel.

Central Mining Research Institute (CMRI) undertook one S&T Project sponsored by the Ministry of Coal and Mines, Govt. of India, to study this sensitive problem in order to evolve suggestive guidelines of design patterns under varying geomining and site constraints. Under such scheme, a total of 180 ring-blasts were conducted at different sites and thoroughly monitored for establishing optimum design patterns. Key parameters like different drill designs, staggered charging patterns, use of shorter and longer spacers while charging, strata monitoring and blast damage index were given special emphasis (CMRI S&T Report of investigation, 2003).

9.2 GENERAL DESCRIPTION

In India, recent estimates indicate that several thousand million tonnes of coal are locked up in underground pillars and a good proportion of which is in seams thicker than 4 m. Extraction of thick seams by conventional hand section is neither productive nor effective from the conservation point of view. The percentage of extraction by hand section mining in thick seams is as low as 25 – 30%. A lot of good grade of coal has been lost and millions of tonnes are standing on pillars. Sand stowing for working of thick seams cannot be considered as an option because the cost is prohibit-ive. Sand has become an increasingly scarce commodity along with timber. At the same time, the coal industry was in search of an economic method for dealing with standing pillars within thick seams. As a solution, Blasting Gallery (BG) method was tried as one of the supplemental measures.

Blasting Gallery method is a French technique for winning of thick coal seam. The Charbonnage de France (CdF) introduced this technique in India for gentle gradient of seams (less than 1 in 6 or 9°) with thickness varying from 6.5–12.0 m. In this method, total seam thickness is extracted in one go by drilling a set of drillholes in a ring pattern. The blasted coal is loaded and hauled out from the goaf by remote controlled LHD and subsequently unloaded to chain conveyor. From the chain conveyor, the coal is then fed into the lump breaker where they are broken into (<) 200 mm sizes. Finally, the coal is transported to the surface from underground through a series of belt conveyors.

In Blasting Gallery method, fragmentation of coal plays a very important role for higher production and better performance of LHDs in the panel. Good fragmentation results in higher production and increased efficiency of LHD's performance. Also more percentage of blasted coal can be recovered from the goaf area, thereby reducing chances of spontaneous heating in the goaf. However, poor fragmentation lowers productivity in that it requires secondary blasting. Also, pneumatic stonebreaker is used to clear the coal lump onto the chain conveyor.

9.3 PREREQUISITES OF BG

- Thickness of coal seam ranging from 6 m to 15 m
- Degree-I gassiness most preferable
- Overlying seams either to be extracted or declared non-vendable
- Sandstone roof preferable
- Gradient should not be steeper than 1 in 6 (9°)

9.4 CONDITIONS BEFORE COMMENCEMENT OF EXTRACTION

- Workings shall be resurveyed and verified for accuracy

- Development shall be done by long roadways just like longwall roadways
- Division into suitable sub-panels in accordance with the incubation period
- Independent ventilation circuit
- Overlying goaf shall be free from water

9.5 ADVANTAGES OF BG METHOD

- Entire thickness of seam can be extracted in single lift
- Percentage of extraction—80 to 90%
- Highly flexible
- Most of the equipment and spares are indigenously available
- Loss of production is minimum while shifting the equipment
- Low investment

9.6 LIMITATIONS

- Gradient steeper than 1 in 6 (9°)
- Gassy seams
- Soft floor

9.7 DRILLING AND BLASTING PRACTICES IN BG METHOD

9.7.1 Drilling

Drilling is done by crawler mounted electro-hydraulic Jumbo drill, which can drill at any angle in the vertical plane. As BG method envisages drilling of holes in a fan cut pattern, length of holes depends on their position and the seam thickness. During experimental trials, it was seen to vary between 2 and 14.5 m. When the top section is already developed, a thin parting of 0.3 to 0.5 m of coal is left during drilling to get better impact of gas energy, which would otherwise puncture the gallery and venting of gases may occur.

The Jumbo drill machine can drill up to 30 m long incline holes with a speed of 1 m per minute. The drill rods are 1.13 m long with connecting features (female types) at the ends. The drill rods are provided with a coaxial central hole to facilitate wet drilling. Water at the rate of 2.5 m^3/hr under a pressure of 8–10 bars is flushed during drilling to clean drill-cuttings out of the hole. This minimizes pollution from coal/rock dust during drilling operation. The diameters of drillrod and drillbit are respectively 32 mm and 34 mm. The finished hole diameter is around 43 mm.

During trials, on an average, 19-24 (for half ring) and 34-39 (for full ring) holes were drilled in a ring pattern (Figure 9.1/b). All the holes were drilled in the roof leaning towards the goaf line at 55°–60°

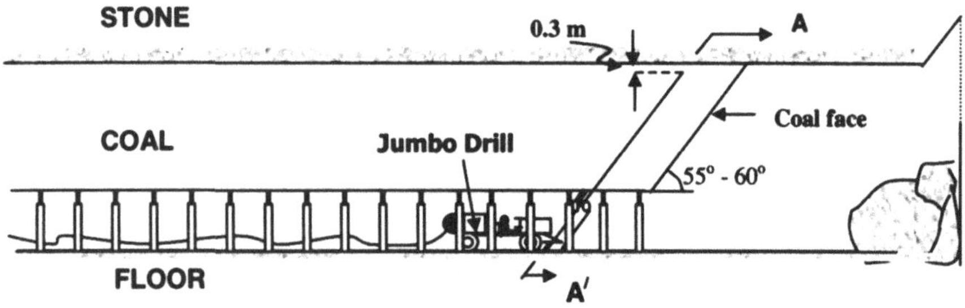

Fig. 9.1(a) Drilling arrangement by Jumbo Drill

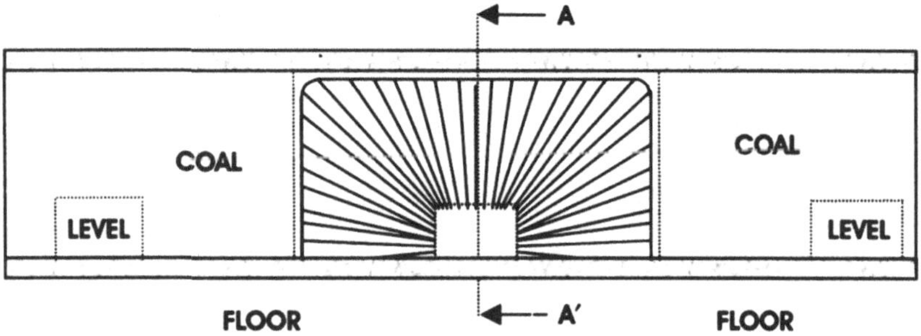

Fig. 9.1(b) Ring pattern of holes in Blasting Gallery (BG) method

(55° average) from horizontal (Figure 9.1/a) and 45° towards the goaf from the pillar side. The holes in the roof were drilled up to the seam thickness leaving 0.3 m parting in the roof. Whereas, holes in the side pillars were drilled in such a manner that they covered nearly half of the pillar on both dip and rise sides from the gallery where Jumbo was operating. A curtain of about 1.5 m in the middle of each pillar was left after such drilling to facilitate safe loading by LHDs. However, it was observed that this curtain falls automatically in successive blasting rounds. The distance between the two consecutive rings was maintained at about 1.5 m.

9.7.2 Ringhole Blasting

After drilling is completed, shot holes are charged with explosive cartridges and spacers are set alternately with interlocking. Belgex Coal-R which is a nitroglycerine based explosive and non-incendive detonating cord viz. G-cord (3.6 gm/m) are used in ringhole blasting in India. These products are duly approved by the Directorate General of Mines Safety (DGMS).

Polyethylene spacers are used between two consecutive cartridges for effective utilization of explosive energy throughout the hole (Figure 9.2). G-cord is inserted in the first explosive cartridge and taped around. It is

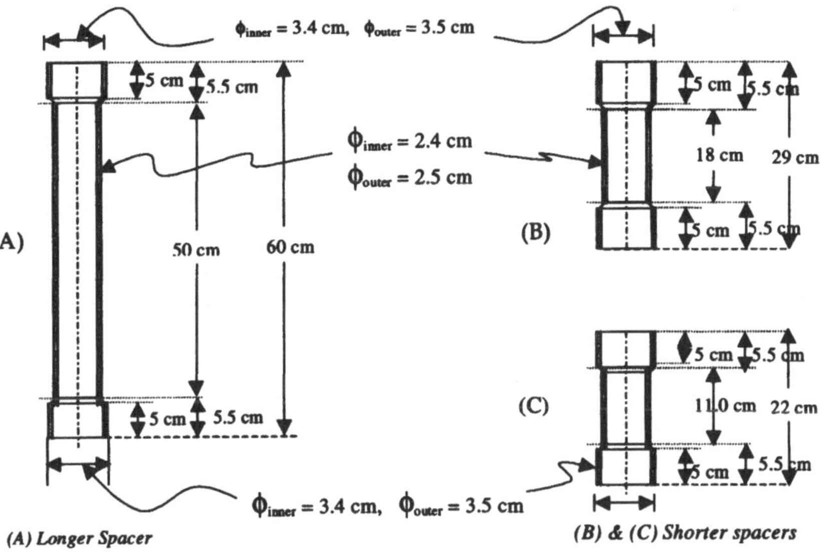

Fig. 9.2 Specification of plastic spacers (longer and shorter types)

then run through the spacers touching each explosive cartridge and again ties in the last cartridge which carries the detonator (Figure 9.3). In the last cartridge, the G-cord and the detonator are inserted obliquely. A portion

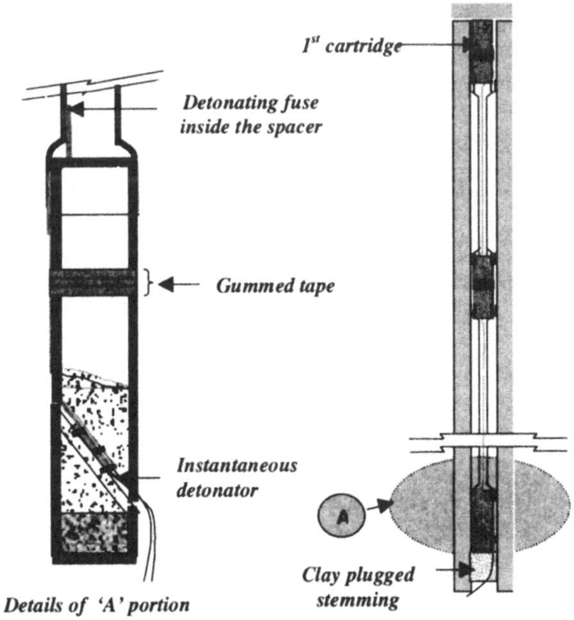

Fig. 9.3 Charging and stemming systems of ringhole using detonating fuse

of the lead wires of the detonator (about 2 m length) and about 0.5-0.6 m uncharged portion of the hole are stemmed with clay. For full ring blast, generally 75-100 kg of explosive is used. Charge per hole generally comes to around 2.0–2.6 kg although, the maximum permissible amount as stipulated by DGMS is 3.00 kg.

After charging and stemming of all the holes, the circuit is connected in series. Before blasting, the last installed roof support at the goaf line is removed and the next support line is reinforced with extra supports. The whole ring is blasted in one round using instantaneous electric detonator. Generally, 250-350 t of coal is produced in one round of ring blast and 500 t in case of junction blast.

9.7.2 Specification of Blasting Materials

9.7.2.1 Explosive

Belgex Coal-R which is a nitroglycerine based explosive composition of M/s Bharat Explosives Ltd. (BEL) is permitted for use in ringhole blasting in BG Panels. It is a special type of newly developed explosive product for which the Directorate General of Mines Safety (DGMS), India, has permitted maximum 3.0 kg of explosive charge per blasthole. The major ingredients of the explosive are:

i. Nitroglycerine;
ii. Nitro-cellulose;
iii. Ammonium nitrate;
iv. Cellulose (woodmeal);
v. Salt.

Table 9.1 Physico-chemical properties of the explosive

Explosive brand and type	*Belgex Coal-R, Bharat Explosives Ltd.; NG-based*
Weight of explosive	185 g/cartridge
Diameter and length of cartridge	32 mm and 200 mm
Density (g/cc)	1.18-1.19 (Avg. 1.185)
VOD (m/s)	2826
Air-gap sensitivity, cm (Unconfined condition)	5.0
Oxygen balance	0.1506535
Detonation pressure (MPa)	2365.93
Explosion pressure (GPa)	1.1823
Explosion energy (MJ/kg)	3.126

9.7.2.2 Oxygen balance

After analysing the composition through 'CHEF' software of Precision Blasting Services of USA, the oxygen balance was calculated as +0.15065350 that indicated an inefficient reaction. This would lead to either yellow or rust coloured fumes.

9.7.2.3 Detonating Cord

Product name and manufacturer	:	G-Cord, Gulf Oil Corp. Ltd., Hyderabad
Weight of ring cord	:	25 g/m
Diameter	:	6 mm
Weight of PERN in G-Cord	:	3.6 g/m

9.7.2.4 Detonators: Instantaneous, No. 6 Strength

9.7.2.5 Spacers

Length of longer spacer	:	600 mm
Length of shorter spacer	:	290 mm

9.8 CASE STUDIES

As mentioned earlier, field investigations were conducted in three mines of M/s SCCL and one mine of M/s ECL. In VK-7 incline (SCCL) and Chora-10 Pit colliery (ECL), the problem of fragmentation was insignificant as compared to GDK-8 Incline and GDK-10 Incline in Ramagundem area. As such, focus was predominantly given to these later two mines to rectify fragmentation problems which were seriously hampering production and productivity.

9.9 FRAGMENTATION STUDY AT GDK-10 INCLINE

9.9.1 Backdrop

GDK-10 Incline is situated in the south-eastern part of Ramagundem coal belt, Andhra Pradesh of the southern part of India. It is nearly 25 km from Ramagundem Railway Station. GDK-10 Incline was opened, for the first time, on November 25, 1976 and production of coal was started from 1978 onwards. Coal extraction by Blasting Gallery method commenced from July 5, 1989 in collaboration with CdF of France. The mine recorded the highest production of 7,135 tonne/day on January 7, 1996 and also received the award for Best Project Implementation from the competent authority for the year 1989-90 and 1994-1995 for BG-I and BG-II panels respectively.

9.9.2 General Geology of the Area

Based on the nature of coal deposits, the mine is divided into three blocks namely, Block-A, Block-B and Block-C. Block-A and Block-C are fairly good patches whereas Block-B contains numerous faults and narrow patches. The GDK-10 Incline is covered by a thin mantle of soil. The stratigraphic succession and sequence of the coal seams are mainly based on the sub-surface data. The Barakar Formation succeeds the Talcher Formation. The Barakar Formation comprises mainly white and greyish, coarse to medium grained fields pathetic sandstone and seven coal seams namely IA, I, II, IIIB, IIIA, III and IV. (in descending order). The coal measure formations trend in E-W direction on the southern side and swing towards N-W in the northern part of the block. The dip of the beds varies from 1 in 8 to 1 in 11. The bottom set of seams namely, IIIA, III and IV are exploited through GDK-10 incline. IIIA seam is having thickness of 1.2 to 1.5 m only, whereas III seam has maximum thickness of 10.5 to 11.0 m. Seam IV is having thickness of 4.2 m. The parting between IIIA and III seam is 25 m whereas the parting between III seam and IV seam is only 4.5 m. Blasting Gallery method is adopted for excavation of III seam. The Rock Mass Rating (RMR) of the immediate coal roof in III seam is 59.90. Compressive strength of the coal roof varies between 221 and 246 kg/cm^2.

The study was conducted at BG Panel No. 2A of Block 'C'. At the time of investigations, nearly 40% of the panel was excavated. Details of the panel are given below—

(a) Name of the panel	-	BG Panel No. 2A (Block 'C')
(b) Size of the panel	-	123 m × 150 m
(c) Seam thickness	-	11.0 m
(d) No. of the pillars	-	12
(e) No. of rooms	-	9
(f) Minimum & Max. depths	-	300 & 320 m
(g) Total coal in the panel	-	2.12 lakh tonnes
(h) Extractable in the panel	-	1.44 lakh tonnes
(i) Panel started on	-	9-12-2001
(j) Gallery dimension	-	4.2 m × 3.0 m (for level and dip galleries)

9.9.3 Causes of Boulder Formations

The main causes of unoptimized fragmentation and formation of boulders, as observed during trial blasts at GDK-10 Incline, are described below:

9.9.3.1 Drilling Error

9.9.3.1a Angle Alignment at the Top of Corner Sides. One of the main reasons for boulder formation during ring blast was improper drilling. As a general practice, it was found that parallel holes were drilled in the dip and rise sides of the gallery. Due to such an irregular act of drilling, a wide gap was created at the corners as depicted in figure 9.4. Boulders generally appeared from these corner gaps.

9.9.3.1b Insufficient Hole Depth. It was a general practice in the mine to drill all side holes including corners, both in dip and rise sides, uniformly

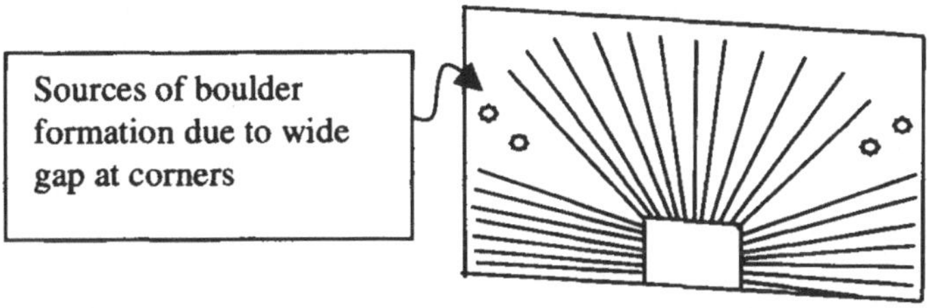

Fig. 9.4 Sectional view of the heading depicting the causes of boulder formation and fragmentation problems

(7 to 9 m). However, for better results, it should have been gradually increased while approaching towards top corners and decreased while away from the top corners. It was seen that the maximum required depth at the side corner was 14.5 m at the rise side and 12.5 m at the dip side. Similarly, for the top portion, hole length should be nearer the corner sides. But, in actual case, it was found that all the top holes were of same length throughout the roof span as indicated in figure 9.5. Because of such inaccuracy in hole depths, boulders used to form in each and every blast. At the same time, severe back cracks used to occur, causing formation of big boulders in subsequent blasts.

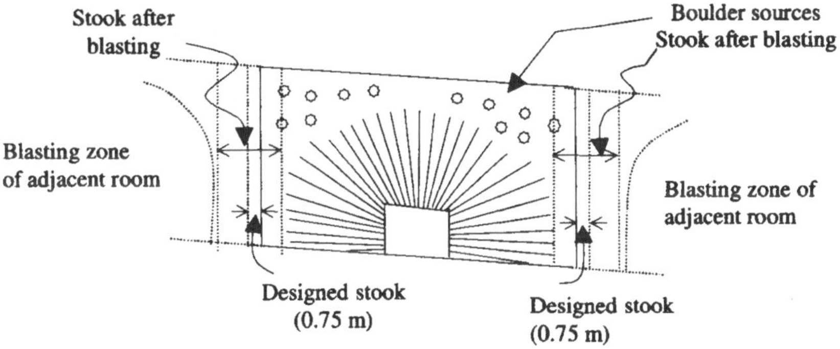

Fig. 9.5 Boulder formation due to insufficient hole depth

Insufficient hole depths at both dip and rise sides of the gallery also resulted in formation of large stook pillar. That case was distinctly observed in 63 level of the panel where big stook was standing in between 63 Level and 63A Level. Such occurrence resulted in poor percentage of coal recovery and more chance of spontaneous heating of coal inside the goaf area.

9.9.3.1c Spacing of Holes. Spacing between holes is an important parameter for controlling fragmentation. Irregular and uneven spacing may lead to boulder formation. The recommended spacing between holes is 30 – 35 cm.

9.9.4 Improper Charging

9.9.4a Misfire

Misfire generally occurred due to improper charging of holes where detonating fuse is not properly in contact with the explosive charge in which the detonator is inserted. Even if, one hole is misfired, it can cause boulder formation. This case was observed in 42 dip in the blasting of 21-02-2002. Sometimes, bending of detonating fuse may result in misfire due to discontinuity of PETN charge inside the fuse. Defection of detonator may also cause misfire.

9.9.4b Blown-out Shot

This is occurred mainly due to insufficient stemming in holes at the bottom of the charge. In case of blown-out shot, some percentage of explosive energy is wasted resulting in formation of unwanted boulders.

9.9.5 Distribution of Explosive Charge

It was seen to be the general practice to put more charge at the top of the hole, especially while charging roof and corner holes and less charge at the bottom of the hole where the burden is usually more. Such acts of charging would always result in boulder formation from the bottom of the hole (Figure 9.6). Moreover, due to restriction of explosive charge per hole by DGMS, sometimes it was found that poor fragmentation was solely due to insufficient explosive charge fired in the blastholes.

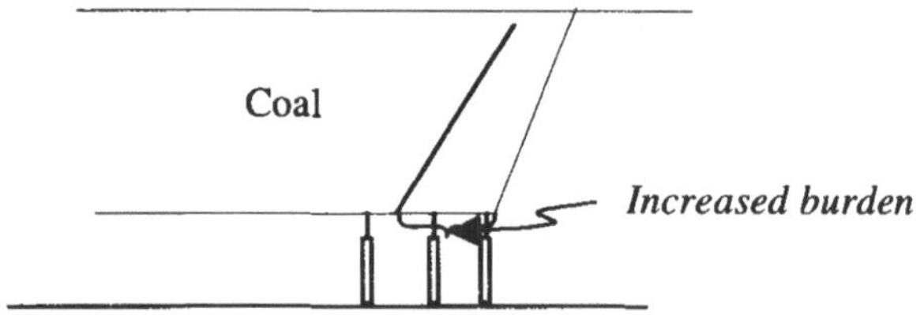

Fig. 9.6 Poor fragmentation due to increased burden and improper charge distribution

9.9.6 Other Causes

9.9.6.1 Hole Jamming Due to Strata Pressure

Sometimes, BG panel is worked out at a depth of 300-320 m from the surface. Under such conditions, more strata pressure is likely to be experienced on the workings, which may result in frequent spalling of coal chunks from the standing pillars. Moreover, during investigation, it was found that the pillars were mostly composed of shale and became weak. Drilling of holes in such pillars always posed problems and usually got jammed due to strata pressure. Due to these problems, proper charging became impossible and ultimately it accounted for boulder formation.

9.9.6.2 Cracks Formation Due to Excessive Charge

It was found that to overcome the problem of strata pressure during charging of blastholes on the side pillars (both dip and rise sides), PVC pipes were inserted immediately after drilling of such holes. Continuous charge was put inside such PVC pipes. Because of this excessive charge, fractures/backbreak used to develop for the subsequent ring blast which indirectly affected blast fragmentation.

9.9.7 Actions Taken to Reduce Boulders

The mine officials of GDK-10 Incline were informed of the causes of boulder formation during ring blasts. The Agent and Manager of the mine jointly convened a meeting of all the Jumbo drill operators and shot-firer groups to make them understand the problem. The following actions were immediately taken:

(i) *Correction of Drilling Errors*: All the Jumbo operators were cautioned that the main reason of bolder formation was due to improper drilling. They were advised to strictly maintain the proper depth of successive holes and angle alignment especially at the corners of the gallery.

(ii) *Charging of Holes*: All shot-firers were advised to charge the shot holes as per guidelines. Like Jumbo operators, they were also cautioned about any sort of negligence.

(iii) *Prevention Against Misfire:* To prevent misfire, an oblique initiation pattern was suggested. The entire detonating cord should remain within the shot hole in-bye of the stemming material. The detonator and the detonating cord should remain in contact throughout the length of the detonator inside the primed cartridge. The detonating cord should be to the explosive cartridges with adhesive tape. Only direct initiation should be used and the detonator should be put in the outermost cartridge.

(iv) *Adjustment of Spacing between Holes:* The spacing between holes at the gallery sides were adjusted to 0.3 m for all the holes.

(v) *Proper Stemming*: Holes should be properly stemmed to prevent any sort of blown-out-shot. At least 0.75 m stemming with clay plugs is required for proper impact.

9.10 CASE STUDY AT GDK-8 INCLINE

During field investigations, it was observed that poor fragmentation was a major problem in BG-II panel of GDK-8 Incline. Generally 5–7 oversize boulders (sometimes more) were found to occur in each ring blast. Secondary blasting was conducted by plaster shooting method, which required more explosives to break the rock at the same time producing lots of fumes and noise. Such act was hampering normal production of the coal.

One of the main reasons for the blast fragmentation problem at this mine was the presence of stone band in the coal seam. There was a 1.5-2.75 m thick shale layer between the developed top section and bottom section in the whole BG-II area as shown in figure 9.7. Due to this, the strata was not uniform and thereby posed fragmentation problems.

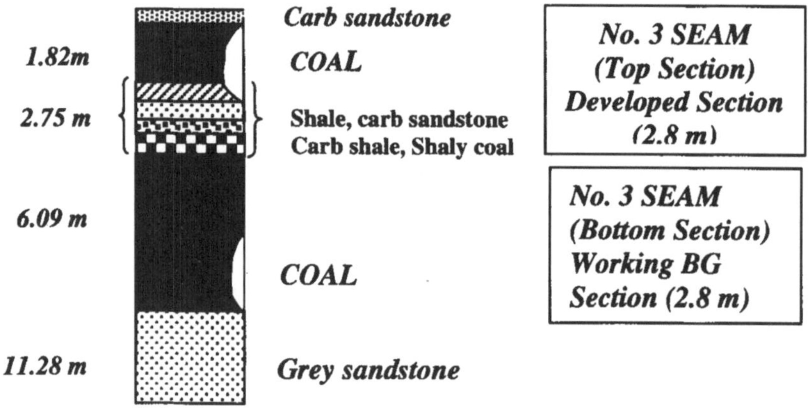

Fig. 9.7 Cross-section of coal seam at BG-II in GDK-8 Incline

In order to minimize this problem, staggered charging patterns using larger (600 mm) and smaller (290 mm) spacers were tried. These patterns are shown in figures 9.8 and 9.9. In such designs, staggered charging was made in the coal portion by using shorter and larger spacers. However, in the shale portion, only shorter spacers were used. The total coal seam, as shown in the figures, was divided into three parts, viz. upper coal portion, shale band and lower coal portion. In the design pattern, the average charge per hole was around 2.58 kg.

9.11 ENERGY-BASED VIBRATION PREDICTOR EQUATIONS

According to extensive mine experiments conducted by Indian and foreign experts, it has been established that the induced particle velocity (PPV) is predominantly related to the emanated seismic energy of a source and the distance from the source. With this concept in mind, nearly 160 experimental blasts were carried out in four BG panels in India under varying geomining conditions. Based on the results, as obtained so far, CMRI has established predictor equations correlating explosive energy

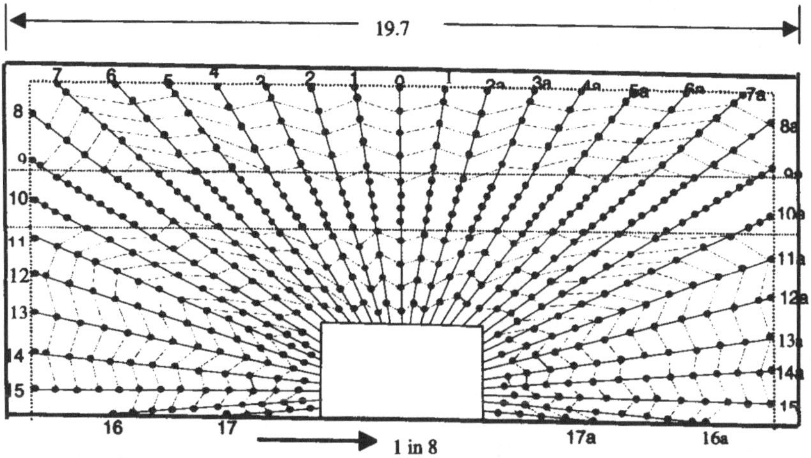

- *Modified staggered pattern established by CMRI*
- *Type of charging: Staggered by using smaller and bigger spacers at the bottom of each hole alternately*
- *In shale/carb-sandstone portion, only smaller spacers were used*
- *Total number of holes is 35 and total explosive charge is around 90 kg*
- *Average charge per hole—2.6 kg*
- *Shorter spacers—110 nos., Longer spacers—389 nos.*

Fig. 9.8 Staggered charging pattern using smaller and bigger spacers in the top and bottom parts of the seam for split level gallery

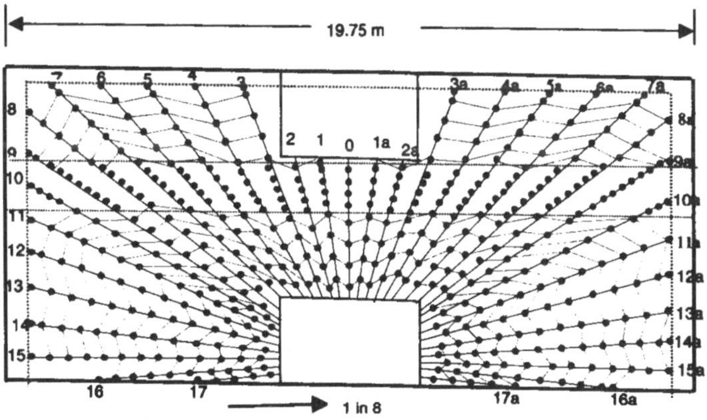

 # Staggered Pattern of charging in coal portion for developed galleries at BG-II, GDK-8 No. Incline, SCCL.
 # Total holes – 35, Total charge – 83.25 kg (450 cart.)
 # Average charge per hole – 2.4 kg
 # Shorter spacers – 108 nos., Longer spacers – 358 nos.

Fig. 9.9 Staggered charging pattern using smaller and bigger spacers in the top and bottom parts of the seam for developed level gallery (in Joules) and distance from the shot to that of induced peak particle velocity received at the sensor point (Pal Roy and Sawmliana, 2003). Good correlation was obtained between the actual and the predicted values. It was realized that these equations might be used as efficient tools to determine the size of a blast in BG panels so that the induced particle velocity did not create any damage in underground workings in respect of the available safety norms.

9.11.1 Energy Based Predictor Equations

When the energy of seismic shocks and the distance from the source can be foreseen, the induced vibration can be better predicted by *EDSD* or Energy Dissipated Scaled Distance [*D*/sqrt. ln (*E*)] rather than conventional scaled distance ($D/Q^{0.5}$ in m/kg$^{0.5}$). Where *E* is the total explosion energy measured in Joules (*J*), *D* (in m) is the distance of the measuring point and *Q* (in kg) is the total explosive charge fired in a round of blast. It is to be noted that in Blasting Gallery (BG) panels, no delay detonator is used i.e. all the holes are fired simultaneously.

It is established that the seismic energy at a source caused by blasting depends on the weight of charge (Kidybinski, 1986). According to mine experiments conducted by Snodgrass and Siskind (1974), induced particle velocity *V* is related to the seismic energy of a source and the distance from the source to the mine opening.

Experiments as well as theoretical calculations using software packages TIGERWIN of BAI and CHEF of PBS, USA have shown that the explosion energy of the Belgex Coal-R was 744.28 Kcal/kg i.e. 3126x10^3 J/kg. Considering such output energy, the following predictor equations were derived separately for roof, pillar and floor vibrations induced by blasting in BG Panels.

Roof: $V_{Roof} = -9.745 + 972.037\,[D/\,(ln\ E)^{0.5}]^{-1},\ r = 0.86$ (9.1)

Floor: $V_{Floor} = -5.206 + 377.196\,[D/\,(ln\ E)^{0.5}]^{-1},\ r = 0.66$ (9.2)

Pillar: $V_{Pillar} = \alpha \times V_{Roof}$ (9.3)

Where, *V* is the peak particle velocity (mm/s), *D* is the distance of the measuring point (m), *E* ($3126 \times 10^3 \times Q$ in Joules) is the explosion energy, *Q* (kg) is the total charge fired in a round of blast, *r* is the correlation coefficient and α is a damping factor whose value lies between 0.45 (fractured, banded and highly jointed deposit) and 0.75 (comparatively compact deposit). When the explosion energy is known, the impact of vibration at a desired distance in underground openings can be estimated using these equations. Figure 9.10 shows the comparison of measured and actual values from 60 experimental blasts whereas figure 9.11 shows the comparison of roof and pillar vibrations of 47 experimental blasts conducted in BG panels of 3 different Inclines of M/s SCCL, Andhra Pradesh, India.

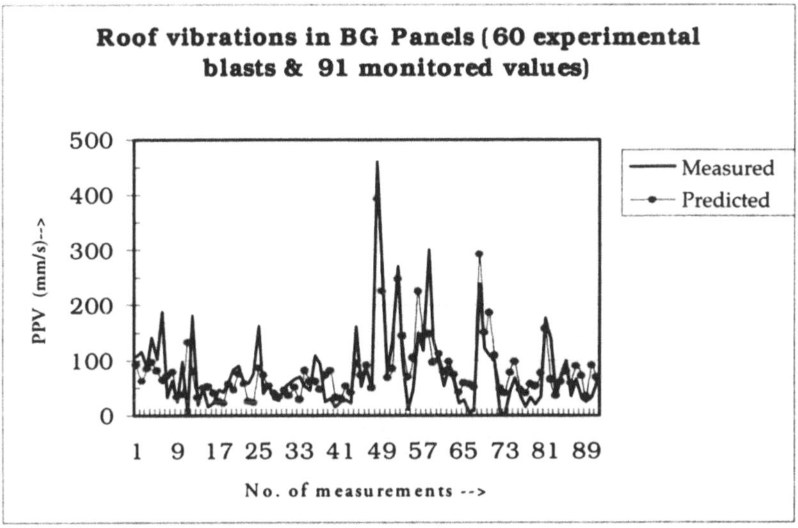

Fig. 9.10 Comparison of predicted (equation 9.1) and measured values of peak particle velocity at the roof of the gallery in BG Panels

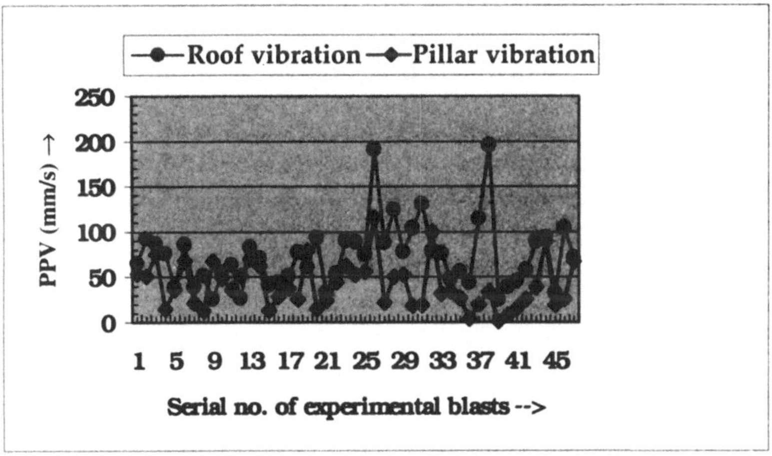

Fig. 9.11 Comparison of roof and pillar vibrations at the same location in BG Panels as measured in 47 experimental blasts

9.11.2 Monitoring Stations

Vibration study was conducted in BG Panels of VK-7, GDK-8, GDK-10 Incline mines of SCCL (Singareni Collieries Company Ltd.) and Chora 10-Pit Colliery of ECL (Eastern Coalfields Ltd.) for safety of the underground workings. It was also intended to develop general guidelines which can be implemented in different panels at different areas under varying geomining conditions. More than 160 ringhole blasts were thoroughly monitored.

Sophisticated instruments like Sinco-6 (USA make), Blastmate II (Canada make), MiniMate-077 (Canada make), convergence indicator (Rheostat type), load cells (Mechanical type), data logger and gas sensing devices were used during monitoring of various parameters (See Plate 8.5 of Chapter 8). A special type of non-ferric sensor attachments were fabricated as per international standards for use in electrodynamic sensors for vibration measurements in roof and pillars (See Plate 8.1 of Chapter 8).

In order to identify the most vulnerable locations, monitoring points were selected at the junctions of the galleries as well as in the middle level gallery between two rise galleries. Load cells and convergence indicators were also installed near vibration monitoring stations with a view to simulate the results for better interpretations of data. A full fledged package study which included gas survey, vibration measurements, strata load and convergence before and after blasting (both for ringhole and induced) enabled to establish the optimum design patterns. Distance of the vibration monitoring stations from the blasting source varied between 10 and 120 m and the consequent recorded vibrations in terms of Peak Particle Velocity (PPV) ranged between 460.8 and 0.921 mm/s. A few solid blasting records during developmental work of BG Panel, movement of LHD and strata movement which resulted bumping sound were also recorded for comparison. A few recorded readings of convergence indicator and load cell are shown diagrammatically in figures 9.12 and 9.13.

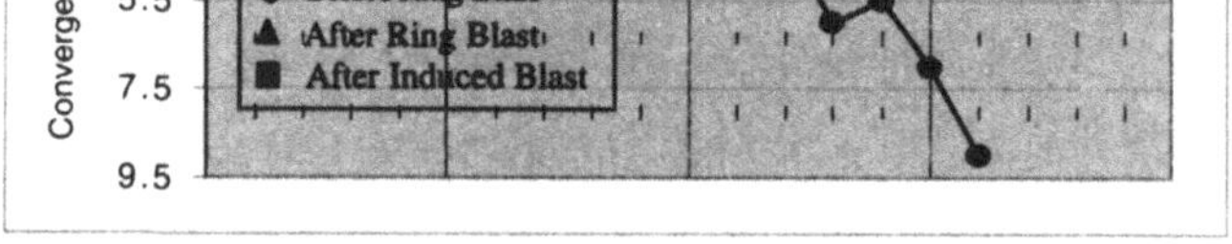

Fig. 9.12 Convergence observation at 81 ½ LN/32 Dip of BG-II, Panel-2 of GDK-8 Incline, SCCL

9.12 CRITICAL CONVERGENCE

Critical convergence is defined as the convergence beyond which the immediate roof in developed galleries may pose stability problem in normal conditions. Critical convergence can be determined from the following formula developed by CMRI (Dhar and Sheorey, 1991):

$$C_m = 40 \times \sqrt{W} \times \rho_{rr}^{1/3} \times (1 - RMR/100)^3 \tag{9.4}$$

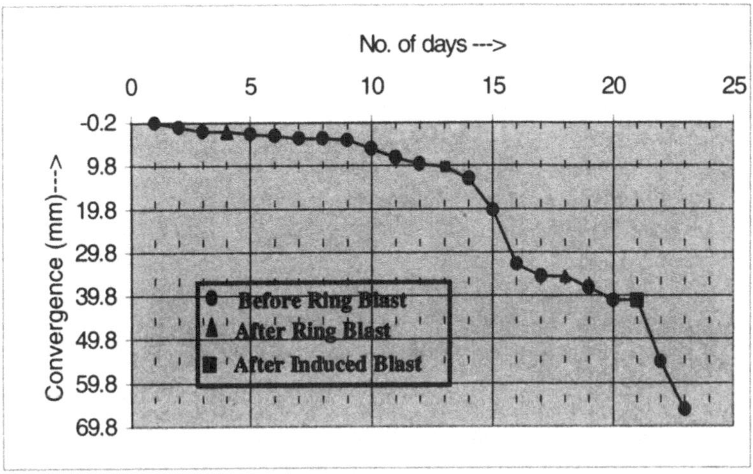

Fig. 9.13 Convergence observation at 82 ½ LN/32 Dip of BG-II, Panel-2 of GDK-8 Incline, SCCL

Where, C_m = Critical convergence (mm)
W = Gallery width (m)
ρ_{rr} = Density of roof rock (g/cc)
RMR = Rock Mass Rating developed by CMRI

9.12a Critical Convergence for Level Gallery

Width of gallery	:	4.2 m
Density of roof rock	:	2.06 g/cc
Rock Mass Rating	:	62.0

Using equation (9.4), critical convergence for level gallery was obtained as 5.72 mm which was subsequently verified during investigations.

9.12b Critical Convergence for Main Dip Gallery

Width of gallery	:	4.8 m
Density of roof rock	:	2.06 g/cc
Rock Mass Rating	:	62.0

Using equation (9.4), critical convergence for main dip gallery was obtained as 6.206 mm. This value was found to be near optimum during trials in the mine.

9.13 PREDICTION OF ROOF FALL

An alternate method was suggested by CMRI (Maity and Prasad, 1999) to predict roof fall, in advance, on the basis of roof closer data. The equation, in an empirical formula form, is given below:

$$\mu = \frac{C_n - C_{n-4}}{C_{n-4} - C_0} \times \frac{n-4}{4} \geq \text{cut-off value} \quad (9.5)$$

Where, C_n = Last reading before fall (mm)

C_{n-4} = Reading of the fourth day prior to last reading (mm)

C_o = Initial reading (mm)

n = Total number of days of observation.

When μ goes beyond the cut-off value, i.e. 2.0, the chance of roof fall within 4 days becomes prominent and precautions are taken accordingly. Case studies conducted by CMRI in different mines showed that in 80% cases, equation (9.5) predicted correctly. Results and recommendations on gas and strata monitoring are given in Tables 9.2 and 9.3.

Table 9.2 Presence of gas before and after ring-blast

Name of gas	*Before blast*	*15 minutes after blast*	*30 minutes after blast*	*40 minutes after blast*	*1 hour after blast*
O_2	19.8 %	19.2 %	19.3 %	19.6 %	19.4 %
CO_2	0.1 %	0.1 %	0.2 %	0.3 %	0.2 %
CH_4	0.0 %	0.0 %	0.0 %	0.0 %	0.0 %
CO	0.3 ppm	51.6 ppm	18.6 ppm	9.6 ppm	8.6 ppm
Total Explosives fired		56.4 kg	96.2 kg	96.2 kg	96.2 kg

Table 9.3 Results of strata monitoring and their impacts

Sl. No.	*Important Findings*
1.	No significant changes in roof convergence and strata load were noticed just before and after ring blasting.
2.	When roof convergence was 4 mm or more and the corresponding strata load was increased by 2 tonne and more within 24 hrs time interval, the situation might be considered as alarming. Under such conditions, to prevent air blast, induced blasting should be conducted.
3.	After induced blasting and local/main fall, strata load was released and convergence remained constant for few days and then gradually increased.
4.	An increasing trend of strata load was observed when the face was approaching towards the monitoring point. This load attained its maximum value at the goaf-edge. Similar phenomena also occurred for roof convergence.
5.	For better results, it was suggested to fix convergence points in the stone roof along with the coal roof for comparison.

9.14 BLAST DAMAGE INDEX (BDI)

Blast Damage Index (BDI) denoted by 'D_{ib}' and defined by the ratio of the induced stress due to blasting to the dynamic tensile strength of the rock mass being blasted (Yu and Vongpaisal, 1996). It is a dimensionless indicator analogous to the reciprocal of safety factor and may vary between

0–3.5. It can be used as an index, to quantify the damage of roof and pillars due to underground blasting, for the safety of man and machinery. The redefined Blast Damage Index (D_{ib}) is represented as follows:

$$D_{ib} = \frac{3.6 \times 10^{-3} \cdot V_R \cdot \rho_r \cdot C_P}{K \cdot U_{CS}} \qquad (9.6)$$

Where,

V_R = Vector sum of peak particle velocities in three orthogonal directions (mm/s)

ρ_r = Density of rock mass (g/cc)

C_p = P-wave velocity (m/s)

K = Site quality constant=RMR (20 -100)

U_{CS} = Uniaxial compressive strength of rock mass (MPa).

Unlike Yu and Vongpaisal's work, the present study involved optimization of blast design parameters for an entirely different situation where extraction of thick coal seams were carried out by Blasting Gallery (BG) method of working with induced caving at periodic time interval. Thus, a different BDI was required to establish the damage potential of underground roof and pillars under such circumstances.

In order to determine the value of BDI, the parametric values of density, uniaxial compressive strength and RMR of immediate strata were evaluated from the field and laboratory studies. Weighted values were considered for such calculations. A Software Package CMRI-ROCK developed by the CMRI Regional Centre, Nagpur was used to calculate the RMR from the base-data with due consideration of the support systems in the underground developed galleries (Chakraborty, 2003).

Based on data monitored in a substantial number of blasts, two different damage-quantifying indices, of which one depended on BDI and the other on Scaled Distance (SD), are given in Table 9.4. Figure 9.14 shows the plotting of some BDI values in different damage zones.

Table 9.4 Damage quantifying indices based on BDI and SD

Classification of zones	*Range of BDI values*	*Range of S.D.**
Safe zone	< 1.0	> 2.5
Alarming zone	1.0 – 1.75	> 2.0 – 2.5
Minor damage zone	> 1.75 – 2.50	1.50 – 2.0
Major damage zone	> 2.50 – 3.50	< 1.5

*S.D. = Scaled Distance

9.15 SUGGESTED DESIGN PATTERNS AND GUIDELINES

Based on the scientific approaches to the problems of Blasting Gallery method of extraction in Indian geomining conditions and consequent extensive field trials and laboratory tests in all existing BG panels of India, CMRI has released a handy BG-Folder and 'INCAB' (INduced

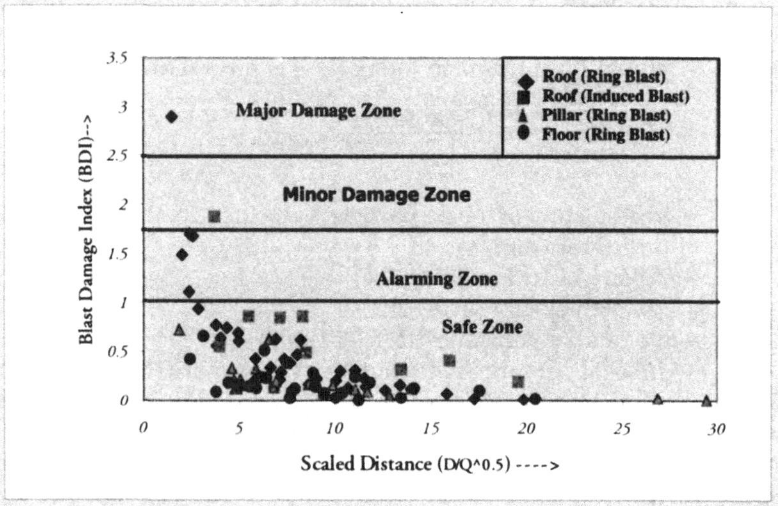

Fig. 9.14 Plot of Blast Damage Index (BDI) at different Scaled Distance (SD) for characterizing damage zones

CAving by Blasting) software for field use (Pal Roy and Sawmliana, 2003). Tables 9.5 and 9.6 show the design guidelines proposed by CMRI for bottom section developed and also for top and bottom both sections developed seam.

Table 9.5 Design guidelines for bottom section developed seam

Seam thickness	7 m–8 m	9.5 m–10.0 m	10.5 m–11.0 m
Size of gallery	4.2 m × 3.0 m	4.2 m × 3.0 m	4.2 m × 3.0 m
No. of holes	30	35	37
Diameter of hole	43 mm	43 mm	43 mm
Hole depth	2.5 m–9.0 m	3.5 m–12.5 m	3.5 m–14.5 m
Spacing between holes	0.3 m–0.4 m	0.3 m	0.3 m
Average charge/hole	1.6 kg	2.0 kg	2.4 kg
Total charge	45 kg–50 kg	60 kg–75 kg	80 kg–95 kg
Hole inclination	45° (Side pillars) 60° (Roof)	45° (Side pillars) 60° (Roof)	45° (Side pillars) 60° (Roof)
Spacer length (end to end)	0.29 m–0.6 m	0.29 m–0.6 m	0.29 m - 0.6 m
Explosive type	Belgex Coal-R 32 mm diameter 185 gm/cartridge	Belgex Coal-R 32 mm diameter 185 gm/cartridge	Belgex Coal-R 32 mm diameter 185 gm/cartridge
Detonating Fuse	G-Cord (GOCL) Diameter – 6 mm PETN – 3.6 g/m	G-Cor (GOCL) Diameter – 6 mm PETN – 3.6 g/m	G-Cord (GOCL) Diameter – 6 mm PETN – 3.6 g/m
Detonator	Copper coated instantaneous No. 6, Range-1	Copper coated instantaneous No. 6, Range-1	Copper coated instantaneous No. 6, Range-1

Table 9.6 Design guidelines for both top and bottom sections developed seam

Seam thickness	7 m–8 m	9.5 m–10.0 m	10.5 m–11.0 m
Size of gallery	4.2 m × 2.8 m	4.2 m × 3.0 m	4.2 m × 3.0 m
No. of holes	30	35	37
Diameter of hole	43 mm	43 mm	43 mm
Hole depth	2.5 m–9.0 m	3.5 m–12.5 m	3.5 m–14.5 m
Spacing between holes	0.3 - 0.4 m	0.3 m	0.3 m
Average charge/hole	1.4 kg	1.6 kg	2.0 kg
Total charge	40 kg–45 kg	50 kg–60 kg	70 kg–75 kg
Hole inclination	45° (Side pillars) 60° (Roof)	45° (Side pillars) 60° (Roof)	45° (Side pillars) 60° (Roof)
Spacer length (end to end)	0.29 m - 0.6 m	0.29 m - 0.6 m	0.29 m - 0.6 m
Explosive type	Belgex Coal-R 32 mm diameter 185 gm/cartridge	Belgex Coal-R 32 mm diameter 185 gm/cartridge	Belgex Coal-R 32 mm diameter 185 gm/cartridge
Detonating Fuse	G-Cord (GOCL) Diameter - 6 mm PETN - 3.6 g/m	G-Cord (GOCL) Diameter - 6 mm PETN - 3.6 g/m	G-Cord (GOCL) Diameter - 6 mm PETN - 3.6 g/m
Detonator	Copper coated instantaneous No. 6, Range-1	Copper coated instantaneous No. 6, Range-1	Copper coated instantaneous No. 6, Range-1

Note: Hole depth and total charge in Tables 9.5 and 9.6 will vary depending on the pillar size and coal strength

9.16 CONCLUSIONS

Based on the field observations and laboratory studies, one of the main reasons for poor fragmentation during ring blast in Blasting Gallery method was determined as the improper drilling and charging of holes. Improper charging of holes can result in misfire and consequent generation of oversize boulders. Due to restriction of explosive charge per hole (maximum limit of 3.0 kg), it may not always be possible to achieve best fragmentation for reasonably thicker coal seam. However, if the modifications and design patterns described in this chapter are followed, overall better fragmentation may be achieved. The presence of stone band in the coal seam usually gives rise to the problem of fragmentation. Staggered pattern of charging using smaller (290 mm) and larger (600 mm) spacers was found to produce excellent fragmentation.

REFERENCES

Chakraborty, A. K, 2003. Personal Communication. CMRI-Regional Centre, Nagpur.

CMRI Interim Report (GAP/30/MT/MOC/01-02), 2003: Investigations to optimize blast design and charge loading parameters in coal for ringhole blasting and in stone for induced blasting in degree-I seams for blasting gallery method in underground mines of The Singareni Collieries Company Limited (SCCL) (Project leaders: P. Pal Roy and C. Sawmliana), Ministry of Coal and Mines, Government of India, New Delhi, February, 2003, p. 89.

CMRI BG-Folder, 2003. Achievements and recommendations of Coal S&T Research Project (GAP/30/MT/MOC/01-02), (Project Leaders: P. Pal Roy and C. Sawmliana), CMRI Publication, p. 10.

Dhar, B. B. and Sheorey, P. R. 1991. Strata control aspects of the blasting gallery method of depillaring thick seam, Journal of Mines, Metals and Fuels, Nov.-Dec., pp. 398-401.

Kidybinski, Anteni 1986. Design criteria for roadway supports to resist dynamic loads, International Journal of Mining and Geological Engineering, Vol. 4, pp. 91-109.

Maity, S. N. and Prasad, S. 1999. Implementation of BG method for extraction of thick seam at Singareni Collieries and strata control investigations to improve productivity, Proc. Platinum Jubilee Symposium on Productivity Improvement in Indian Mining Industries, January 21-30, IT, BHU, Varanasi, India, pp. 175-191.

Pal Roy, P. and Sawmliana, C. 2003. Energy-based vibration predictor equations and blast damage index for blasting gallery panels. J. of Explosives Engineering, Vol. 20, No. 5, September/October, pp. 34-39.

Snodgrass, J. J., Siskind, D. E. 1974. Vibrations from underground blasting, U.S. Bureau of Mines, RI 7937, p. 31.

Yu, T. R. and Vongpaisal, S. 1996. New blast damage criteria for underground blasting, CIM Bulletin, 89(998), Canada, pp. 139-145.

Note: In recent times, in place of Belgex Coal-R new emulsion-based product Powerring and in place of G-Cord new product powercord are in use in Blasting Gallery panels in India.

Properties

(a) Powerring : Density = 1.15 ± 0.05 g/cc
RWS = 67%
RBS = 92% (Compared to ANFO@0.8 g/cc)
VOD = 3200 ± 400 m/s

(b) Powercord : Colour = Red
PETN grammage = 9.0–9.5 g/m
Diameter = 4.0–4.5 mm
VOD = 6000+ m/s

10

Rock Fragmentation and Assessment

10.1 INTRODUCTION

The term fragmentation refers to the post blast size distribution of the rock mass. Till date there is no entirely acceptable way of measuring fragmentation. In many cases reference is made to the average size of the broken rock and sometime it refers to the largest boulders or the number of oversize boulders produced from a blast. The generally accepted terms like 'better fragmentation' or 'poor fragmentation' are used in practical situations where the primary requirement of a good blast is the production of desired fragmentation by using a proper blast design and a matching explosive combination.

Fragment sizes may acceptably be controlled by using the proper amount of energy at strategic locations within the rock and its subsequent release at a precise time to allow proper interactions to occur. In most of the situations, the optimum fragment size depends on the requirements of the operation including those of loading equipment and crusher size.

Quantitative assessment of fragmentation at a larger scale, till date, is the most difficult task. It is highly impractical to screen the whole fragmented materials at the production scale. Worldwide researchers are now employing various numerical and instrumental techniques to arrive at a plausible solution. In the context of CMRI investigations, the present chapter gives an overview of the fragmentation principles, prediction and assessment methodologies.

10.2 INDIAN COAL MINING SCENARIO

The increasing demand of coal in India has thrown a big challenge to its mining industry for developing simple operational methods of production and fruitful utilization of available machinery. Industry-wise explosive demand and global hard coal production of major countries are shown in figures 10.1 and 10.2. To meet the growing demand of coal and mineral productions in the country, it is now a challenge for researchers to establish standard blast parameters such as explosive combination, blast geometry and rock properties for the safe and productive extraction of coal and minerals.

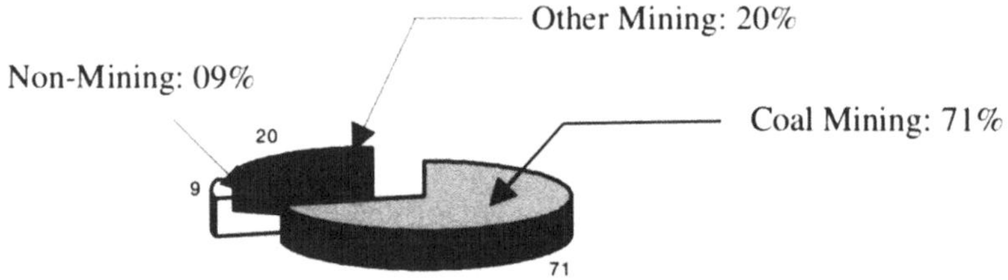

Fig. 10.1 Industry-wise explosive demand

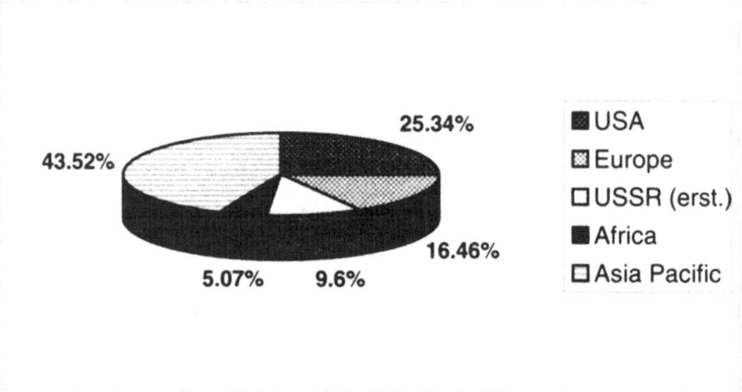

Fig. 10.2 World coal production scenario (in percentage)

10.3 FRAGMENTATION ASSESSMENT METHODS

Researchers, worldwide, have tried with different tools and methods to assess the post-blast fragment size distributions. The commonly employed methods, which are used in fragmentation assessment, may be classified into the following segments (Singh and Pal Roy, 1993).

Assessment of Fragmentation

* Visual analysis
* Photographic method
* Conventional and high-speed photogrammetry
* Counting oversize boulders
* Explosive consumption in secondary blasting
* Shovel loading rate
* Bridging delays at the crusher
* Screening

10.3.1 Visual Analysis

In this method the post blast muck is viewed immediately after blasting and a subjective assessment is made. This technique is not dependable, as the superficial view cannot tell anything about the unseen portion.

10.3.2 Photographic Method

This technique evaluates fragmentation through various analyses based on photographic presentation of the blasted muck. This technique is very much time-consuming and fails to quantify the nature of fragmentation.

10.3.3 High-speed Photography

It is used to evaluate the fragmentation size distribution with respect to time, even when the fragments are not in contact with the original block or other fragments (Figure 10.3). The limitations of the technique become evident when gas venting, smoke and dust obscure crack formation and muck displacements.

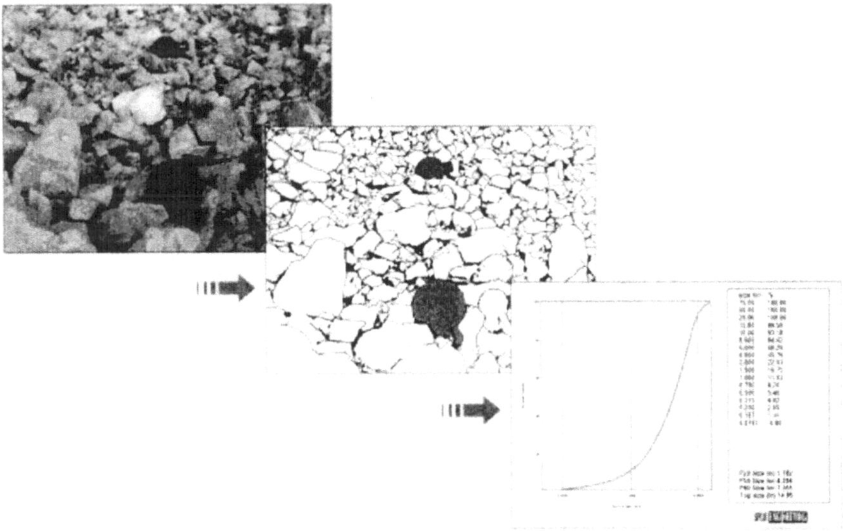

Fig. 10.3 Fragmentation image analysis and size distribution (Courtesy: BAI)

10.3.4 Conventional and High-speed Photogrammetry

This is more reliable and accurate than photography. It can provide three dimensional measurements and thereby helps in fragmentation volume calculations.

10.3.5 Counting of Oversize Boulders

Manual counting of oversize boulders, which cannot be handled by the shovel, may direct an over-size index with respect to the total *in situ* rock blasted. This is a very popular method for estimation of post-blast fragmentation.

10.3.6 Explosive Consumption in Secondary Blasting

In this method, an index regarding the consumption of explosives in secondary blasting by either pop shooting or by plaster shooting is used for comparing the degree of fragmentation of a group of blasts.

10.3.7 Shovel Loading Rate

This technique assumes that the faster the mucking, the better the fragmentation. It can be used more accurately for a comparative account of the nature of fragmentation of a group of blasts.

10.3.8 Bridging Delays at the Crusher

This method is usually preferable in a smaller production site rather than in large scale blasting situation. Bridging delays at a crusher mainly attributes to oversize boulders.

10.3.9 Screening

This technique directs the accurate way of quantifying a post-blast fragmentation. It is economical in case of only small-scale blast. For bigger production blast, it is very cumbersome and impractical.

10.4 RESULTS OF CMRI INVESTIGATIONS

Based on the study carried out by CMRI, Rock Quality Designation (RQD) and charge factor are correlated as per Table 10.1. Such a correlation may be useful for controlling fragmentation (Singh, 1991).

Table 10.1 Charge factor against various RQD levels

RQD (in per cent)	*Charge factor (kg/m^3)*
10-50	0.3
50-70	0.6-0.7
70-100	0.8-0.9

In bedded rocks where thickness of bedding planes is more than 1.8 m, distribution of explosive in each layer is essential for satisfactory breakage. When shooting direction is across the major joint set, side break may extend even more than 5 m.

If the same explosive is used, increased fragmentation can be obtained by reducing either spacing or burden or both by increasing the diameter of the borehole. These techniques increase the charge factor (kg/m^3). One can resort to an explosive of higher strength to improve fragmentation. However, in a highly jointed and thinly bedded deposit as viewed in Plate 10.1, even smaller amounts of explosive charge does not pose any problem in fragmentation.

Plate 10.1 View of a highly jointed bedded deposit

Experiments conducted by CMRI, in iron ore mines with explosive combinations of three different categories like (i) LOX special (A1 powder) and LOX ordinary (loxite power); (ii) ANFO with Slurry booster or ANFO with NG-based booster and (iii) Slurry as column charge and Slurry with base charge, revealed that the explosive combinations of either ANFO with Emulsion or ANFO with NG-based booster performed best from fragmentation point of view (Pal Roy and Dhar, 1993).

10.5 CHARGE FACTOR

While designing any production blast, charge factor is considered as a basic parameter. It is the weight of explosive to be detonated in a blast per unit volume of rock. The unit of charge factor is kg/m^3. As charge factor increases, the average fragment size decreases when the burden remains constant. In bench blasting, charge factor usually varies from 0.1 to 0.7 depending on the strength of the rock blasted. Tables 10.2 and 10.3 show different ranges of charge factor of explosives loaded with ANFO for surface coal and metal mining, considering there is no vibration problem arising due to blasting (Bhandari, 1997). Table 10.4, lists the various charge factors used by CMRI in different rock conditions.

Table 10.2 Charge factors for surface coal mining

Major removal equipment	*Predominate geological unit fragmented*	*Charge factor (kg/m^3)*	*Bench height (m)*
Large dragline	Shale	0.30	15
	Shale	0.35–0.40	20-30
	Sandstone	0.35	18
Small dragline	Shale	0.20	6
	Shale	0.50	7.5
	Shale	0.35	18
	Sandstone	0.65	26
Front-end loader	Shale	0.35	9
	Sandstone	0.65	16
	Sandstone	0.4-0.5	20

Table 10.3 Charge factors for surface metal mining

Rock Type	*Charge factor (kg/m^3)*
Weathered limestone	0.30
Weathered porphyry	0.35
Rhyolite breccia	0.45
Monzonite porphyry	0.60
Quartz sericite porphyry	0.60
Fresh limestone	0.60
Massive jasperoid	0.75

Table 10.4 Charge factor recommended by CMRI in different rock deposits

Sl. No.	*Rock mass type*	*Charge factor (kg/m^3)*
1.	Sandstone—alluvium	0.40-0.55
2.	Metasedimentary limestone and dolomite	0.40
3.	Strong coarse grained sandstone and hard sandy shale	0.50-0.65
4.	Lateritic ores, limonitic ores laminated ores	0.70-0.75
5.	Gneisses and schists and their variants	0.25-0.30
6.	Massive sandstone intercalated either with sandy shale or shale layers	0.30-0.35
7.	Compact ferruginous lateritic	0.60
8.	Massive weathered sandstone	0.30
9.	Massive bedded sandstone	0.35
10.	Massive, hard, thickly bedded limestone	0.33-0.35

(Contd.)

Table 10.4 (*Contd.*)

Sl. No.	*Rock mass type*	*Charge factor (kg/m³)*
11.	Fire affected hard massive sandstone	0.40-0.44
12.	Sedimentary blocky limestone	0.35-0.40
13.	Iron ore (a) soft rock	0.44
	(b) blue dust little bit friable and hard one	0.65
	(c) hard and little bit soft mixed	0.71
	(d) hard ore	0.75
14.	Weathered, highly fractured jointed lateritic mantle	0.25
15.	Jointed, fissured, calcitic dolomite	0.40-0.45
16.	Coarse to pebbly felspathic sandstones with clays	0.30-0.35

10.6 BLAST DESIGN FOR HETEROGENEOUS STRATA

In case of heterogeneous strata of different physical and geological characteristics, precious blast designs are of utmost important for better fragmentation and reduction of boulders which generally occur from the top strata (Plate 10.2). To deal with such problems, two different situations are considered.

(i) Rock strata consisting of two different physical and geological characteristics (6.0-6.5 m bench height);

(ii) Rock strata consisting of three or more different physical and geological characteristics (11.0-14.0 m bench height).

Plate 10.2 View of fragmentation obtained from heterogeneous strata containing oversize boulders from the top hard deposit

10.6.1 Case Studies

The following are the results of case studies conducted at X-Seam Area of Pundi Quarry-E, Sarna Area of Quarry-E and 6-Overburden of Quarry-A of West Bokaro Collieries of M/s Tata Iron and Steel Company Ltd. (TISCO).

10.6.1a Strata Consisting of Hard and Massive Layered Formation at the Top (6 - 6.5 m bench height)

These types of rock strata were encountered in 7-O/B, X-Seam Area of Pundi Quarry-E. Bench height varied between 6.0 and 6.2 m. The top layer was made-up of coarse-grained, massive sandstone with average thickness of 1.8 m (varying from 1.5 to 2.3 m). Boulder formation frequently occurred from this massive sandstone. Average point load index was 2.0 MPa and corresponding compressive strength was 48 MPa. The bottom strata were heavily jointed and fractured shale with thin shaly coal layer of 0.2 m. As the nature of joints and structure of shaly coal were almost similar to that of shale structure, they were taken as the same strata for the study. There were two types of major joint sets namely J_1 and J_2 present in the bottom strata. J_1 was having 80° dip in the direction of N 5°E and was more prominent than J_2. While, J_2 was having 80° dip in the direction of N 280°E. Layer thickness varied between 5 and 30 cm. Average point load index for this strata was 1.8 MPa with average compressive strength of 45 MPa.

To reduce boulder formation from the top zone consisting of massive and coarse-grained sandstone, pilot holes were drilled in between the production holes along the rows (Figure 10.4). Drill depth for production holes varied between 6.2 and 6.5 m and pilot holes were 1.6 to 2.2 m depth with 150 mm hole diameter for both. Burden and spacing for both production and pilot holes were 3.0 m x 3.0 m and the holes were drilled in staggered pattern. Production (or main) holes were charged with 65 kg of Heavy ANFO/SME explosives. Pocket charge (5 kg) was used in each production hole. Pilot holes were charged with 5.0 kg explosive each (Figure 10.4). Detonating fuse was used for initiating the charge with Primex booster (100 gm). Cord relay (25 ms) was used in conjunction

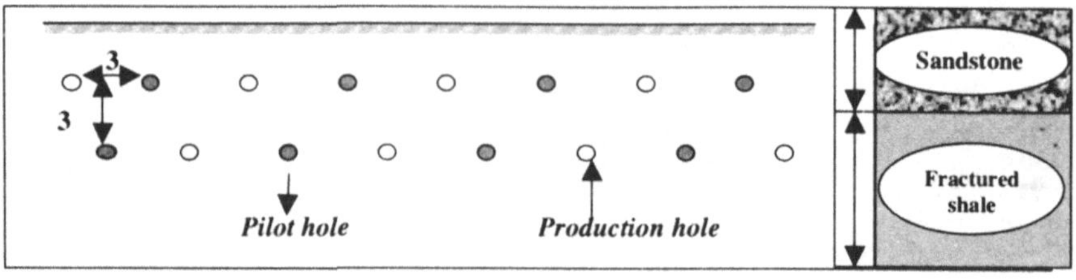

Fig. 10.4 Drilling pattern for two different strata—massive at the top whereas heavily fractured and jointed at the bottom

with detonating fuse during surface connection. Firing pattern was row-to-row delay.

10.6.1b Three or more Rock Strata with Massive Thickly Bedded Layered Strata at Top (11m-14 m bench height)

These types of rock strata were found in 6-O/B Sarna Area of Quarry-E and 6-O/B of Quarry-A. In 6-O/B Sarna Area of Quarry E, three different types of rock strata were found. The top stratum was coarse-grained sandstone, blocky in nature. The middle portion consisted of jointed shale and bottom part was blocky medium grained sandstone. 6-O/B-1 of Quarry-A was also had almost the same type of rock properties and blocky sandstone at the top, but had more layers than Sarna Area of Quarry-E. For such type of rock strata, the established design pattern was a bit different from that of section 10.6.1a.

Burden and spacing were both 6.0 m and the holes were drilled in rectangular pattern. 3.0 m deep (150 mm diameter) auxiliary (Pilot) holes were drilled in between two production holes to reduce boulder generation from the top hard strata (Figure 10.5). Inter hole delay was 17 ms and row-to-row was 42 ms. Production holes (13 m depth) in the front row were charged with 190 kg explosives whereas rest of the holes in subsequent rows (13 m depth) were charged with 260 kg explosives each. Pilot holes were charged with 15.0 kg of explosives each (Figure 10.5).

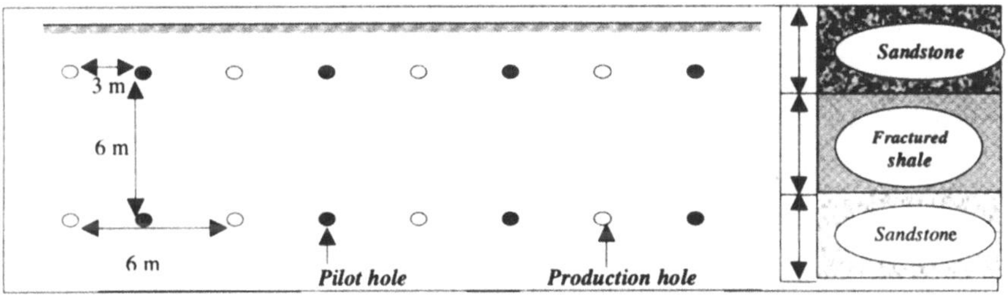

Fig. 10.5 Drilling pattern for three or more rock strata of different characteristics

10.7 FRAGMENTATION PREDICTION MODELS

The fragment size in any blast depends on two major principles namely (a) proper explosive energy input at strategic locations within the rock mass and (b) its proper time of release for optimum interaction. Some of the very standard equations for the determination of fragment size distribution are as follows (Cunningham, 1983)—

10.7.1 Kuznetsov Equation

$$\overline{X} = A \cdot (V/Q)^{0.8} \cdot Q^{0.167} \tag{10.1}$$

Where,

$\overline{X}$ = mean fragment size (cm)
A = rock factor = 1: for extremely weak rock
7: for medium hard rock
10: for hard, highly fissured rock
13: for hard, weakly fissured rock
V = rock volume associated per blasthole (m^3)
Q = mass of TNT which is equivalent in energy to that of the explosive charge in each blasthole (kg)

10.7.2 Rosin-Rammler Equation

$$R = e^{-(X/X_C)^N} \tag{10.2}$$

Where,
R = proportion of material retained on screen
X = screen size
$X_c = \overline{X}/(0.653)^{1/N}$
N = index of uniformity

'N' is called the fragmentation index or the index of uniformity. The value of 'N' may be determined from the following equation—

$$N = (2.2 - 14.B/d)\cdot(1 - W/B)\cdot[1 + (A-1)/2]\cdot(L/H) \tag{10.3}$$

Where,
d = charge diameter (mm)
B = Burden (m)
W = standard deviation of drilling accuracy (m)
A = spacing/burden ratio
L = charge length above grade level (m)
H = bench height (m).

10.7.3 Cunningham Modification of Kuznetsov Equation

$$\overline{X} = A\cdot(V/Q)^{0.8}\cdot Q^{0.17}\cdot(E/115)^{-0.63} \tag{10.4}$$

Where, 'E' is relative weight strength in terms of the actual explosive considering 'ANFO' as 100 while the relative weight strength of 'TNT' is 115. 'N' is related to the potential damage of the following nature.

(i) Index of Uniformity (IU) or fragmentation Index (FI): 'N' $<1.0 \rightarrow$ should be avoided
(ii) $1 \le$ 'N' $< 1.3 \rightarrow$ poor wall control
(iii) $1.3 \le$ 'N' $\rightarrow$ good wall control.

10.7.4 SveDeFo Model

The Swedish Detonic Research Foundation (SveDeFo) proposed the following model for the fragment size distribution quantified by mesh sizes with 50% of the mass passing (k_{50}). The model was actually a modified version of the original equation developed by Larsson in the year 1973.

$$k_{50} = s \cdot \left[1 + 4.67\left(\frac{T}{L}\right)^{2.5}\right] \cdot e^{0.29 \ln B^2 \sqrt{\frac{S}{1.25}} - 1.18 \ln\left[\frac{CE}{c}\right]^{-0.82}} \quad \text{...(10.5)}$$

Where,

B = Burden (m); CE = specific charge (kg/m^3); c = rock constant (it is the charge factor of gelatin dynamite necessary to break the rock, 0.3-0.5 kg/m^3); S = Blastibility index (for very jointed and fissured rock = 0.60, jointed rock = 0.55, relatively homogeneous rock = 0.45 and homogeneous rock = 0.40), T = stemming length (m), L = depth of blasthole (m).

10.8 DRILLING AND BLASTING COST ASSESSMENT

Estimation of drilling and blasting costs are important to the management and their contractors. Mining companies need accurate financial evaluation of ongoing projects to see the status of the project, make managerial decisions for betterment and to assess bids from contractors. Contractors on the other hand need accurate cost estimation to win contract awards. Any gross underestimates can result in heavy financial loss while overestimates will probably result in the loss of the contract to a competitor.

Every mining project has their own system of evaluation which mostly directs the management to take their decisions regarding excavation, maintenance, day-to-day operations and productions (Figure 10.6). But, unfortunately, the system being followed in the industry is mostly rough estimation taking anticipated and average values and to some extent unscientific. Moreover, it does not take into account important parameters.

In mining, although all unit operations are interrelated, it is however accepted that drilling and blasting are of utmost importance. Drilling pattern and blasting techniques are planned to break the rock efficiently while avoiding undesirable side effects and unstable conditions of the mine. Desired fragmentation from production blasts makes loading, hauling and crushing operations easy or less burdensome. The economics of these interrelated procedures are strongly dependent on drilling and blasting operations, which in turn are directly responsible for providing the desired rock fragmentation. It is thus critical to analyze the economics involved in drilling and blasting operations.

The three basic elements governing the economics of drilling and blasting designs are (i) rock geotechnical parameters; (ii) explosive properties; and (iii) characteristics of drilling equipment. Figure 10.6 shows the flowchart to select an optimized drilling and blasting design. It is interesting to note that the important dimensions of significance in the design of blast parameters are burden and spacing. However, there are a wide range of other variables like sequence of firing, stemming, charge loading and available machinery that have direct and indirect impacts on drilling and blasting performances.

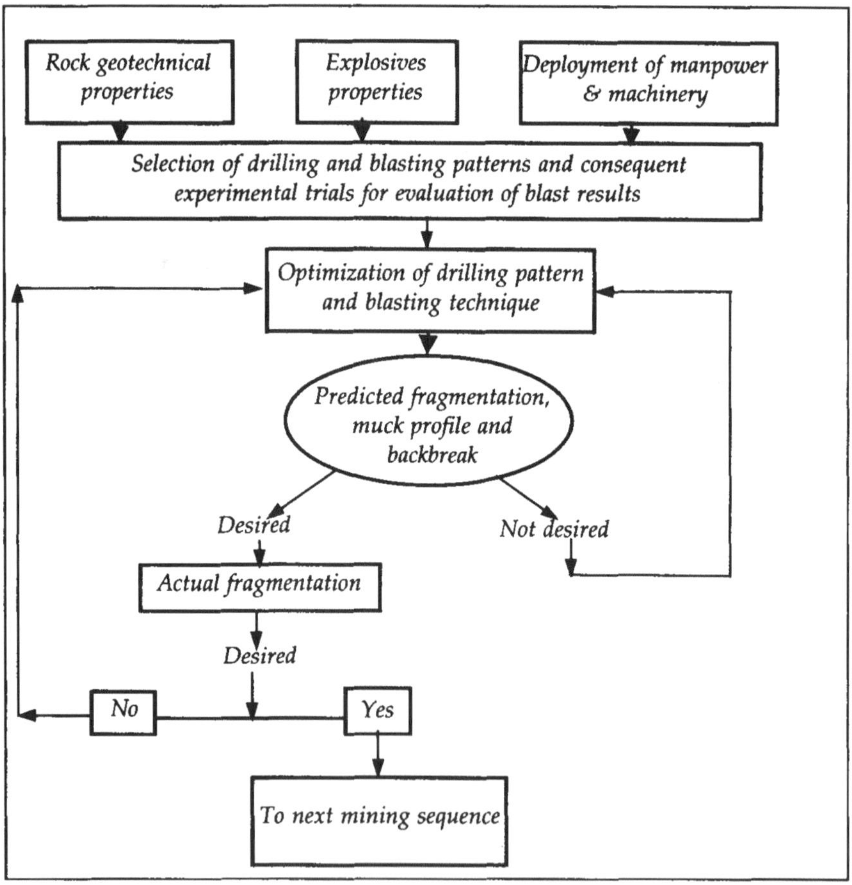

Fig. 10.6 Stages of operations to achieve drilling and blasting cost optimization

10.8.1 Basic Equations for Cost Calculations

Jimeno et al. (1995) have suggested the following basic equation to calculate the drilling cost per metre of drill (C_T).

$$C_T = [(C_A + C_I + C_M + C_O + C_E + C_L) / D_R] + C_B \quad \text{...(10.6)}$$

Where,

Indirect costs: → C_A = Depreciation (Rs./h)
C_I = Interest rates, taxes and insurance (Rs./h)

Direct costs: → C_M = Maintenance (Rs./h)
C_O = Labour costs (Rs./h)
C_E = Power costs (Rs./h)
C_L = Grease and lubrication (Rs./h)
C_B = Bit, stabiliser and drill pipe (Rs./h)
D_R = Average drilling rate (m/h).

10.8.2 Depreciation

The depreciation cost may be calculated from the following equation:

C_A = *(Purchase price -Residual value) / (Hours of operative life)* ...(10.7)

10.8.3 Interest rates, Insurance and Taxes

When the machinery is purchased with borrowed money, then its interest must be taken into account in association with insurance costs and taxes. In order to calculate the amount of interest, equation (10.8) may be used. It should be noted that equation (10.8) is a direct cost and should be taken into account at the end of any cost calculation (CMRI S&T Report of Investigations, 1998).

$$C_I = [\{(N + 1)/2N\} \times \text{Purchase price} \times \%\ (\text{Interest} + \text{Insurance} + \text{Taxes})] / [\text{Work hours per year}] \quad \text{...(10.8)}$$

where, N = Number of years of service life.

However, in most of the mines, the management calculates this cost using the following equation—

$$C_1 = \frac{\{\text{Factory price including freight cost of the drill machine}\} \times \text{Rate of interest (say 18\%)}}{\text{Economic life of the drill (hours)} \times \text{Penetration rate (m/h)}} \quad \text{...(10.9)}$$

With the help of equation (10.9), for an IDM-70E drill machine whose cost is Rs. 1,01,70,889.00 (including freight cost) and economic life is 12 yrs or 16,000 hours, the value of C_I (Rs./h) will be Rs. 5.721 per metre of drill when the penetration rate of the drill machine is 20 m/h.

10.8.4 Bit, Stabiliser and Drill Pipe

Generally it is difficult to get details of previous information from the working technicians on the performance of drill bit, stabiliser and drill pipe. But it is very critical to analyze this particular area although it has a tremendous influence on the overall drilling cost. This cost may vary from 15 to 40% of the total drilling cost depending upon the hardness of the rock.

The service life of a tricone bit can be estimated from the following equation (Jimeno et al., 1995)—

$$Life(m) = \frac{[84420 \times V_p \cdot D^{1.55} \cdot E^{-1.67}]}{N_r} \qquad \ldots(10.10)$$

Where: D = Diameter (in inch)
E = Pulldown force on the bit (thousands of pounds)
N_r = Rotary speed (r/min)
V_P = Penetration rate (m/h).

For a standard company's product, the drill pipes and stabiliser usually have an average life of 30,000 and 11,000 m, respectively.

10.8.5 Relation Between Penetration Rate and Drilling Rate

It is very easy to determine the drilling rate (D_R, m/h) at a particular location for a particular brand of drilling machine operating in a mine. The following formula may be used to convert the drilling rate to the penetration rate (V_P, m/h).

$$V_P = [D_R / 2]^{1.54} \qquad \ldots(10.11)$$

10.8.6 New Drilling Index

Jimeno and Hevia (1987) developed the following drilling index (I_P) for the design of bench blasting—

$$I_P = V_P / (W \cdot V_r / d^2) \qquad \ldots(10.12)$$

Where,
V_P = Penetration rate (m/h)
W = Pulldown weight on bit (thousands of pounds)
V_r = RPM of the drill bit (r.p.m.)
d = Diameter of the drill hole or bit (inches).

The following three points must be taken care of—

— Use of tricon bit for drilling.
— The air flow is sufficient to clear the cuttings from the blasthole.
— The down times are removed, i.e. to measure the net penetration.

When I_P is known, then the following equation would determine the powder factor ($P.F.$)—

$$P.F. = 1.124 \times e^{-0.5727\, xI_P} \qquad \ldots(10.13)$$

Where, $P.F.$ = Powder factor in kg ANFO/m^3
I_P = Drilling index
r = Correlation coefficient = 0.92

Equation (10.13) is a powerful and effective tool for the design of blasting since it is useful for:

— Calculating the drill pattern.
— Estimating the optimum amount for each blasthole in accordance with fixed pattern.
— Developing a cost model for bench blasting.

10.9 CASE STUDIES

Central Mining Research Institute (CMRI) carried out intensive studies over a period of 2 years in 9 different mines (both in Shovel-Dumper and Dragline faces), and evolved various cost factors and their interrelationships. These results are summarized in Tables 10.5 to 10.13. The study also categorically identified the factors like cost of explosive, drilling rate, selection of drill machine, cost of drill bit, cost of drilling rod, non-electric initiation systems, selection of spacing and burden, as the major parameters to be properly implemented or controlled in order to optimize the cost associated with drilling and blasting operations.

Table 10.5 Shovel-Dumper faces (≤ 10 m High Benches) [Lingaraj OCP, MCL]

(i) Drilling cost	(0.23 - 0.30) × Explosive cost
(ii) Cost of [DTH + D.F. + Relay]	(2.0 - 2.12) × Cost of [D.F. + Relay]
(iii) Cost of [DTH + NTD]	(2.70 - 2.85) × Cost of [D.F. + Relay]
(iv) Drilling cost/length	Stone = 0.83 × O.B.*/Coal=0.69 x O.B.*
(v) Explosive cost/volume	(0.53 - 0.68) × [Drilling & blasting cost/vol.]
(vi) Drilling & Blasting cost/volume	(9.10 - 9.80) Rs./m^3

* O.B. Stands for drilling cost per length in overburden.

Table 10.6 Shovel-Dumper faces (≤ 10 m High Benches) [Lakhanpur OCP, MCL]

(i) Total Drilling cost	(0.46 - 0.51) x Total Explosive cost
(ii) Cost of [DTH + TLD]	(4.3 - 5.7) x Cost of [D.F. + Relay + Delay] Note: Because of decking in DTH/TLD system, its cost became much higher than that of conventional system of blasting
(iii) Cost of [D.F., Relay & Delay]	4% of the total drilling & blasting costs
(iv) Cost of DTH & TLD System	17.55% of the total drilling & blasting costs Note: This was optimized to 12% after experiments
(iv) Drilling cost/length	53.90 - 61.25 Rs./m
(v) Explosive cost/volume	6.65 - 10.50 Rs./m^3
(vi) Drilling & blasting cost/volume	(0.53 - 0.65)x [drilling & blasting cost/vol.]
(vii) Total Drilling & Blasting cost/vol.	(10.50 - 16.10) Rs./m^3

Table 10.7 Shovel-Dumper Faces (≤10 m Bench Height) & Dragline faces (>10 m Bench Height) [BELPAHAR OCP, MCL]

(i) Drilling cost	(0.20 - 0.25) × Total drilling & blasting cost → ≤ 10 m B.H.
	(0.15 - 0.17) × Total drilling & blasting cost → > 10 m B.H.
(ii) Explosive cost	(0.67 - 0.70) × Total drilling & blasting cost → ≤ 10 m B.H.
	(0.77 - 0.81) × Total drilling & blasting cost → > 10 m B.H.
(iii) Drilling cost/length [250 mm drill dia.]	(1.98 - 2.60) × Drilling cost/length [150 mm drill dia.]
(iv) Explosive cost/volume	(6.60 - 7.50) Rs./m^3 → ≤ 10 m Bench Height (B.H.)
	(9.0 - 10.0) Rs./m^3 → > 10 m Bench Height

Table 10.8 Shovel-Dumper Faces (>10 m High Benches) [Amlai OCP, SECL]

(i) Drilling cost	0.25 × [drilling & blasting cost]
(ii) Drilling cost/volume	(2.10 - 3.50) Rs./m^3
(iii) Explosive cost/volume	(8.20 - 10.15) Rs./m^3
(iv) Drilling & Blasting cost/volume	(10.85 - 13.62) Rs./m^3
(v) Drilling cost/length	(104.30 - 141.40) Rs./m
(vi) Drilling cost/hour	(2827.30 - 3129.00) Rs./h

Table 10.9 Dragline Faces (25-35 m High Benches) [Dhanpuri OCP, SECL]

(i) Explosive cost:	(4.21 - 5.37) × Drilling cost
(ii) Initiator cost:	0.02 × Total explosive cost
(iii) Drilling & blasting cost/vol.	(5.30 - 6.50) × Drilling cost/volume
(iv) Drilling cost/length	(76.83 - 88.31) Rs./m
(v) Drilling cost/hour	(2925.40 - 3268.69) Rs./h
(vi) Drilling cost/length [250 mm]	Rs. 4.66 + Drilling cost/length [270 mm]→D_R=35m/h
	Rs. 5.02 + Drilling cost/length [270 mm]→D_R=40m/h

Table 10.10 Dragline faces (25 m Bench Height) [Sonepur Bazari OCP, ECL]

(i) Drilling cost	(0.15 - 0.16) × [Drilling & blasting cost]
(ii) Explosive cost	(0.77 - 0.79) × [Drilling & blasting cost]
(iii) Cost of [DTH + TLD]	(0.060 - 0.065) × [Drilling & blasting cost]
(iv) Cost of [DTH + TLD]	(0.060 - 0.084) × [Explosive cost]
(v) Drilling cost/length	(61.60 - 63.35) (Rs./m)
(vi) Charge factor	[0.29 - 0.36] (kg/m^3)
(vii) Explosive cost/volume	[4.80 - 5.99] (Rs./m^3)
(viii) Drilling & blasting cost/volume	[6.23 - 7.63] (Rs./m^3)
(ix) Drilling & blasting cost/wt.	[2.83 - 3.39] (Rs./t)

Table 10.11 Dragline faces (28-30 m Bench Height) [Sasti OCM, WCL]

(i) Drilling cost	(0.10 - 0.12) x [Drilling & blasting cost]
(ii) Explosive cost	(0.86 - 0.88) x [Drilling & blasting cost]
(iii) Cost of [DTH + TLD]	2.44 × Cost of [D.F. + Relay + Delay]
(iv) Drilling cost/length	(80.85 - 83.65) (Rs./m)
(v) Charge factor	[0.33 - 0.40] (kg/m^3)
(vi) Explosive cost/volume	[5.60 - 7.00] (Rs./m^3)
(vii) Drilling & blasting cost/volume	[6.65 - 8.05] (Rs./m^3)
(viii) Drilling & blasting cost/wt.	[3.15 - 3.50] (Rs./t)

Table 10.12 Shovel-Dumper faces (10-13 m Bench Height) [West Mudidih Opencast Project, BCCL]

(i) Drilling cost	(0.43-0.50) × [Drilling & blasting cost] → 150 mm Dia. (0.32-0.33) × [Drilling & blasting cost] → 250 mm Dia.
(ii) Explosive cost	(0.40-0.51) × [Drilling & blasting cost] → 150 mm Dia. (0.65) × [Drilling & blasting cost] → 250 mm Dia.
(iii) Cost of [DTH + TLD]	5.56 × Cost of [D.F. + Relay + Delay]
(iv) Drilling cost/length	(117.95 - 126.70) (Rs./m) → 150 mm Drillhole Dia. (125.65 - 133.10) (Rs./m) → 250 mm Drillhole Dia.
(v) Charge factor	[0.56 - 0.71] (kg/m^3) → 150 mm Drillhole Dia. [0.88 - 0.97] (kg/m^3) → 250 mm Drillhole Dia.
(vi) Explosive cost/volume	[8.75 - 11.20] (Rs./m^3) → 150 mm Dia. [13.65 - 16.80] (Rs./m^3) → 250 mm Dia.
(vii) Drilling & blasting cost/vol.	[18.55 - 28.35] (Rs./m^3)
(viii) Drilling & blasting cost/wt.	[7.35 - 11.20] (Rs./t)

Table 10.13 Shovel-Dumper faces (6-12 m Bench Height) [Jambad Opencast Project, ECL]

(i) Drilling cost	(0.26-0.41) x [Drilling & blasting cost] → O.B. & COAL
(ii) Explosive cost	(0.47-0.67) x [Drilling & blasting cost] → O.B. (0.40) x [Drilling & blasting cost] → COAL
(iii) Cost of [DTH + TLD]	4.40 x Cost of [D.F. + Relay + Delay] → Deck charging 3.00 x Cost of [D.F. + Relay + Delay] → Direct charging
(iv) Drilling cost/length	(78.86 - 83.95) (Rs./m) → O.B. (41.34) (Rs./m) → COAL
(v) Charge factor	[0.46 - 0.69] (kg/m^3) → O.B. [0.30] (kg/m^3) → COAL
(vi) Expl. cost/volume	[8.27 - 12.08] (Rs./m^3) → O.B. [5.41] (Rs./m^3) → COAL
(vii) Drilling & blasting cost/ volume	[13.35 - 5.76] (Rs./m^3)

10.10 FRAGMENTATION COST ASSESSMENT

In a bid to determine the optimum fragmentation cost, Nielson (1983) used a computer model to obtain an optimum cost from four different mining operations namely drilling and blasting (D&B), loading, hauling and crushing (Figure 10.7). Another study by Nielsen depicted in figure 10.8, revealed that the optimum cost for the entire mining operation resulted when the charge factor was substantially increased over the standard one. This is likely to be the case in most operations, rather than the exception, when all costs are carefully analysed and accounted.

Da Gama and Jimeno (1993) carried out an investigation to give an account of the average production costs in hard rock open-pit mines. Their investigation revealed the following cost breakage:

Drilling and blasting	30%
Loading	17%
Crushing	20%
Hauling	33%

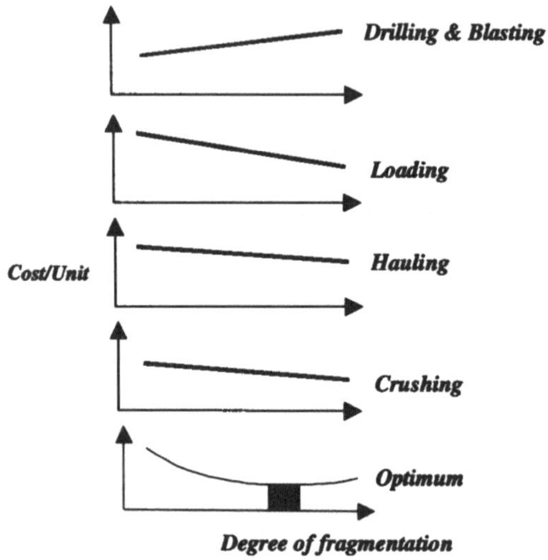

Fig. 10.7 Evaluation of optimum blasting (after Nielsen, 1983)

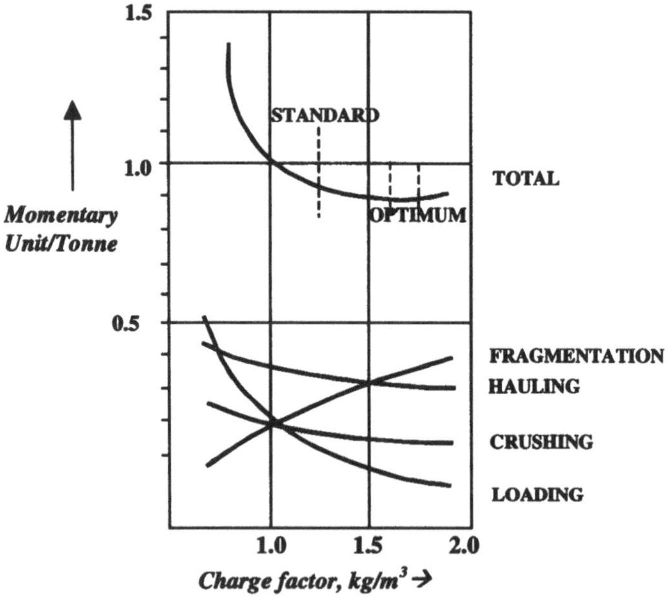

Fig. 10.8 Influence of powder factor on cost of mining operation (after Nielsen, 1983)

They also pointed out that the percentage of costs corresponding to drilling and blasting may even increase up to an additional 15%, in the case when large boulders are produced, thus requiring secondary blasting.

10.11 EFFECT OF INITIATION TIMING

The selection of improper initiation time periods between rows in a blast is a frequent cause of backbreak and poor fragmentation as well. Over-confinement may cause considerable uplift in the stemming zone. This happens more in the back rows. Figure 10.9 shows how over-confinement which affects uplift, also affects the piling or scattering of the blasted rock (Konya, 1987).

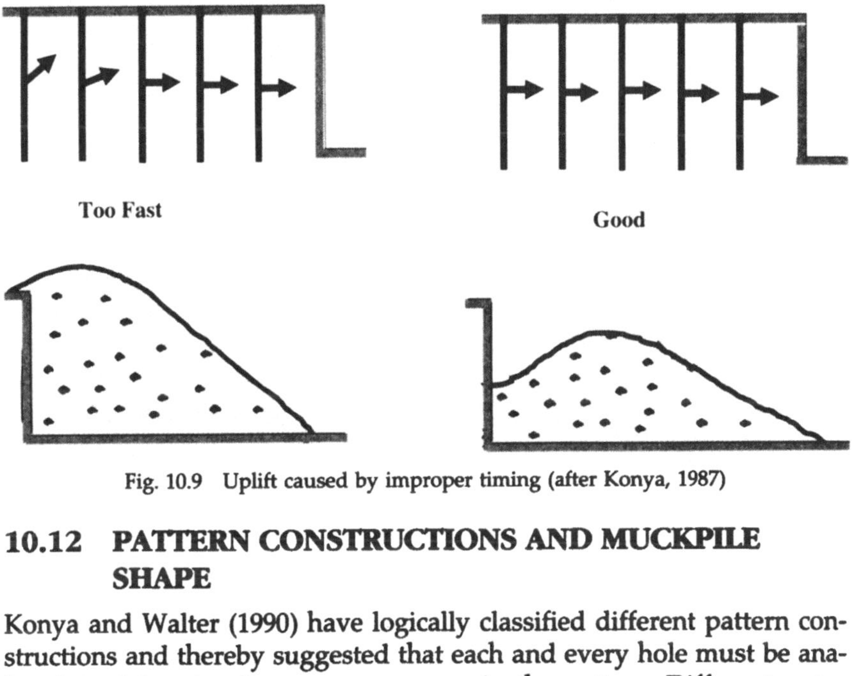

Fig. 10.9 Uplift caused by improper timing (after Konya, 1987)

10.12 PATTERN CONSTRUCTIONS AND MUCKPILE SHAPE

Konya and Walter (1990) have logically classified different pattern constructions and thereby suggested that each and every hole must be analysed to determine its proper response in the pattern. Different crater shapes will be created from firing of independent holes depending upon the relationship between the blasthole and free face. While explaining the theory, they assumed that the breakage angle between the burden line and the edge of the crater was approximately 45°. The plan view of the different crater-forms under varying conditions are shown in figure 10.10.

A group of scientists, of the CMRI Regional Centre at Nagpur, recommended that the muckpile is quite dependent on the blast design, especially initiation timing, sequence and type of initiation and cut configuration (CMRI S&T Report: Chakraborty et al., 2002). They also schematically demonstrated probable muck profile and fragmentation with various blast designs conceptually developed by Konya and Walter (1990). All these patterns and their consequent effects are shown in figures 10.11 to 10.14.

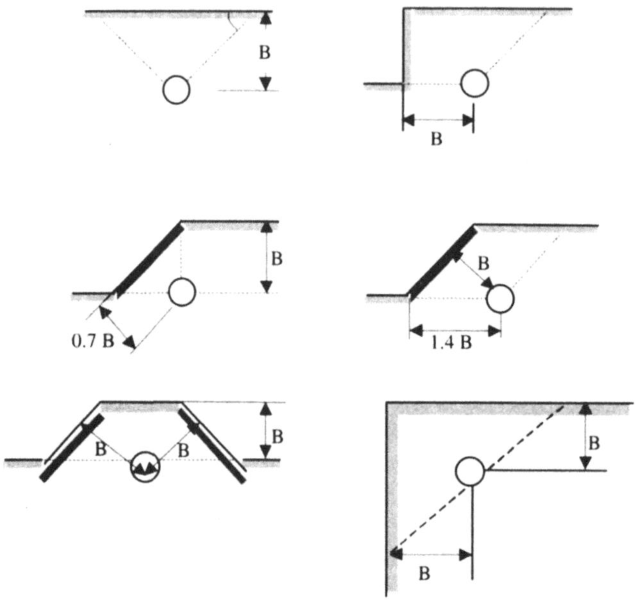

Fig. 10.10 Crater forms which result from borehole orientation to face(s) (after Konya and Walter, 1990)

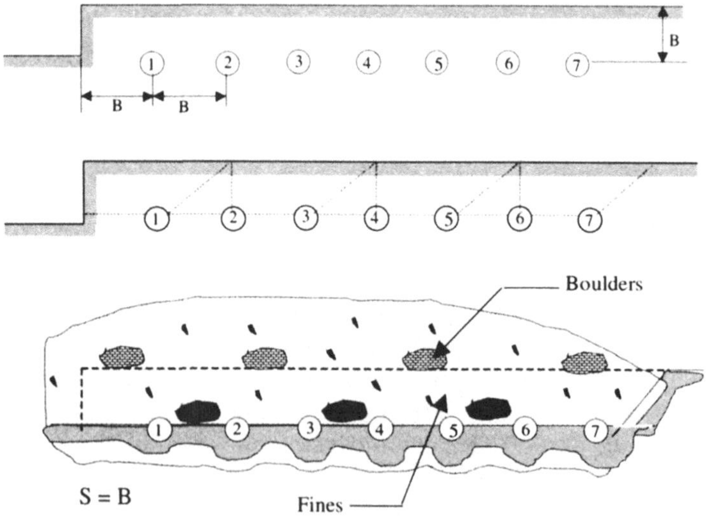

Fig. 10.11 Muck profile and fragmentation under condition 'A'

<u>*Condition 'A':*</u> *B (burden) = S (spacing)*
Initiation: Progressive MS delays; B_h *(bench height) / B > 4*
Vibration level: Low
Fragmentation: Fines + Boulder

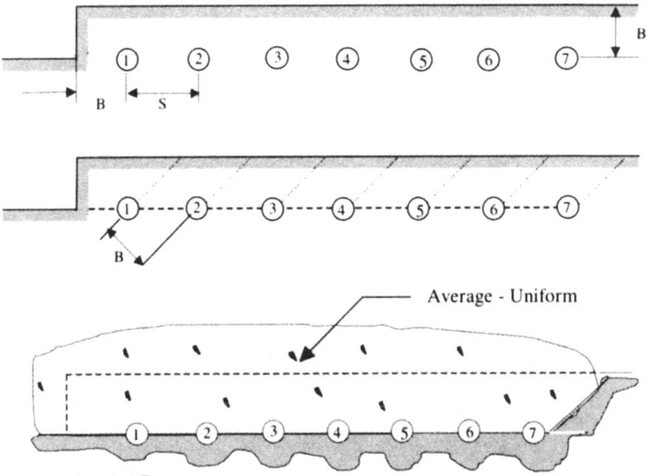

S = 1.4 B

Fig. 10.12 Muck profile and fragmentation under condition 'B'

<u>Condition 'B'</u>: *S = 1.4 B*
Initiation: Progressive MS delays; B_h */ B > 4*
Vibration level: Low
Fragmentation: Uniform

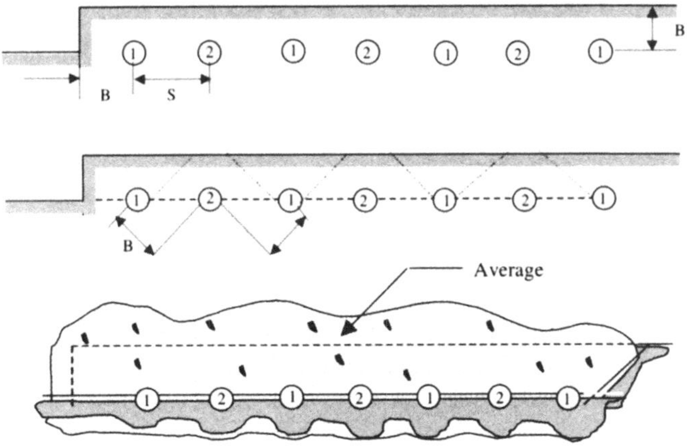

S = 1.4 B

Fig. 10.13 Muck profile and fragmentation under condition 'C'

<u>Condition 'C'</u>: *S = 1.4 B*
Initiation: Alternate MS delays; B_h */ B > 4*
Vibration level: High
Fragmentation: Uniform but in two different sizes

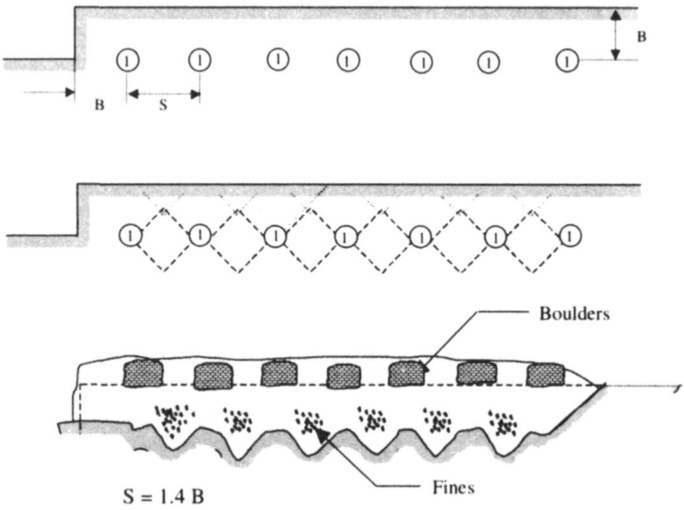

Fig. 10.14 Muck profile and fragmentation under condition 'D'

Condition 'D': *S = 1.4 B*
Initiation: Instantaneous; $B_h / B > 4$
Vibration level: High
Fragmentation: Uneven

10.13 EFFECTS OF FRAGMENTATION ON SUBSYSTEMS

Nielson (1983) evaluated the optimum powder factor (kg/m^3) in Sydarager's open-pit iron ore mines in Norway by considering its influence on different subsystems of mining. It is clearly seen from figure 10.8 that the powder factor is optimum when its value approaches as high as 1.6 to 1.7.

A thorough scientific study conducted by CMRI (Chakraborty et al., 2002) categorically classified that every blast design leads to a distinct size distribution of fragments, which influences the productivity of the subsequent operations in the following manner—

(i) Increasing fragmentation requires higher drilling and explosive costs,
(ii) Optimum utilization of loading equipment is related to the fragment size,
(iii) Truck or dumper cycle time is reduced because of less waiting period and loading time in case of good fragmentation,
(iv) Better fragmentation reduces secondary blasting, and operational cost in crushing is reduced if the fragment size is smaller.

Jimeno et al. (1995) recommended that the maximum fragment size for the crusher is 80% of the maximum permissible size in the crusher. They also mentioned that the maximum recommended fragment size for loading a bucket is 0.7 times the bucket size. However, recommended optimum fragment size for a shovel should be (0.125-0.166) times the shovel bucket size. Plate 10.3 shows an overview of optimum fragmentation obtained in a large size dragline bench blast.

Plate 10.3 View of fragmentation in a large size dragline bench blast

10.14 DETERMINATION OF FACE ORIENTATION

CMRI-evolved fragmentation analyzing scale has proved to be an effective tool for selection of face orientation during opening of new opencast mines so that better fragmentation and wall control can be achieved during blasting (Pal Roy, 1995). The scale, as depicted in Figure 10.15, indicates that when strike of the joints forms an acute angle (below 25°) with the strike of the face then both breakage and wall stability problems would occur during blasting. In such cases, blocky fragmentation would result from the burden and the back wall would be shattered. When joints make angles between 25° and 65° with the face, then endbreak would be within controllable limits and good fragmentation would appear i.e. there would be no oversize boulders. However, when the joints make angles between 65° and 90° with the face, possibility of blocky fragmentation i.e. poor fragmentation would always be there. As such, thorough geological mapping of the exposed area can help in determining the face orientation for better fragmentation when combined with good blasting practices.

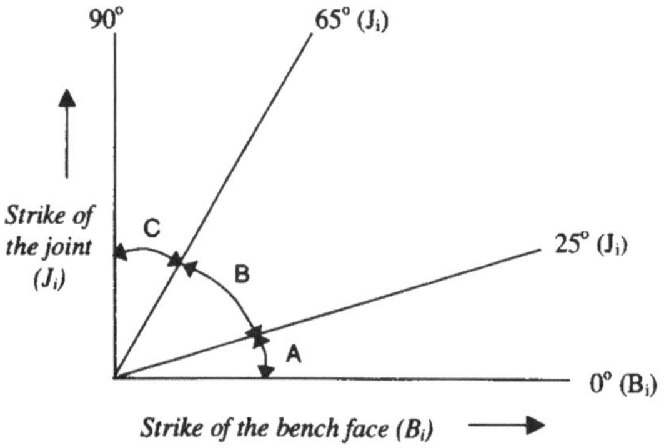

A: Breakage and wall stability problems
B: Good fragmentation. Endbreak and backbreak within controllable limits
C: Little endbreak but significant backbreak. Blocky fragmentation

Fig. 10.15 CMRI—evolved fragmentation analyzing scale

10.14.1 Case Study

The fragmentation analyzing scale, as depicted in figure 10.15, was used to design the face orientation of Turamdih West/Banduhurang opencast mines of M/s Uranium Corporation Limited (UCIL). The geological mapping of the proposed area indicated the presence of three major joint sets. Table 10.14 represents these sets with their respective dip direction and amount. With the help of figure 10.15 and the results obtained in Table 10.14, the orientation of different faces were derived along the direction of E-W with strike approximately N80°. It was aimed so that the strike directions of all major joints would make angles between 25° and 65° with the strike of the face (Table 10.15). Figure 10.16 depicts the diagrammatic view of the proposition.

Table 10.14 Physical characterization of major joint sets

Joint sets	*Mean orientation of joint sets*		*Persistence (m)*	*Spacing (m)*
	Dip direction (Degree)	*Dip amount (Degree)*		
J_1	N028	50	5.00	0.10
J_2	N118	65	1.00	1.00
J_3	N297	81	0.90	3.00

Table 10.15 Angle between strike of the face and joints

Joint sets	*Strike*	*Strike of the face E-W*	
		N90°	*N80°*
J_1	N118°	28°	38°
J_2	N028°	62°	52°
J_3	N027°	63°	53°

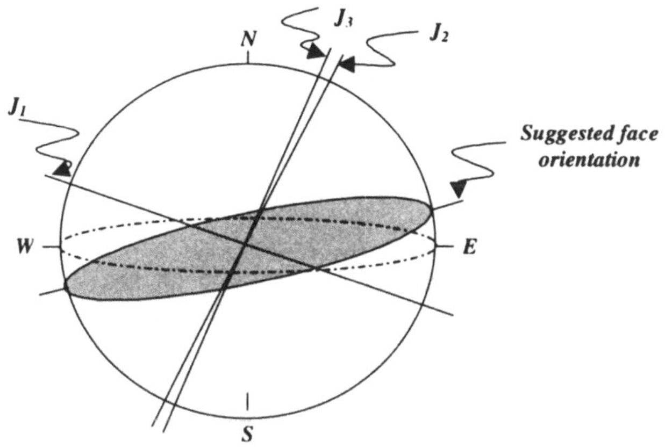

Fig. 10.16 Design of face orientation in consideration to various joint sets for optimum blast results

10.15 DESIGN PATTERNS FOR NON-COAL DEPOSITS

The following are a few echelon and box-cut type patterns generally seen to be effective in non-coal deposits (limestone, dolomite, granite, iron and copper ores etc.). These patterns are shown in figures 10.18 to 10.21. The two important aspects considered while implementing such design patterns are (i) Maximum charge per delay (i.e. charges fired within 15 ms of time interval) should not exceed the prescribed limit (ii) Total firing time of holes should be restricted to 1000 ms. The maximum charge per delay (fired within 15 ms of time interval) plays an important role in the generation of ground vibration as depicted in figure 10.17. Plates 10.4

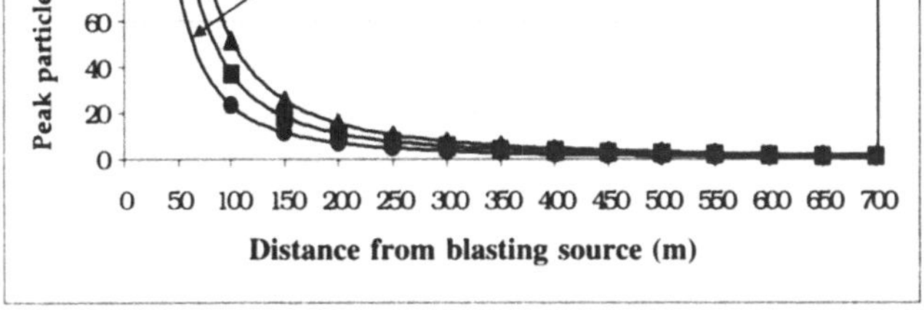

Fig. 10.17 Variation of peak particle velocity with distance for different charge/delay (charges fired within 15 ms time interval)

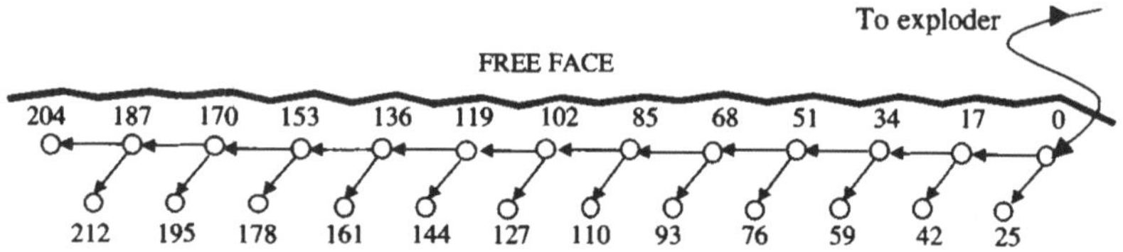

Fig. 10.18 Control-row or echelon type of connection of holes in two rows

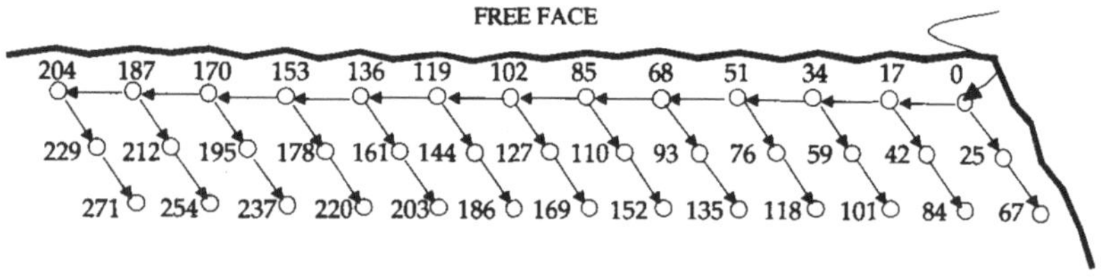

Fig. 10.19 Control-row or echelon type of connection of holes in three rows

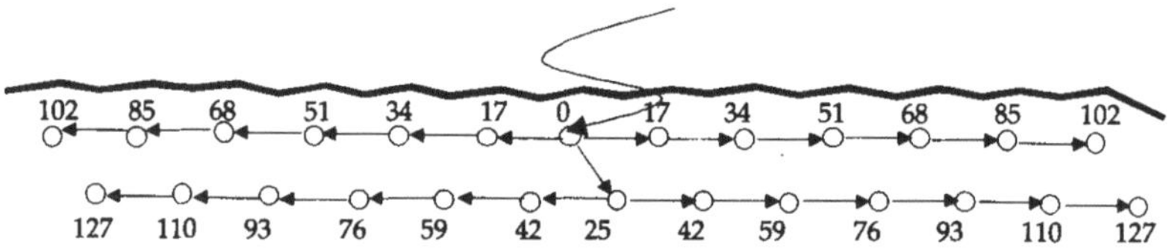

Fig. 10.20 Box-cut type of connection of holes in two rows

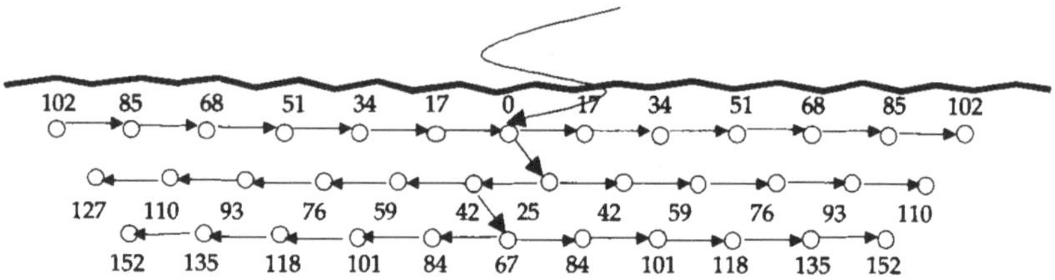

Fig. 10.21 Box-cut type of connection of holes in three rows

to 10.7 show four typical blasting events carried out at Pundi Opencast Mine of West Bokaro Collieries, TISCO.

10.16 COMPUTER MODELS OF FRAGMENTATION PREDICTION

The ICI's computer blasting model 'SABREX' is based on the fact that both shock waves and gas pressure are important in describing a blast. 'SABREX' is a blasting programme which can deal with many complexities of blast design and predict blast results in terms of fragment size distribution, muck pile profile, grade level fragmentation, collar back

Plate 10.4 View of stemming ejection in coal blasting

Plate 10.5 View of fine fragmentation in OB blasting

Plate 10.6 View of an optimally thrown blasted materials

Plate 10.7 Ejection of coloured fumes from 2 different explosives

fragmentation, flyrock control, bedrock damage and costs (drilling, accessories and explosives). The predictions are realistic and sensitive to all factors controlled by a blasting engineer (Kirby et al., 1987).

The PBS (Precision Blasting Services, USA) blasting model 'BREAKER' predicts and compares relative differences in fragmentation as affected by many of the normal production blasting variables, such as burden, spacing, bench height, rock properties and explosive properties. It also takes into account the rock geologic and structural parameters. It calculates a single size distribution based on the input from one blast and presents both numerically and graphically. Comparative assessment of fragment size distributions of different blasts is also possible by this model (Pal Roy, 1992).

'SAROBLAST' developed by Swedish researchers (Shaoquan and Rustan, 1993), is an IBM microcomputer based programme. It was developed in turbo pascal language for the design and prediction of the fragmentation and muckpile profile. It may be simulated to examine three most important fragmentation prediction formulae, namely the Kuz Ram, The SveDeFo and the composite formula.

'FRAGSCAN' developed in France, is basically an image-processing tool for continuous scanning of fragmentation after blasting is performed. It provides the fragmentation distribution and then rebuilds the whole pile dumper, and not directly on the whole pile, in order to reduce the problem of surface measurement validity (Cheimanoff, et al., 1993).

'FRAGALYST' jointly developed by CMRI Regional Centre, Nagpur and M/s Wavelet Group, Pune is an image analysis system for scientific evaluation of blast fragmentation. The system consists of video photographs of the muckpile, downloading the photographs to the computer, converting the images to greyscale, image enhancement, calibration and bolb (grain) analysis. With the aid of menu-driven software, it is possible to determine the area, size, shape and sphericity of the fragments in a muckpile/grain aggregates on the basis of greyscale difference (Raina et al., 2002).

'WipFrag' system consists of a Pentium computer equipped with dedicated video board, frame grabber, additional RAM memory, camera, suitable monitor and PostScript compatible printer. The muckpile is photographed or videotaped and the image is transferred to the WipWare system. The broken rock image is transformed into a particle map or network. Network areas are converted into volumes and weights and the resulting data is displayed as a graph. Throughout the globe, 'WipFrag' is the most widely used system for fragmentation assessment (Palangio et al., 1995).

JKMRC, Australia has developed interactive graphics blast design editors for surface open cut, underground development and ring blasts. The open cut version was first released as a commercial software package, known as 3x3o-PRO. The programme provides a flexible blast design editor combined with a collection of practical blast analysis techniques aimed directly at developing and evaluating new blast design options, not just repeating an existing formula (Scott et al., 1996). Besides many specific features, 3x3o-PRO enables the derivation of the vibration amplification function at any design location for a selected frequency range. There are two animation modes namely dynamic and kinematic, available in the package. 3x3o-PRO has the ability to generate plots and reports for use by drill and blast crews and the design engineer.

10.17 CONCLUSION

Whatever be the geo-mining conditions, good fragmentation results from the perfect choice of explosives, proper selection of blast design parameters, implementation of flexible and relatively simple techniques and optimum utilization of existing operating conditions. In order to maximize fragmentation and minimize unwanted side effects from blasting, the blast design variables and timing must be selected such that all variables work mutually. The behaviour of the explosives poured into the holes also plays an important role in controlling fragmentation. When delays are used adjacent holes are affected due to high-stress levels from earlier firing holes. These high-stress levels in some cases can cause pre-compression or dead pressing of the main charge and may result in insufficient explosive firing and release only a small fraction of their available energy.

Geologic conditions also play significant roles in controlling fragmentation. The single most important geologic consideration is geologic structure. The joints, dip and strike of bedding planes, seam conditions can have serious influence on fragmentation.

Though considerable work has been done worldwide regarding rock fragmentation, a perfect technique for assessing post-blast fragmentation is yet to develop (Cunningham, 1983; Da gama et al., 1993 and Jimeno et al., 1995). Out of the existing methods, site screening and photographic techniques have appeared to be somewhat satisfactory.

REFERENCES

Bhandari, S. 1997. Engineering rock blasting operations, A. A. Balkema, Rotterdam, Netherlands, p. 375.

Cheimanoff, N. M., Chavez, R. and Schliefer, J. 1993. 'FRAGSCAN'—a scanning tool for fragmentation after blasting, abatement. Proc. 4th International Symposium on Rock Fragmentation by Blasting, H. P. Rossmanith (ed.), Technical University, Vienna, Austria, 5-8 July, A. A. Balkema, Rotterdam, Netherlands, pp. 325-330.

CMRI S&T Report of Investigations, 1998. Techno-economic assessment of drilling and blasting parameters vis-à-vis establishment of cost effective optimum blast pattern for opencast mines (P. Pal Roy et al.), Funded by the Ministry of Coal, Govt. of India, GAP/2/Blasting/MOC/96, p. 88.

CMRI Report of Investigations, 2002. Development of innovative models for optimization of blast fragmentation and muck profile applying image analysis technique and subsystems utilization concept in Indian surface coal mining regime (A. K. Chakraborty et al.), Coal S&T Project No. MT/103 of the Ministry of Coal, Govt. of India, January, p. 125.

Cunningham, C. V. B. 1983. The KUZ-RAM model for prediction of fragmentation from blasting, Proc. 1st International Symposium on Rock Fragmentation by Blasting, August, Lulea, Sweden, pp. 653-664.

Da gama, C. Danis and Jimeno, C. Lopez, 1993. Rock fragmentation control for blasting cost minimization and environmental impact abatement, Procs. 4th International Symposium on Rock Fragmentation by Blasting, H. P. Rossmanith (ed.), Technical University, Vienna, Austria, 5-8 July, pp. 273-280.

Jimeno, E. L. and Hevia, Muniz, E. 1987. A new method for the design of bench blasting, Proc. 2nd International Symposium on Rock Fragmentation by Blasting, August 23-26, Keystone, Colorado, USA, pp. 302-309.

Jimeno, C. L., Jimeno, E. L. and Carcedo, F. J. A. 1995. Drilling and blasting of rocks, A. A. Balkema, Rotterdam, Netherlands, p. 391.

Kirby, I. J., Harries, G. H. and Tidman, J.P. 1987. ICI's computer blasting model 'SABREX'-basic principles and capabilities, Proc. Inter. Soc. of Explosive Engineers (ISEE), Annual Meeting, February 1-6, Florida, USA, pp. 184-197.

Konya, C. J. 1987. Controlled backbreak with proper borehole timing, Proc. 13th Annual Conference on Explosives and Blasting Technique, International Society of Explosives Engineers (ISEE), Florida, USA, February 1-6, pp. 49-59.

Konya, C. J. and Walter E. J. 1990. Surface blast design, Prentice-Hall, Inc., New Jersey, USA, p. 303.

Nielson, Kai, 1983. Optimization of open-pit bench blasting, Proc. 1st International Symposium on rock fragmentation by blasting, August, Lulea, Sweden, pp. 653-664.

Pal Roy, P. 1992. Computer applications in blasting research—Capabilities of the Central Mining Research Institute, IM & EJ, October, India, pp. 5-9.

Pal Roy, P. and Dhar, B. B. 1993. Comparative assessment of ground vibration and fragmentation in three iron ore deposits in India. Proc. 4th Inter. Symposium on Rock Fragmentation by Blasting, Vienna, Austria, H. P. Rossmanith (ed.), Technical University, Vienna, Austria, 5-8 July, A. A. Balkema, Rotterdam, Netherlands, pp. 287-294.

Pal Roy, P. 1995. Breakage assessment through cluster analysis of joint set orientations of exposed benches of opencast mines, Geotechnical and Geological Engineering, U.K., Vol. 13, pp. 79-92.

Palangio, Tom C., Fanklin, John, and Maerz, Nobert, H. 1995. WipFrag® photoanalysis fragmentation measurement, Paper presented at the Explosives Technology Course, Department of Mining Engineering, Queen's University, Kingston, Ontario, Canada, 25-29 September, p. 32.

Raina, A. K., Choudhury, P. B., Ramlu, M., Chakraborty, A. K., Dudhankar, A. S., Udpikar, V., Ghatpande and Misra, D. D. 2002. FRAGALYST—An indigenous digital image analysis system for grain size measurements in mines, Journal of Geological Society of India, June, Vol. 59, pp. 561-569.

Scott, A., Cocker, A., Djordjevic, N., Higgins, M., Rosa La, D., Sarma, K. S. and Wedmaier, R. 1996. Open pit blast design analysis and optimization, Julius Kruttschnitt Mineral Research Centre, The University of Queensland, Australia, p. 338.

Shaoquan, Kou and Rustan, P. A. 1993. Computerized design and result prediction of bench blasting, abatement, Proc. 4th Int. Symposium on rock fragmentation by blasting, Vienna, Austria, H. P. Rossmanith (ed.), Technical University, Vienna, July, 5-8 A. A. Balkema, Rotterdam pp. 263-271.

Singh, M. M. 1991. Field investigations on the effect of structural discontinuities on rock blasting, Unpublished Ph.D. Thesis, IT, Banaras Hindu University, Varanasi, India, p. 180.

Singh, B. and Pal Roy, P. 1993. Blasting in ground excavation and mines, A. A. Balkema, Rotterdam, Netherlands, p. 177.

11

Socio-economic and Environmental Impacts of Blasting

11.1 INTRODUCTION

Research on human response to blasting operations reveals that human beings are much more sensitive to vibrations and noise than the structure. Although blasting vibration and noise are short-term transient phenomena, still mine management faces a lot of practical difficulties in conducting their daily operation. On hearing any loud noise, residents in the vicinity of operating mines feel that if this type of operation continues then one day their houses would fall down. This general psychology plays an important role on the behavioural aspects of the neighbouring villagers.

Continual noise, dust, flyrock and overall vibration from blasting operation may be upsetting and annoying to the residents. This may get to stage of violence if such operations involve any physical or property damage. The overall aggravation may result first to mass complaints, then violent resistance and when combined with political interference may ultimately lead to the closure of blasting operations.

While considering human sensitivity, both ground vibration and air blast must be considered simultaneously. Dowding (1985) has pointed out that ground motion and air blast jointly excites structures, which causes more noise inside a structure than outside. The additional noise level inside the structure may result, due to rattling of things, falling in the low frequency zone of the air blast.

German scientists, Reiher and Meister (1946) concluded from their experiments that at frequencies greater than 10 Hz, continuous vibrations became objectionable and annoying at even 2.5 mm/s. However, ordinary production blast lasting less than one second would result in a different human response. They also concluded from their experimental study that when events occur more than 10 times per day then the annoying situation would come even at a very low order of noise and vibration levels.

Siskind et al. (1980) revealed that at an exposure time of 0.6 second to a single vibration event, it does not become perceptible until the vibration level is 12.5 mm/s. In recent years, in India, human response to blast induced vibrations and air blast has been identified as an important aspect for any successful mining operation. Numerous complaints, arising due to unoptimized blasts, ultimately lead to many unsavoury situations, as a result of which majority of the mines are either running at a loss or approaching to closure. Sensing the importance of the issue, CMRI (Dhar et al., 1993) published a booklet, in the form of standard guidelines, in which a set of human response criteria was established for Indian mining conditions. In addition to that, CMRI also made a thorough study on human response and air blast impacts based on opinion polls conducted in several villages in and around the mining areas. CMRI regional centre at Nagpur, while studying human response in non-coal mines, came to the conclusion that there was no major change in the brain function to the ground vibration and air overpressure impacts generated with pyrotechnic materials, both in sitting and resting positions (Chakraborty et al., 2003). The following sections give a brief account of CMRI's study on socio-economic and environmental impacts of blasting operation.

11.2 HUMAN RESPONSE TO BLAST-INDUCED VIBRATIONS

In spite of all the possible precautionary measures, numerous complaints arise from all residential sectors in the vicinity of operating mines. Human beings have developed a perception system well adapted to environmental disturbances. As a result, the sensitivity of human perceptive system to vibration is very high, similar to the extremely high sensitivity of the ear to sound waves. There were many instances where the affected villagers manhandled and threatened the mine management to vent their grievances, which were sometimes genuine and sometimes motivated.

Human response to vibrations may be distinctly categorized into three different stages namely viz. (a) acceptable limit of tolerance, (b) troublesome and (c) severe. General human factors like age, health, tolerance limit and education all have some direct and indirect effects on human response. Based on the subjective assessment of community response to blast induced vibrations, CMRI published the following chart, as given in Table 11.1, for approximate scaling of human response to vibration.

Table 11.1 Human response to particle velocity (Dhar et al., 1993)

Response	*Particle velocity (mm/s)*
Noticeable	0.5
Troublesome	5.0
Severe	15.0

Plate 11.1 Socio-economic survey being carried out in village-2

11.3 HUMAN REACTION TO THE EFFECT OF BLASTING

The tolerance and reaction of human beings to vibrations are important when standards are based on annoyance, interference, work efficiency and health. Humans notice and react to blast-produced vibrations of levels that are lower than the damage thresholds. The technical problem of quantifying responses is complicated because of the simultaneous presence of both ground vibration and air blast, as well as of the presence of many secondary effects like window and dish rattling etc. Ground vibrations are occasionally blamed for house vibrations when long-range air blasts propagating under favourable weather conditions are responsible. The very infrasonic air blast itself cannot be heard, but the house responds as if subjected to a ground vibration (Singh, 1998).

To get an overall idea about human response to blasting operation and subsequent feelings of the affected villagers, a survey was carried out in five villages affected due to either underground or opencast blasting operations. The village statistics are briefed in Table 11.2.

Statistical analyses were carried out by selecting, at random, 100 adult villagers including a few housewives from each village (in village-4, it was 134). They were then individually asked to give their unbiased opinions about their demand and any specific feeling. On the basis of the opinions received, the following six human reaction criteria were established.

Table 11.2 Village statistics of four mining affected villages (Pal Roy et al., 1993; Dhar et al., 1993)

Village specification	*Population*	*Pacca House (%)*	*Services (%)*	*Business (%)*	*Literacy (%)*
Village-1: Affected by U/G blasting operation. Depth of cover = 86-92 m	10000	70	75	10	50-60
Village-2: Affected by opencast blasting operation. Distance of the village from the stripping boundary of the mine = 150 m	3000	50	7	3	70
Village-3: Affected by U/G blasting operation. Depth of cover = 47-56 m	250	0.8	2.8	0.8	70-80
Village-4: Consisting of three small villages including company staff qrs. Affected by opencast blasting operation. Minimum distance = 100 m	3500	60	40	25	70
Village-5: Affected by opencast blasting operation. Minimum distance = 250 m	300	16.67	7.67	5	45

- Rehabilitation due to heavy damage and unbearable mining hazards
- Compensation due to small or severe damage or any other property loss
- Service against damage and loss of property
- Closure of mining operation as the management did not fulfill past commitments
- Confrontation to keep the management under mental pressure
- Continuation of mining operation keeping villagers' property intact and no disturbance to the villagers.

Plate 11.2 Ground vibration monitoring near the entrance of villge-5

Plate 11.3 Vibration monitoring in a residential house due to continuous traffic movements at a close-in distance to NH39

11.3.1 Analysis of the Results

Figure 11.1 shows the diagrammatic representation of the general feelings of the five villages whose statistics are given in Table 11.2. It is clear from the diagrams that majority of the villagers did not want any direct confrontation with the management. Their main demand was only to protect their hard-earned property. In the case of village-4, majority of the villagers who were questioned were mostly businessmen or company workers including some very poor local residents. The reasons why their claims were biased towards either service to their relatives or to extort some compensation from the mine management. It is interesting to note that the majority of the villagers did not, at heart, want the closure of the blasting operation. In spite of the fact that many labourers were the residents of such neighbouring villages. Besides this, younger generation of such villages was involved in different types of businesses involving coal and other mining properties. As a result, closure of mining operations would lead to an economic crisis for them.

11.3.2 Persistence of Ground Vibrations

Persistence or duration of ground vibrations from a blast was found to be an important aspect of human annoyance. With a view to obtaining the scientific reasoning of the duration of blast at a measuring point on the compacted ground, the author carried out extensive experiments at various sites with due consideration of different blast geometry and initiation sequence. However, no conclusive results supporting any specific mechanism were obtained. Table 11.3 shows a few of the experimented values monitored during detonation of very high, moderately high and low amount of explosive charges fired in different blasting rounds. The results did not signify distinctively the effect of total charge, maximum charge per delay; total firing time and the firing pattern (either by

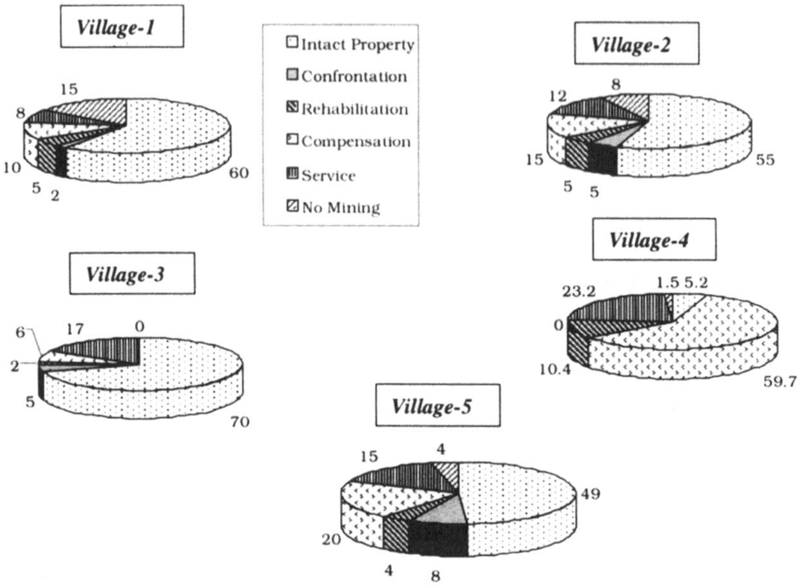

Fig. 11.1 Diagrammatic representation of villagers' opinions

conventional or through shock tube system) as the parameters responsible for the persistence of blast vibrations.

Though there was no distinct parameter seen to be responsible for the persistence of ground movement, yet, in broader perspective, the following three classifications could be drawn from the results obtained.

1. Higher charge fired in a round of blast did not indicate more persistence of ground vibration at a certain measuring point.
2. Higher amplitude did not signify more persistence of ground vibration at a measuring point.
3. Duration of ground vibration received at a measuring point was dependent predominantly on the propagating medium and rock geologic conditions.

11.4 AIR BLAST AND ASSOCIATED ANNOYANCE

When blasting is carried out, it is accompanied by a loud noise called air blast. It is an atmospheric pressure wave consisting of high frequency sound that is audible and low frequency sound or concussion that is subaudible and cannot be heard. Air blast overpressure is measured in decibels (dB) or in pounds per square inch (psi). In many practical situations, it was observed that although the air blast did not create any damage but it created a lot of annoyance which ultimately resulted in direct confrontation between mine management and those affected. When the

Table 11.3 Assessment of the persistence of ground vibrations due to blasting of very high and low charges (Pal Roy and Chakraborty, 2000)

Total charge (kg)	*Maximum charge per delay* (kg)	*Distance* (m)	*PPV* (mm/s)	*Domnt. Freq.* (Hz)	*Total firing time of holes (ms)* [b]	*Duration of vibration (ms)* [a]	*Ratio* (a)/(b)	*Rock type*
a) 222,792.50	18,880.74	350	86.40	15	1225	4050	3.31	MHSS with weak joints
b) 25,761.00	3101.84	500	4.36	20	1025	3984	3.88	-do-
c) 21,600.00	7200.00	1470	13.80	4	100	5025	50.25	MHJSS
d) 20,710.00	4680.50	115	36.90	17	300	1200	4.00	Same as (a)
e) 5600.00	2862.50	1510	3.11	4	50	4050	81.00	Same as (c)
f) 1443.75	593.75	175	9.97	13	75	2516	33.55	SMH-CGSS
g) 1443.75	593.75	100	23.30	14	75	2391	31.88	-do-
h) 1300.00	31.25	100	7.91	28	356	3172	8.90	-do-
i) 1300.00	31.25	175	3.65	23	356	2672	7.50	-do-
j) 1000.00	253.40	268.20	4.10	22	200	891	4.46	-do-
k) 906.25	25.00	294	2.14	34	288	1219	4.23	-do-
l) 500.00	25.00	100	14.70	37	19	1047	5.39	-do-

(a), (c), (e), (f), (g) and (j) -> D.F. + Cord relay System
(b), (d) -> D.T.H. + D.F. & Cord relay
(h), (i), (k) and (l) -> D.T.H. + NTD

MHSS = Massive hard sandstone.
MHJSS = Medium hard jointed sandstone.
SMHCGSS = Soft to medium hard coarse-grained sandstone.

residents hear any loud noise, then psychologically, they think about damage to their property. These problems when aggravated by local leaders then turn to a different direction and sometimes it becomes difficult for the mine management to conduct their day-to-day production blast.

According to Wiss and Linehan (1978), the causes of air blast are the following:

(i) Rock pressure pulse due to detonation of explosive
(ii) Gas ejection from stemming column
(iii) Firing of holes with the help of detonation cord
(iv) Air pressure pulse due to displacement of rock from its original position
(v) Collision of projected rock fragments.

11.4.1 Temperature Inversion

Konya and Walter (1990) reported that under normal conditions, temperature decreases with height in the atmosphere, cooling at the normal lapse rate of 3.5°F for each 300 m of height. However, when it is reversed i.e. the temperature increases with the increase of height, it is termed as inversion. The velocity of sound in air increases when air is warmer and decreases when air is colder. Konya and Walter (1990) also reported that the change of velocity is approximately 0.3 m/s for a temperature change of 1°F. In this context, the five different conditions of figure 11.2 are explained below—

1. *Condition-a*: Ideal condition with isothermal air and zero wind velocity, (although does not exist in reality).
2. *Condition-b*: Normal condition, when the temperature decreases with increase in height. While travelling from warmer to cooler air conditions, the sound velocity decreases causing the waves to curve upward into the atmosphere, -away from the earth following Snell's law (Konya and Walter, 1990).
3. *Condition-c:* When temperature inversion occurs, then the velocity of sound also increases, resulting in the sound waves curve bend downward due to refraction toward the ground with intensity increased by a factor of 2 to 3. Such a condition usually occurs in the early morning, in the evening or at night.
4. *Condition-d*: The situation, when strong wind persists in normal condition-b. In that case, the region of sound concentration would be further away from the blasting source.
5. *Condition-e*: On the upwind side, the wind velocity would oppose the inversion velocity effect and decrease the sound velocity; but on the downwind side, the wind velocity would add up the sound velocity.

11.4.2 Air Blast Measurements

Air blast is used to define air pressure waves from explosive detonations in surface mining. Air blast is an atmospheric pressure wave consisting of audible high frequency sound and sub-audible low frequency concussion. For practical purposes "decibel" unit is used to define the level of air blast. The unit "decibel" is connected to "psi" by the following equation.

$$dB = 20 \log (P/P_0) \tag{11.1}$$

where,

dB = Sound level in decibels

P = Overpressure in psi,

P_0 = Overpressure of the lowest sound that can be heard = 3.0×10^{-9} psi.

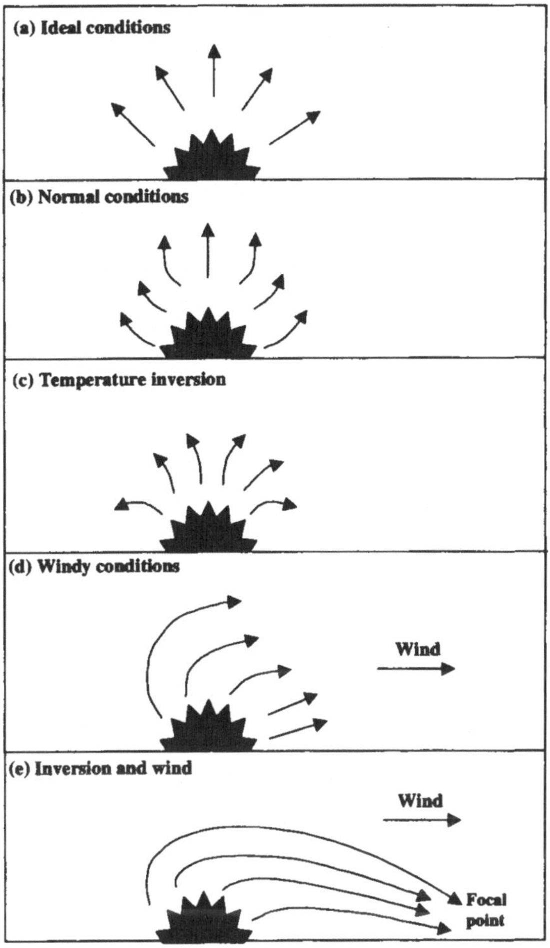

Fig. 11.2 Air blast propagation under different meteorological conditions (after Bhandari, 1997)

Different weighting networks are used to measure sound levels depending on the requirement. The networks are designed as *A, B, C,* linear (L) and differentially in the ability to measure low-frequency sound. A-weighting discriminates against low frequencies, it appropriates the human ear. B-weighting discriminates moderately against low frequencies and C-weighting makes a slight discrimination at the low frequencies. Linear network measures all frequencies. Blast sound is primarily in the low frequency range (Konya and Walter, 1990). In general, *A*-weighted sound level i.e. *dB* (*A*) is used. It is a simple measuring technique keeping high correlation with *N* (Noise Criteria) and PNC (Preferred Noise Criteria). It is suitable for measuring fluctuating noise and is adequate for predicting general public acceptance of most of the community noises.

The lowest frequency which normally gives a sensation of sound for the average individual is about 25 Hz. Hertz (Hz) is the most common unit of sound frequency, defined as the cycle per second. The ear of a normal adult human has the ability to hear a very broad spectrum of frequencies from 25 to 16,000 Hz. Beyond 16,000 Hz, sounds are known as ultrasonic. Some typical sound level threshold values are shown in Table 11.4. Table 11.5 shows some specific sources of industrial noise due to opencast mining, while Table 11.6 shows a comparative assessment of the values of air overpressure and noise levels monitored at the same place but in different locations of an opencast mine. All such measurements were made using Bruel and Kjaer make sound level meter.

Table 11.4 Typical sound level damage threshold values (Dhar et al., 1993; Konya and Walter, 1990)

Type of damage	*Values in dB-L*
Structural damage	180
Plaster cracks	176
Loose windows sash rattles	140
Failure of badly installed window panes	140-145
Failure of correctly installed window panes	Over 168
All window panes fall	176

Table 11.5 Noise levels from different sources of opencast mining

Equipment	*Noise level, dB-A*	*Measurement location*
Compressed air	110-120	1 m
Rock drill	100	12 m
Large portable compressor	85	6 m
7cu m dragline	90-93	Operator's position
Diesel trucks	75-110	Driver's position
Shovels	85-87	Driver's position
	90-95	At 12m distance, while loading
Dumper	88-90	Neutral position of the engine in idle condition
	95	With full accelerator
	105	Without load, down gradient
	98-102	Operator's position with load (uphill)
	87	Same as above (4 m)
	83	Same as above (12 m)
Rotary drills	75-100	Operator's position
Front end loader	90-92	Operator's position
	96	Near motor
	102	While loading (at 1 m distance)
	97	While unloading (at 10 m distance)
	94	Same as above (at 12 m)
	95	Free movement with load at 12 m
	89	Free movement without load at 12 m
Scrapers	90-105	Operator's position

Table 11.6 Values of air overpressure and noise levels monitored at the same place in different locations

Distance from the source (m)	*Total charge fired (kg)*	*Air overpressure (dB-L)*	*Noise (dB-A)*
68	212.44	145	113
72	197.96	141.2	92
75	273.80	140	109
78	196.00	140	106
90	260.13	138	107
98	600.00	142	104
102	380.19	141	108
120	230.25	138	95
125	218.75	135	85
130	531.25	136	94
140	283.50	140	96
145	275.93	140	106
155	180.90	146.6	105
165	255.90	143.1	103

Reed et al. (1968) derived the following empirical equation which may be used to describe window breakage:

$$D = 3.71 \times 10^{-4} \times A^{1.22} \times P_0^{2.78} \tag{11.2}$$

where, D = damage probability in per cent

A = window area in square feet

P_0 = peak overpressure in millibars (1 psi = 69.4 millibars)

In environmentally conscious places, air blast is seldom a problem in normal blasting operations. In most places, it creates psychological fear rather than any probable damage. The random nature of the process involved in the generation and propagation of air blast, together with the many variables which determine the strength of a window pane, breakage prediction of any window pane is extremely uncertain. However, the above equation may give some probability information of potential window damage. The following precautionary measures may be taken while blasting in opencast projects to reduce noise and air overpressure levels.

11.4.3 Control Measures

The following control measures will reduce the impact of air blast and noise arising due to opencast blasting.

(a) Use long stemming column in blast holes
(b) Use shock tubes rather than detonating fuse as trunkline
(c) Use 150 mm thick cover of sand or drill cuttings to cover trunkline detonating fuse

(d) Avoid blasting in cloudy weather
(e) Carry out blasting at midday
(f) Avoid blasting when strong winds are blowing towards the residences.

11.5 FLYROCK —OCCURRENCE AND CONTROL

Any rock fragments thrown unpredictably from a blasting site is known as flyrock. Flyrock may occur in situations where there is insufficient stemming, too small a burden because of overbreak from the preceding or proceeding blast, planes of weaknesses in rock which reduce resistance to blasting, and finally existence of loose rock fragment on top of the bench.

Flyrock is produced when explosive energy is excessive for the amount of burden, stemming is inadequate or the energy is too rapidly vented through a zone of weakness in the rock. When this happens, much of the explosive energy is used to throw rock rather than to create fragmented rock and controlled displacement of the burden material. Geology, improper blast design, or carelessness can cause unwanted flyrock. Other factors such as powder factor, pattern layout, delay sequence, stemming height, drilling accuracy, and charge loading method should also be evaluated if any excessive flyrock is generated. Plates 11.4, 11.5 and 11.6 show different views of practical situations which generally happen in actual field conditions.

Plate 11.4 View of clear bench faces in a limestone mine

Plate 11.5 Optimum fragmentation and desired muck profile obtained from the 2nd bench of the mine shown in plate 11.4

(a) Uncontrolled blast

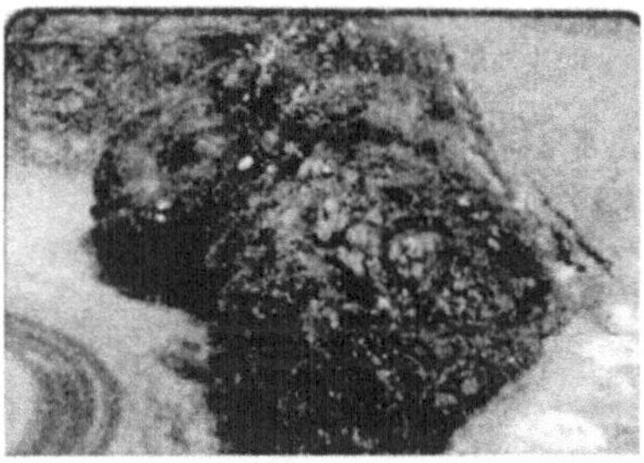

(b) Controlled blast

Plate 11.6 Views of uncontrolled and controlled blasts (Courtesy: BAI)

All blast can be expected to generate some flyrock. The distance that flyrock travels can range from a few metres to more than 1 km with poorly controlled shots. In surface mine blast, there is a desired muck pile displacement or cast. There is always a normal flyrock range, and a secured blast area. The normal flyrock range is about 2.5 times the bench height and that the secured area is about 4 times of the bench height. Flyrock beyond the secured blast area is termed as excessive flyrock. Excessive flyrock is a major cause of injury and damage to mine equipment. Table 11.7 gives a few measurements of flyrock in a limestone mine of the eastern part of India. Following are the effective measures to control flyrock problems—

1. In loose and fractured rock mass, explosive added with additives may be used.
2. Proper burden and spacing should be maintained according to the bench height.
3. Length of stemming column should be greater than 0.6 times the burden. For better protection, it may be taken as greater than or equal to the burden.

4. Angular hole may be drilled in conformity with the slope of the bench.
5. As far as possible, holes should be located beyond weak zones.
6. Burden and spacing should be marked by a responsible person with the help of a surveyor, taking the condition of the face and fractures into consideration.
7. Before loading, blasting officials should always check the hole depths and ensure that the holes are drilled as per the blast design.
8. All loose pieces of rock from the blasting site should be cleared before charging.
9. Stemming of holes is very important. All the holes should be properly stemmed using 6-8 mm size stone chips along with drill cuttings.

Table 11.7 Measurements of flyrock and maximum throw of blasted materials at the back side and front side of the face

Location	*(a) Charge per hole (b) Maximum charge per delay (c) Total charge fired in the round (all in kg)*	*Flyrock distance at the back side of the face (m)*	*Maximum throw at the front side (m)*
1st bench, right	(a) 60.00 (b) 155.257 (c) 1950.00	12	20
2nd bench, left	(a) 70.20 (b) 173.125 (c) 2204.60	15	25
1st bench, right	(a) 76.05 (b) 171.06 (c) 2206.00	15	25
Mixed bench, front	(a) 40.95 (b) 98.70 (c) 1404.60	20	30
1st bench, left	(a) 87.75 (b) 189.70 (c) 1917.95	10	25
1st bench, left	(a) 87.75 (b) 169.72 (c) 877.80	15	20
2nd bench, left	(a) 64.35 (b) 169.72 (c) 151.85	10	30
1st bench, left	(a) 76.86 (b) 189.36 (c) 1504.25	15	25
1st bench, left	(a) 82.35 (b) 266.68 (c) 1653.00	10	20

11.6 BLASTING IN UNDERGROUND COAL MINES

For underground coal mine blasting, the various classes of permitted explosives are summarized below—

P_1 group— Permitted explosives used in drifting, simultaneous firing in coal, and ripping;

P_2 group— Sheathed permitted explosives used in simultaneous firing in coal, and ripping (replaced by P_3 group);

P_3 group— Equivalent to sheathed permitted explosives used in places where P_2 group is approved;

P_4 group— Used for delay firing in ripping, undercut coal and cross measure drifts;

P_5 group— Used for delay firing in solid coal.

Different types of permitted explosives as shown in Table 11.8 may be used in various categories of coal mines which have been classified in terms of potential gas hazards.

Table 11.8 Types of permitted explosives for different degrees of gassiness

Degree of gassiness of coal seam	*Classification of gassiness*	*Type of permitted explosives*
I	< 0.1% of gas in the general body of air and rate of emission of such gas is less than 1 m^3/t of coal production.	$P_1/P_3/P_5$
II	> 0.1% of gas in the general body of air and rate of emission of such gas is greater than 1 but less than 10 m^3/t of coal production.	P_3/P_5
III	Rate of emission of the gas is greater than 10 m^3/t of coal production	P_3/P_5

11.7 BLASTING PROCEDURES IN HOT HOLES

CMRI carried out substantial experiments to obtain a general guideline of blasting in hot holes to be used to acquire huge amounts of coal, which would otherwise be lost or unrecovered. The following stepwise approach may be followed for successful extraction of coal from a fire zone—

Step-1: Select the number of holes properly so that the total blasting operation should not exceed 2 h from the charging of first hole

Step-2: Measure the temperature of the holes

Step-3: Use water at least 12 h before blasting to flush hot holes till the temperature comes down below 80°C.

Step-4: Record the temperature of holes at a regular interval of time

Step-5: Use a mixture of bentonite, sodium silica and water in holes, which do not retain water, to seal micro-fractures and cracks. Guar-gum up to 5% may also be used for the same purpose.

Step-6: Check before charging whether the detonating cord is detached from the main reel or not, if not, then detach it immediately before charging operation starts.

Step-7: Stemming operation should be done after charging of all the holes.

11.8 Measurement of Dust Clouds

In a joint collaborative study conducted by CMRI (presently renamed as Central Institute of Mining and Fuel Research or CIMFR), CMPDIL (Central Mine Planning and Design Institute Ltd.), WCL (Western Coalfields Ltd.), DGMS (Directorate General of Mines Safety) and DRDO (Defence Research Development Organisation), observations were made regarding dust clouds generated during blasting in dragline and shovel-dumper faces of Ghugus Opencast Project of WCL. The measurements of SPM levels were carried out using 47 mm open-ended air samplers (controlled valve) downstream of the sample filter in the suction line.

The sampling rate was 25, 30 and 50 lpm respectively at three locations and sampling time was 5 to 50 minutes to cover the drift of the dust cloud. From the background data collected, it was seen that even during the non-working period, the SPM level of 749 $\mu g/m^3$ in the mine area exceeded the 500 $\mu g/m^3$ specified limit set by CPCB (Central Pollution Control Board, Govt. of India) for industrial area (Kapoor et al., 1998). Table 11.9 shows the maximum permissible SPM limits established by CPCB for different areas.

The solution of this major problem can effectively be controlled by time-bound water spraying before and after blast, especially in blasting area and nearby haulage roads. The use of DTH (down-the-hole delay) and NTD/TLD (trunk-line-delay) systems (or in other words shock-tube system) would reduce the dust and noise hazards by nearly 30 to 40%. Proper planning, time-bound water spraying and plantations are the general ways to reduce the environmental hazards arising due to blasting.

Table 11.9 Maximum permissible SPM limits established by CPCB

Type of Area	*Concentration in ($\mu g/m^3$)*
Industrial Area	500
Residential Area	200
Sensitive Area	100

11.9 REMEDIAL MEASURES: GENERAL GUIDELINES

As a general guideline, the following precautionary measures may be taken to minimize the blasting hazards in sensitive regions.

11.9.1 Non-technical Sides

- Good liaison with the villagers through the 'Sarpanch' or other influential persons
- Blasting team should include one or two persons who are working in the mine from the neighbouring village
- Villagers are to be alerted well before blasting which would minimize their sudden psychological impact.

11.9.2 Technical Sides

- Muffle blasting should be done wherever blasting is to be carried out within 100 m or less of any structure
- Thorough inspection of the site of blasting is necessary before marking of the holes
- For controlling air blast impacts, suggestions made in section 11.4.3 are to be followed
- For controlling flyrock, suggestions made in section 11.5 are to be followed.

REFERENCES

Bhandari, S. 1997. Engineering rock blasting operations, A.A. Balkema, Rotterdam, Netherlands, p. 375.

CMRI-RC Interim Report, 2003. Evaluation of human response to vibration and air overpressure due to blasting in opencast non-coal mines in India, (Chakraborty et al.), S&T project report submitted to the Ministry of Environment and Forests, Government of India, p. 11.

Dhar, Bharat B., Pal Roy P. and Singh, R. B. 1993. Optimum blasting for Indian geomining condition—suggestive standards and guidelines, CMRI publication, September, p.40.

Dowding, C. H. 1985. Blast vibration monitoring and control, Prentice-Hall International Inc., USA, p. 287.

Kapoor, J. C., Singh, A. P., Pandit, M. and Bhalla, A. K. 1998. Measurement of suspended particulate matter (SPM) and noise levels generated during blast trials at Ghugus Opencast Mine of Western Coalfields Ltd., Interim Report No. DRDO/CEES/NS/106/98, June, p. 11.

Konya, C. J. and Walter, E. J. 1990. Surface blast design, Prentice-Hall International Inc., USA, p. 298.

Pal Roy, P., Singh, R. B., Bagchi, A. and Barman, B. K. 1993. Investigation into the causes of 'cut-off' and remedies thereof and to establish optimum safe blast design for better fragmentation in Chasnalla OCP, SAIL, EC/64/92, May, p. 29.

Pal Roy, P. and Chakraborty, A. K. 2000. Critical evaluation of ground vibrations and consequent structural damage under varying blast configurations, Proc. 1st World Conference on Explosive and Blasting Technique, Munich, Germany, 6-8 September, Holmberg (ed.), pp. 71-78.

Reed, J. W., Pape, B. J., Minor, J. E. and Dehart, R. C. 1968. Evaluation of window pane damage intensity in San Antonio resulting from Medina facility explosion on November 3, 1963, Annals of the New York, Academy of Sciences, Vol. 1521, October, pp. 566-584.

Reiher, H. and Meister, F. J. 1946. The effect of vibration on people, Forschung aut dem Gebeite des Ingenieurwesens, Vol. 2, No. 11, Translated by HQ, Air Materials Command, Wright Field, Ohio, USA, Report F-TSL-616-RE.

Singh, P. K. 1998. A study of ground vibrations due to rock blasting, Ph. D. Thesis submitted to the Technical University of Clausthal, Germany, February, p. 210.

Siskind D. E., Stachura, V. J., Stagg, M. S. and Kopp, J. W. 1980. Structure response and damage produced by airblast from surface mining, U.S. Bureau of Mines, R.I. 8485, p. 111.

Wiss, J. E. and Linehan, P. 1978. Control of vibration and blast noise from surface coal mining, Wiss, Janey, Elstner and Associates Report for the U.S. Bureau of Mines, Contract JO255022, 2, p. 111.

12

Emerging Blasting Techniques

12.1 AIR-DECKING PRINCIPLE

12.1.1 Introduction

The method of air-decking has become an increasingly popular technique in production and presplit blasting. In the blasthole, an explosive column is combined with an air-chamber. The main aim of the technique is to control breakage through effective distribution of explosive energy and thereby enhance uniform rock fragmentation. An air column lowers the initial pressure of the explosion products and simultaneously increases the time of interaction for the detonation front with the surrounding rock, which enhances rock breakage.

To study air-deck blasting, CMRI initially conducted experiments using indigenously developed, low-cost wooden spacers of different sizes and strengths. Air-deck blasting was first tried in 1993 at Jambad OCP of ECL and Chasnalla OCP of IISCO (Indian Iron and Steel Company Ltd.). The experiments were extended further in 1994 at Bermo Mine of DVC (Damodar Valley Corporation). The results were subsequently published by Pal Roy et al., in 1995. It is worth mentioning that the performance of air-deck blasting during that period was only compared with conventional blasting practices without going into detail of the theory behind the process.

To establish the scientific principles of air-decking, a series of experiments were conducted by CMRI at Parej East Opencast Mine of CCL, Murli Pahari Limestone Mine of KCL (Kalyanpur Cement Ltd.), Chasnalla OCP of IISCO and Nigahi OCP of NCL using a high-speed video camera, VOD measuring instrument and BOSTECH Gasbags. The results were analysed to evaluate the effectiveness of air columns in comparison to material decking using explosives like slurry (cartridge), emulsion (cartridge), ANFO and SME (site-mixed emulsion). After analysing various field results, the theory and principle of air-deck blasting and its breakage mechanism under varying charge loading and air-column configurations were established (Pal Roy and Singh, 2001).

12.1.2 Theory of Air-deck Blasting

In air-deck blasting, the products of explosion expand into the air gap decreasing the overall explosion pressure. Expanding explosion products generate a shock wave during expansions. Expansion from two charges, separated by an air gap, generates two shock waves that collide in the air gap. This results in a high-pressure zone in the air gap. While in the case of material decking, the detonation fronts from the adjacent explosive columns attenuate due to the resistance of drill cuttings and pebbles, and although there is interaction in the decking column, their intensity is too low to propagate through the decking materials. Consequently, the latter provides little energy output to break the rock adequately.

12.1.2.1 Measurements of VOD

Using Instantel's VOD probe, the detonating velocity of the explosive was measured to examine the efficiency of the explosion; via fibre optic probes inserted into the explosive at predetermined distances from each other (Figure 12.1). On detonation, the emitted light strikes each probe at a different instant. The time difference, together with the known probe separation was then used to compute the velocity of detonation. Two typical measurement set-ups are shown in figures 12.2a and 12.2b. With bulk explosive used, the probes were inserted inside a fibre optic insertion guide and taped with the explosive cartridge before lowering into the borehole. The VOD was measured by the instrument by taking the ratio of probe separation distance and the elapsed time.

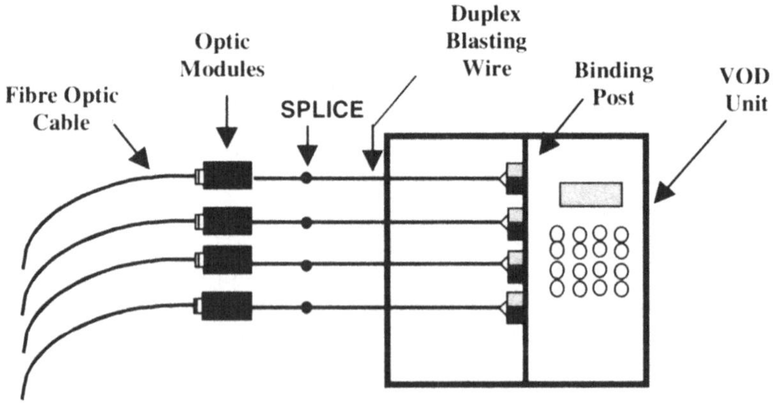

Fig. 12.1 Diagrammatic view of Instantel's VOD measuring instrument

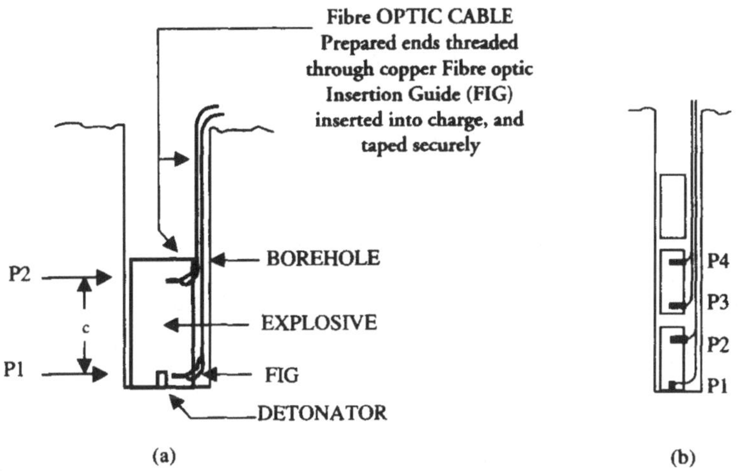

Fig. 12.2 Typical fibre optic cable placement for VOD measurement

During field trials, different experimental set-ups, as shown in figure 12.3, were made in regard to the placement of gasbags, non-electric initiation systems and explosive charge to monitor the velocity of detonation-fronts in the explosive, air and material decked columns. The main emphasis of such arrangements were to ascertain:

- the mobility and power of the original detonation front while traversing an air column inside a blasthole;
- the reduction of the original detonation front while traversing through material decking in the blasthole;
- the VOD values for different blasthole diameters; and
- the performance of explosives used in the test blasts.

Waveform and frequency distributions of the seismic pulses were recorded at varying distances from the blasting source. A steeply deepened (high attenuation) waveform was observed. A high-speed video camera was used to observe the rock throw in the air-deck column. The results of all the VOD tests are listed in Table 12.1.

The results obtained in cases 1 and 4 clearly indicated that the detonation fronts from opposite explosive columns interacted in the air column separating them and created a new front which was comparable to 80% of the original VOD of the explosive being used.

When the decking comprised drill cuttings, the VOD in the decked portion abruptly reduced (cases 2 and 3). The reduction corresponded to 1/10 to 1/13 of the original value, thus yielding very little energy output to break the surrounding rock. If low-strength explosives are used, then material decking would provide an ineffective breakage zone.

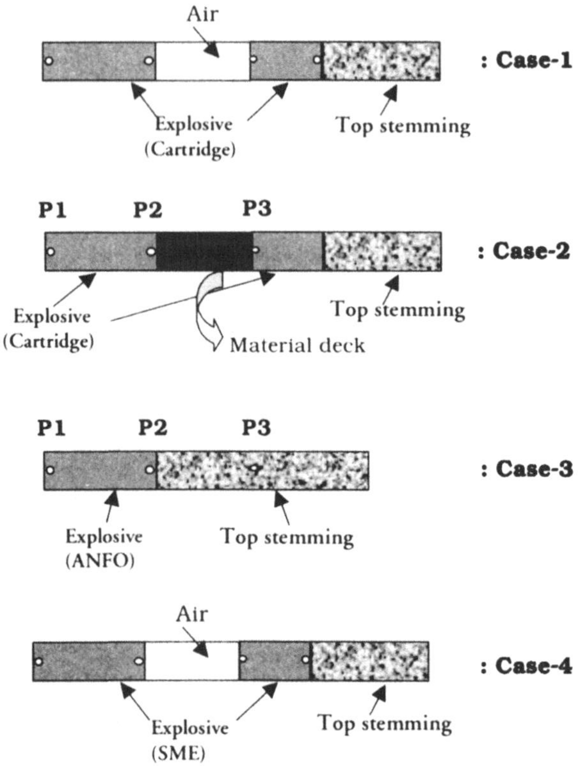

Fig. 12.3 Different experimental set-up for VOD measurements

12.1.2.2 Accessories Used in Air-deck Blasting

(a) Wooden Spacers. To popularise air-deck technique in Indian market, CMRI designed different kinds of wooden spacers which are economical and easy to use. Figure 12.4 shows the specifications proposed by CMRI for 150 mm (≅6″) drillhole diameter. Laboratory test results show that the spacers of 0.5 m length made of Murga wood in the middle and hard wood at the top could withstand safe loads of approximately 2000 kg and yield loads of around 2200 kg. For 1.0 m length spacers made of Hollock wood, the optimum load was 1000 kg both for 150 mm and 100 mm drillhole diameters. In the case of 250 mm drillhole diameter, and a 1.25 m length spacer, the load threshold was found to be 600 kg.

(b) BOSTECH Gasbags. BOSTECH gasbags, manufactured by the BOS Technologies (India) Pvt. Ltd., Bangalore in collaboration with S.P.A. Pvt. Ltd. NSW, Australia, marketed by SUA Explosives and Accessories Ltd., Bangalore, are self-inflating units via an aerosol can inside producing the inflating gas (Figure 12.5). The Gasbag is inflated by depressing the trigger at the top of the aerosol can. When the trigger is thus locked down, it

Table 12.1 VOD measurements results

Case-1: Air-decking (mid-column)	*Remarks*
(150 mm drillhole diameter) V1 to V2: 5268 m/s V2 to V4: 4656 m/s V1 to V4: 4883 m/s V1 to V3: 4701 m/s V3 to V4: 5464 m/s V2 to V3: 4163 m/s	1. For air-decking BOSTECH gasbags were used. 2. Raydet shock tube systems (DTH and TLD) were used in the hole and on the surface. 3. Toeblast and Supergel Explosives (M/s Gulf Oil Corporation Ltd.) were used. 4. The value of V2 to V3 emphasizes the effectiveness of air column in the blasthole. 5. Effectiveness of air-decking is more pronounced and profitable when the length of hole is more than 10 m and the diameter is ≥ 150 mm (6 in.)
Case-2: Material-decking (mid-column)	*Remarks*
(250 mm diameter drillhole diameter) V1 to V2: 5298 m/s V2 to V3: 0399 m/s V1 to V3: 0612 m/s	1. Drill-cuttings were used for material decking. 2. EXEL non-electric initiating devices (DTH and NTD) were used in the hole for single-hole firing. 3. Bharat Toe and Bharat Base (M/s Navbharat Explosives Co. (P) Ltd.) were used. 4. The values of V2 to V3 and V1 to V3 show that the detonation fronts dampen when passing or intercolliding in the decking material in the hole. 5. No changes in VOD were observed due to a change in hole diameter from 150 to 250 mm.
Case-3: Stemmed hole and no air gap	*Remarks*
(115 mm drillhole diameter) V1 to V2: 3706.23 m/s V2 to V3: No signal passed	1. Drill-cuttings were used for material decking. 2. Electric detonators were used at the bottom of hole to initiate firing. 3. Blasting was done using ANFO. 4. No signal was received at point P3 and therefore no VOD record was available. The reason could be due to comparatively lesser value of VOD of ANFO dampened sharply before reaching any electrical pulse to P3.
Case-4: Air-Decking [mid-column]	*Remarks*
(259 mm drillhole diameter) V2—V3: 3017 m/s (air column)	1. BOSTECH gasbags were used for air-decking. 2. EXEL non-electric initiating devices were used down-the-hole for a series of 39 holes of 15.5-16 m depth. 3. Site-mixed Emulsion (SME) was used with 0.2% cast booster. 4. SME had VOD in the range of 4200–4500 m/s.

releases the contents of the aerosol can into the gasbag and inflates it to the correct pressure to lock the bag in the borehole. Gasbags are available for a variety of drillhole diameters. One of its many interesting and useful

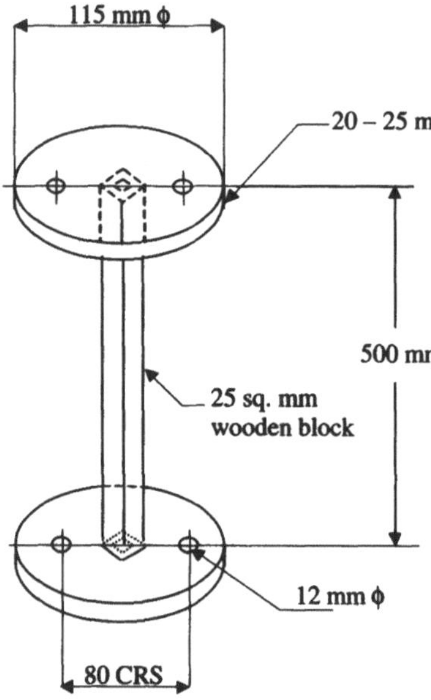

Fig. 12.4 Specifications of wooden spacer for 150 mm (≅6 in.) drillhole diameter

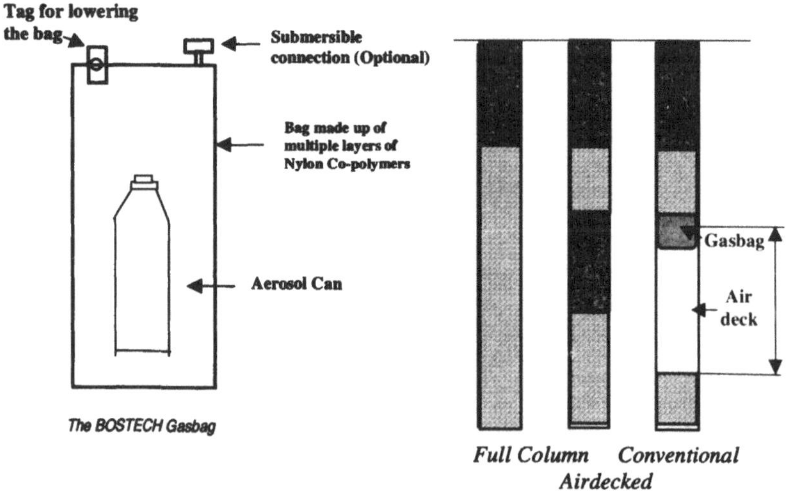

Fig. 12.5 Configuration and usage of a BOSTECH gasbag

features is that it can seal off water at the bottom of a borehole, enabling ANFO to be used instead of slurry/emulsion explosives. It can also be used to check explosive wastage through fissures and voids where SMS, SME or Bulk explosives are used. While wooden spacers are cheaper, there are restricted dry-hole conditions, whereas, gasbags may be used in any condition, while they are found to perform well even in watery holes and holes on the top of old underground workings.

(c) Non-electric initiating devices. Non-electric initiating devices allow bottom initiation without desensitisation or side initiation of non-capsensitive explosives in the borehole. They consist of a plastic shock tube with an outside diameter of 3 mm to 4 mm. The inside walls are coated with a fine layer of explosive dust (generally 10 to 15 mg/m length). The shock tube initiating system comprises a length of shock tube sealed at one end with a detonator crimped at the other. When the tube is fired from the sealed end, a mild shock wave propagates inside the tube which in turn initiates other accessories or explosives primed with it. DTH (down-the-hole) delays can be used in conjunction with surface NTD or TLD (noiseless or trunk-line-delay). In India, shock tube systems are available in the brand name of EXEL (product of Orica-Indian Explosives Ltd.) and Raydet (Product of IDL-Gulf Oil Corporation Ltd.).

12.1.3 Breakage Process in Air-deck Blasting

The principle of air-deck blasting relies on the superimposition of opposite directional detonation fronts in the air-column, followed by the breakage of explosion products. As a result of this superimposition, new sources of compressive shock air waves are formed. The direction of maximum compression depends on the type of charging, blasting accessories and initiation sequence. Figure 12.6 depicts the ideal situation where the direction of maximum compression is perpendicular to the axis of the hole. This case is appropriate for massive to medium-hard strata. The

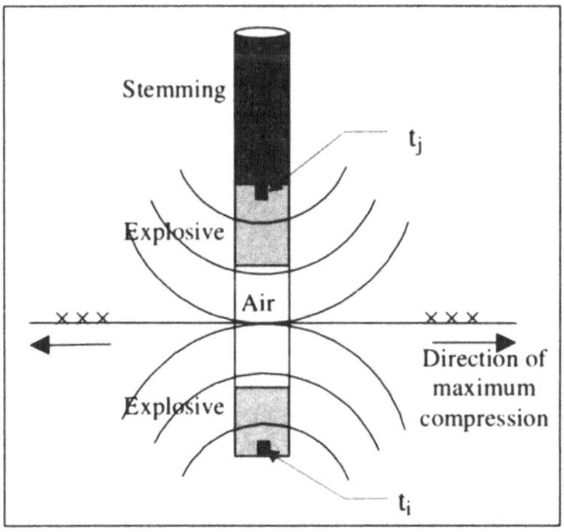

Fig. 12.6 Ideal case of maximum compression in the air column perpendicular to the borehole (in-hole delays of equal timings t_i=t_j).

case of figure 12.7 may be useful when the top strata are slightly harder than the bottom set. Figure 12.8 shows the case where detonating fuse replaces the in-hole delay. This case may be effective when a toe problem is encountered in the mine.

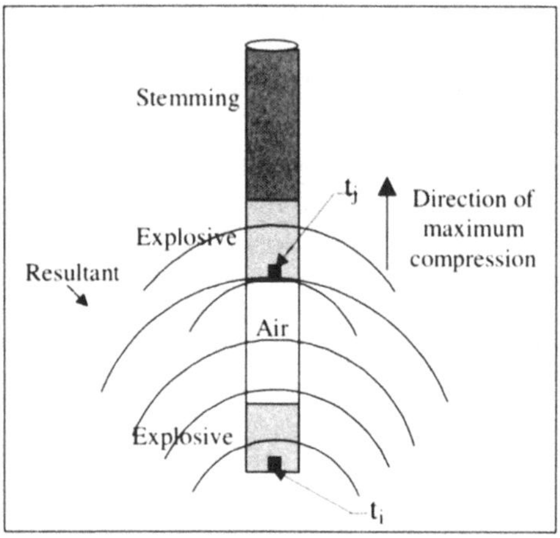

Fig. 12.7 In-hole delays of unequal timings ($t_i < t_j$).

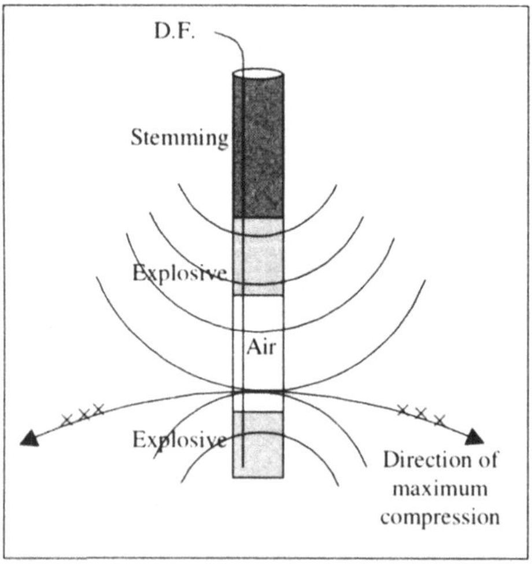

Fig. 12.8 Conventional charging, using detonating fuse.

12.1.4 Usage of Air-decking in Presplit Blasting

An air-decking principle can be used in presplit blasting to enhance more effective distribution of shock wave power derived from explosive energy. As reported by Chironis (1987), Atlas Power Co. developed a technique of trapping air columns in presplit blastholes. It was actually a variation of open-hole shooting where no stemming was used and the holes were left open at the top. Open-hole shooting has little potential for use in mines because it generates a high-noise level leading to adverse situations. The key difference, therefore, in the air-shock presplitting technique of the Atlas Power Co. was the use of a rugged, inflatable plug as stemming to bottle up the blastholes.

Bussey and Borg (1995) outlined the concept of loading a bottom charge in an otherwise empty borehole, which was then stemmed at a desired elevation with the aid of a special device. They used this technique for presplitting in surface coal mines to produce highwall conditions for greater safety, lower cost of explosives, labour and drilling. The concept leaves a competent highwall providing greater safety for personnel working in the vicinity of such structures.

The theory behind the above studies is based on the concept that when the holes are fired, the peak borehole pressure is lowered due to expansion of gases in the empty space. The crushing and cracking immediately around the holes are reduced, but other effects are sufficient to start the cracking process from the hole but lack sufficient energy to cause breakage beyond desired limits.

12.1.5 Conclusions

The experiments conducted by CMRI confirm the effectiveness of air columns in comparison to conventional material decking approaches. The measured VOD values in the air column approached 80% of the original explosive VOD used in the borehole. This reduction was observed to be similar for 150 mm and 250 mm diameter drillholes. However, in the case of material decking, the reduction was nearly 1/10 to 1/13 of the original explosive VOD. It has been shown that the direction of compressive shock pressure pulse produced due to the superimposition of detonation fronts in air columns, depends on the type of air-decking and initiation sequence. This technique has significant potential advantage such as the reduction of explosive weight by nearly 20% without the sacrifice of fragmentation. It can be used very effectively in presplit blasting for wall control, with greater safety and lower cost outcomes.

12.2 BABY-DECKING TECHNIQUE

12.2.1 Introduction

Winning of coal in opencast mines is generally carried out by removal of overburden by blasting. In bench blasting, sub-grade drilling (nearly 5-10% of the bench height) is generally recommended for removal of toe problem. However, due to sub-grade drilling and conventional blasting patterns, the underlying coal seam gets fractured or crushed due to blasting (along the separation line between overburden and coal) for both dragline and shovel benches. This damaged coal is predominantly mixed with overburden during bench blasting and ultimately causes loss of substantial amount of coal. Patrik et al. (1988) reported a loss of about 20,000 tonnes from one cast blasting while Wedmair and Scott (1995) estimated that such losses would reduce the mine revenue by more than US dollar 1 million (approximately Rs. 4.70 crore). In India, such type of losses in dragline blasting is reported by Kalam and Singh (2000). They specifically pointed out coal bed deterioration (overbreak) in deep-hole NONEL blasts. It is thus essential for the practicing mining engineers to protect such loss of coal during overburden bench blasting.

12.2.2 Scientific Identification of the Problem

There are number of factors affecting the degree of coal damage. Amongst them, some factors cannot be controlled, such as local geology, whereas, others can be controlled or altered by the blasting engineers. Chiappetta and Postupack (1995) made a number of practical and economic recommendations for the entire coal and blasting industry, especially commercial operations whose economic growth relies heavily upon recovering an undamaged coal seam. The authors classified the problems through various diagrammatic representations. A few of such are shown in figures 12.9 to 12.11.

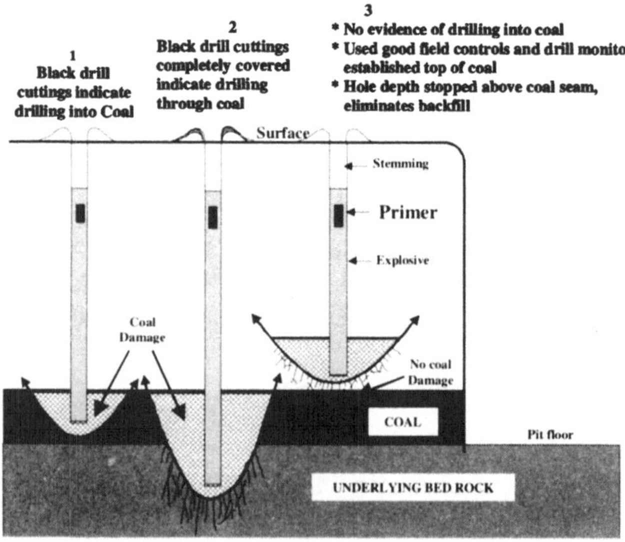

Fig. 12.9 Faulty drilling due to lack of field control (after Chiappetta and Postupack, 1995)

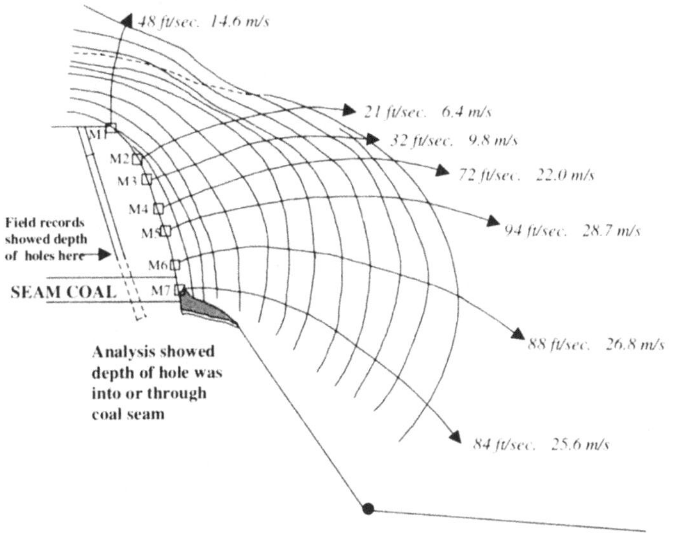

Fig. 12.10 High-speed photography results for inclined holes (after Chiappetta and Postupack, 1995)

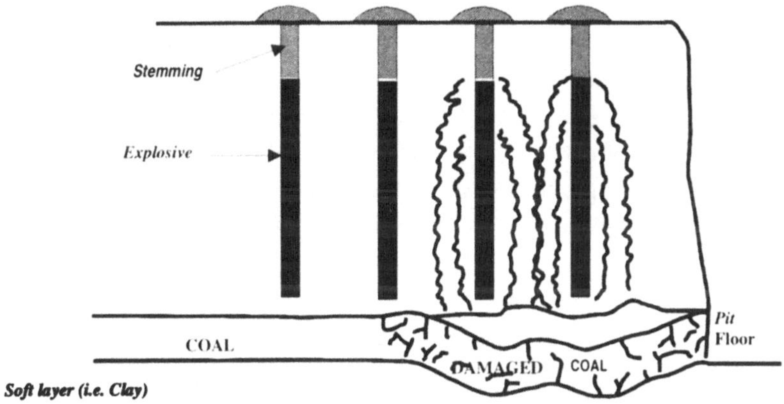

Fig. 12.11 Coal damage resulting from soft material underlying coal seam (after Chiappetta and Postupack, 1995)

The degree of explosive confinement at the toe is one of the most important controllable factors that affect coal damage during blasting. Sarma and Scott (2000) studied the influence of toe confinement on coal damage during cast blasting. They introduced the baby-decking technique to minimize coal losses and its damage during cast blasting. The main purpose of the baby-decking technique is to reduce explosive confinement at the toe in order to prevent coal damage due to overbreak. In baby decking, the total explosive column is divided into two decks, a small baby-deck and a main deck. The primers and timing of down-the-hole delays are designed in such a way that the baby-deck is initiated 25 ms after the main deck. The amount of explosive charge in baby-deck would depend on the hole depth, hole diameter, explosive type and rock properties.

Basically, damage to underlying coal seam occurs due to the following reasons—

- Stress wave induced damage;
- Damage due to gas penetration; and
- Damage due to overburden movement.

Sarma and Scott (1999) used inert materials at the bottom of blastholes. They reported significant reduction of coal damage without affecting the casting of the material. However, they did not mention the possibility of toe problem due to insufficient explosive charge at the bottom and/or desensitisation of the bottom column by the main deck.

In India, no such improvised pattern has been implemented so far in coal mining industries although similar damage and losses of coal in deep hole dragline blasting are observed (Kalam and Singh, 2000).

The economic viability and saving of cost are briefed in the following example—

Box-cut = 60 m × 40 m (dragline face)

Coal damaged = 1 m (say)

Loss of coal = 60 × 40 × 1 × 1.5 = 3600 tonnes

Unit price of coal = Rs. 800/tonne (approximate)

Loss of revenue = Rs. 800 × 3600 = Rs. 28.8 lakh.

Therefore, saving even of 10% of damaged coal would earn a revenue of Rs. 28.8 lakh (for a cut of 60 m by 40 m dragline blast).

12.2.3 Baby-decking

- In the baby-decking technique, the explosive near the toe is initiated after the main deck and also less amount of explosive quantity is placed in the baby-deck. The stress wave generated by baby-deck may attenuate more quickly because the surrounding rock has already been fractured by the main deck.
- The explosion gases from the main deck extend the fractures further into the burden rock which in turn offer a path of least resistance for the explosion gases from the baby-deck. The explosion gases from the baby-deck would try to penetrate into the fractured rock above rather than penetrating into the coal seam below. That is why the gas-induced damage to the coal seam from the baby decking should be less.
- The initial burden movement in baby-deck priming starts considerably high from the coal seam (depending on the charging

pattern) because it is driven by the main deck. When baby-deck is initiated (after 25 ms), the burden rock in front of the main deck is already in motion. This discontinuous burden-movement induces less shear stress on the coal seam resulting less damage.

12.2.4 Conclusion

The main purpose of baby-decking is to reduce explosive confinement at the toe in order to prevent coal damage due to overbreak. In baby-decking, the total explosive column is divided into two decks, a small baby-deck and a main deck. The primers and timing of down-the-hole delays are designed in such a way that the baby-deck is initiated 25 ms after the main deck. The amount of explosive charge in baby-deck would depend on the hole depth, hole diameter, explosive type and rock properties.

12.3 SEGREGATION BLASTING

12.3.1 Introduction

It is generally observed that a thin layer of coal parting (0.5 to 1.2 m) sandwiched between overburden rocks is generally lost in the process of blasting the overburden benches for extracting the main coal seam(s). Extracting such coal in small working benches is time consuming and non-optimistic from the economic point of view. At the same time, if not extracted, such coal may be a source of spontaneous heating besides creating danger to unauthorized manual pickers.

Segregation blasting technique encounters the difference in densities of coal and overburden and their impact sensitivity at different energy levels so that when the composite medium is blasted, coal and overburden are separated and thrown away at different distances.

12.3.2 Mechanism of Segregation Blasting

The literature surveyed has indicated that the basic principle involved in segregation blasting is the concept of projectiles of the blasted materials which firmly depend on energy utilization at different rock layers of varying densities.

Russian scientists have proposed mathematical approaches to explain the phenomenon of single-hole blast separation. Chernigovskii (1985) concluded that the degree of retarding effect of the medium on a fragment in motion depends on its linear size, density and speed of fragmentation. Borovikov and Vanyagin (1995) showed from vector analysis and analysis of projectile movement, expressed in quadratic form, that the horizontal displacement of rock material depended on its density and initial velocity.

Hence rock, being denser and bigger in fragment size than coal, encounters greater air resistance and would therefore, require a greater initial velocity for the same displacement as coal. Therefore if the optimum explosive is distributed along the blastholes for proper fragmentation, it would be observed that rock is thrown at a shorter distance than coal. Keeping this concept in view, CMRI scientists (Mandal et al., 2001) derived a few mathematical equations for determining horizontal displacement 'X' and acceleration 'a' due to particle motion.

The basic idea of segregation blasting relies on distributed charge at specified locations in the blastholes. The advantage of distributed charge over concentrated buried charge is that the former helps rock blasting in directivity and accuracy of throw. The distributed charges move the entire volume of rock with controlled speed except at the corners whereas the chamber charges have maximum velocity at the centre and dies down towards the corner.

The movement of any blasted volume of rock is subjected to three motions viz. a) uniform and rectilinear in the direction of motion, b) uniformly accelerated motion under the action of gravitational force and c) the gravitational force. For uniformly accelerated motion, the maximum height traversed can be determined from equation (12.1) and standard velocity-displacement equation (12.2). Eliminating 't' from equations (12.1) and (12.2), the required initial velocity responsible for the movement of flyrock to a pre-determined location can be determined by equation (12.3).

$$l = gt^2/2 \quad (12.1)$$

$$V_0 = D_s/t \quad (12.2)$$

$$V_0 = D_s/\sqrt{(2l/g)} \quad (12.3)$$

Where, D_s = Slant distance, (m)
l = Slant height gradient (m)
t = Time in seconds

Moreover from the conservation of energy, the following equation (12.4) holds good. Where, the left side denotes the kinetic energy per cubic metre of rock and the right side the blast energy required on throwing a cubic metre of rock in forward direction.

$$\rho V_0^2/2 = q e_{exp} \eta \quad (12.4)$$

Where, ρ = density of rock to be blasted, (kg/m^3)
V_o = initial velocity required, (m/s)
q = specific charge consumption, (kg/m^3),
e_{exp} = specific energy of explosives (J/kg), and
η = blast efficiency.

12.3.3 Air-Drag

It has been observed that air-drag exerts a considerable influence on bodies of small linear dimensions as they move in the air at high speed. In the case of a rock fragment of diameter 0.08 m, the horizontal range in comparison to the movement in vacuum decreases by 4.5 times. Similarly, for rock diameter of 0.5 m it reduces to almost half of its value than in vacuum. Therefore, it can be concluded that the degree of retarding effect of the air on a fragment in motion depends on its linear size, density and speed of fragmentation.

12.3.4 Law of Resistance

All bodies moving in gaseous or a fluid medium encounter a resistance to their movement. This resistance depends on the viscosity of the medium as well as the pressure developed on the surface of the body due to deflection and retardation of the flow of the medium. This resistance on account of viscosity of air is known as drag. Air-drag is expressed by the following mathematical equation (12.5).

$$F_{air} = C_x S_f (\rho_{air} V^2)/2 \qquad (12.5)$$

Where, C_x = drag coefficient,

S_f = area of the middle section of the fragment perpendicular to the velocity vector (m^2)

ρ_{air} = density of air (kg/m^3)

V = velocity of the particle in motion in m/s.

The drag factor b_d depends on the shape and the mass of the fragment [equation (12.6)]. It has also been observed that the lesser quantity of j_{air}, the greater will be the mass of the rock fragment for the same dimension i.e., the greater will be its density [equation (12.7)].

$$j_{air} = b_d V^2 \qquad (12.6)$$

Where, j_{air} = deceleration due to air-drag.

$$b_d = C_x (\rho_{air}/2)(S_f/M) \qquad (12.7)$$

12.3.5 Correction for Air-drag

Therefore, taking air-drag into account and to achieve a projection range equal to the range in vacuum, the required specific charge consumption of explosive (Q) can be derived from equation (12.8).

$$Q = \rho . V_0 / (2g\eta \cdot e_{exp}) \qquad (12.8)$$

Therefore, any particle in motion containing initial motion ($mv^2/2$) is subjected to the action of gravitational forces and air resistance. Therefore, considering the opposing forces, as indicated above, the system in motion can also be described by differential equations along the rectangular co-ordinates *x-y* (Borovikov and Vanyagin; 1995). They are:

$$m\frac{d^2x}{dt^2} = mg\sin\alpha - \phi\cdot\vec{V}^2\cos\theta \tag{12.9}$$

$$m\frac{d^2y}{dt^2} = -mg\cos\alpha - \phi\cdot\vec{V}^2\sin\theta \tag{12.10}$$

$$\vec{V}^2 = \left[\frac{dx}{dt}\right]^2 + \left[\frac{dy}{dt}\right]^2; \ \cos\theta = dx/(\vec{V}dt);\ \sin\theta = dy/(\vec{V}dt) \tag{12.11}$$

φ = Coefficient of air resistance
α = Angle of inclination of bench toe to the horizontal plane (deg.)
θ = Angle of velocity vector with the horizontal axis (deg.)
m = Mass of the elemental particle in motion (kg).

Putting the values of $m = \rho_r W S$, and $\varphi = C_s \rho_{air} S/2$, can be written as:

$$\frac{d^2x}{dt^2} = g\sin\alpha + \left[\frac{C_s\rho_{air}}{2\rho_r W}\vec{V}\frac{dx}{dt}\right] \tag{12.12}$$

$$\frac{d^2y}{dt^2} = -g\cos\alpha + \left[\frac{C_s\rho_{air}}{2\rho_r W}\vec{V}\frac{dy}{dt}\right] \tag{12.13}$$

Where, ρ_r = density of rock, (kg/m^3)
ρ_{air} = density of air, (kg/m^3) and
S = a cross-section of the elementary mass m, (m^2)

From the above five equations (12.9, 12.10, 12.11, 12.12 and 12.13) and earlier equations (12.1, 12.3, 12.4, 12.5, 12.6, 12.7 and 12.8), it can be concluded that the horizontal displacement of rock material due to blasting depends upon the density and the initial velocity. Knowing the rock to be denser and greater in fragment size than coal, due to blasting, it will be exerted upon greater air resistance and hence will require more initial velocity for the same displacement. Therefore if optimum explosive is distributed along the blast holes for proper fragmentation, it can be presumed that the higher density rock will be thrown at a smaller distance than that of the lower density coal.

12.3.6 Case Study

Good quality coking coal occurring at shallow depths was worked extensively in the early days of coal mining in India, but in combination with the propensity of the seams to spontaneous heating this unplanned mining has made fires, a threat to most of the mines of the Jharia coalfield. Lakshmi opencast mine of the Bhowra Area of Bharat Coking Coal Limited

(BCCL), a subsidiary of Coal Indian Limited (CIL), fell victim to such ill-organized private operations and had to be excavated to protect existing mines and other important structures that surround the fire area.

The Lakshmi mine of Bararee Colliery in Bhowra Area was formerly worked in two lifts by underground method and is at present worked by opencast method to protect the surrounding operations and other structures from fire due to spontaneous heating in the underground workings. The overburden comprised 2.5-4 m of alluvial soil at the top and 9-10 m massive sandstone with intercalations of shale bands. Within the 10 m of overburden, 3.0-3.5 m below the massive sandstone, was the 1.2 m thick seam XIIIA, which was a steel-grade coking coal. Below the composite strata, about 14 m below surface, seams XI and XII combined to form a 14 m thick single seam.

The top alluvium was excavated without blasting by backhoe and the composite strata of sandstone and coal were removed by drilling and blasting. Since the area was on fire, a continuous flow of water from the nearby surface-mine water drainage canal was used for quenching. The area where the fire was exposed, blasted muck was spread to minimize contact with the atmosphere, and within the coal galleries artificial smothering was employed. Pumps and flexible ducts provided a water supply from the canal.

Drilling was undertaken with a 110 mm bit. The blasts were designed with 2.2 m of burden and 9.5-10.0 m depth with a hole spacing of 2.8 m. Powergel-1 and 2 were selected as the explosives with due regard for their thermal behaviour and were charged in 83 mm diameter, 2.78 kg cartridges. The area to be drilled and blasted was saturated with water for about 24 h before drilling and for at least 12 h before charging of the holes and also during charging. The hole temperatures were measured at 30 minutes intervals both before and during charging. Care was taken to ensure that after fully quenching with water the temperature of the holes did not exceed 80°C; if, however, the temperature remained higher, the holes were left uncharged. The maximum number of holes charged was limited to 15 so that the charging and blasting operation could be completed within 2 h. The massive sandstone was heavily fractured as a result of the contraction and expansion caused by repeated heating and cooling with water. Because of this, optimum fragmentation was produced at a low charge factor, which varied between 0.35 and 0.40 kg/m^3 (Mandal et al., 2001).

12.3.7 Experimental Trials

The experimental trials were designed so as to take into account the basic parameters of detonation, shock or stress wave propagation, gas pressure expansion, mass movement during fragmentation and displacement of material. Attention was given to previous work on the role of the decking length in softer strata in the optimization of rock mass fragmentation and on the movement of softer strata caused by venting of gaseous energy in deep-hole blasting. On the basis of these previous findings, and the theory of blasting projectiles, several trials were conducted in which the explosive and decking lengths in the blastholes were varied so as to arrive at the desired projectile movement of the rock masses. The differential projectile movement was predicted from the difference in the specific gravities of the two types of materials—rocks mass, 2.2-2.8; and coal seam, 1.4-1.6. The experimental trials on segregation of the splinter seam were conducted by single-shot blasting of the composite strata (Mandal et al., 2001).

By proper placement of a specific quantity of explosive in the blasthole and decking within the softer layers to confine the gases for better borehole pressure, successful segregation of the splinter seam was obtained, as shown in Plate 12.1. Placement of the explosive above and beneath the coal seam possibly increased utilization of the explosive energy at the point of collision of the two stress-strain waves, causing maximum throw of the coal with optimum fragmentation of the waste. The charging pattern was followed for proper segregation, except when the splinter seam was 3.5 m below surface, an explosive quantity of 1.85 kg at the floor of the splinter seam was found to give the best segregation. It was also observed that about 60% segregation of the sandwiched coal by single-shot blasting was achieved in single-row blast, but this was reduced to about 40% in double-row blasts. This was possibly because the throw of coal from the second row was on the waste of the first row, thereby mixing the coal from the second row with the waste of the first.

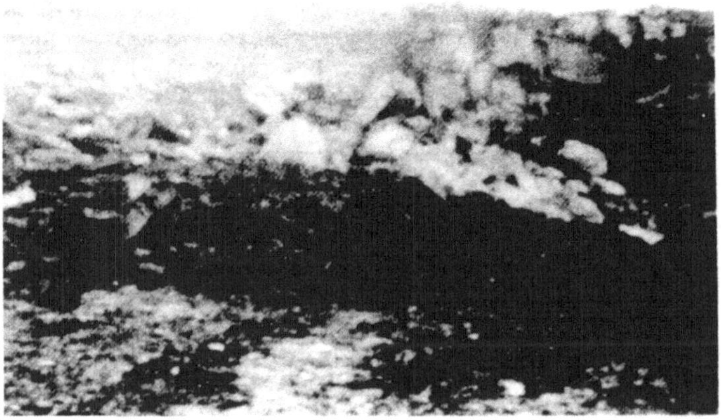

Plate 12.1 A view of segregation of coal obtained in Lakshmi opencast mine of BCCL (after Mandal et al., 2001)

12.3.8 Conclusion

Considering the densities of rock and coal fragments as the former being greater—and that rock fragments are larger than coal fragments, the gravitational force for rock fragments would be greater than for coal fragments. Similarly, when air resistance is considered as the retarding force against the projectile movement of the fragments, it is clear from equation (12.7) that rock fragments, having a greater surface area, will be subjected to greater air resistance than coal fragments. However, this particular area requires more attention and extensive field trials in varying rock-geologic conditions to achieve a greater throw of coal from multiple-row blasts with ease and much more confidence. High-speed videographic analysis and laser profiling system can be effective tools for such study.

12.4 ADDITIVES IN ANFO FOR OPTIMUM BLASTING

12.4.1 Introduction

Field investigations have revealed that sawdust can be used as an efficient additive to ANFO for carrying out production blasts in limestone mines. Sawdust can be used at a reasonably reduced cost without sacrificing fragmentation, muck profile and wall control. Such additives lower the detonation pressure of ANFO by acting as low-grade fuel that reacts or burns more slowly in the presence of ammonium nitrate than does fuel oil. On detonation, it eventually forms near-ideal oxygen balance that helps in liberating maximum energy to break the surrounding rock. It also forms a low-density stable mixture that helps in reducing loading density (kg/m) of explosive in the blasthole. Experiments performed at Hirmi Limestone Mine, M/s Larsen & Toubro Ltd. and at the Lambasta Limestone Mines of M/s Jaypee Rewa Cement, showed excellent acceptance of the technique in a highly competitive market where cost reduction has been a major area of thrust. Successful adaptation of the technique in Kota Stone mining sector of India also shows its importance.

12.4.2 Scientific Background

Hard and massive rock types require the creation of new fractures for adequate fragmentation. These new fractures are generated through shock energies associated with high explosion pressures (velocity of detonation and high density), whereas rock types that are already cracked depend more on the heaving/displacing action provided by the gas pressure for breaking. This may be best accomplished by an explosive with a lower detonation velocity and density (Hustrulid, 1999). As a general rule, rock that is relatively not very hard or that exhibits natural cracks and bedding planes, a low-order velocity of detonation (VOD) may be preferred which would yield less shock energy but greater heave energy. Usually, the strain component of the shock energy produces the pressure pulse that causes fracture and cracking. The gas energy produces effective work in propagating the cracks and fragmenting the rock.

The colour of post detonation fumes can be a good measure of energy release of an explosive product. Light grey indicates near-ideal oxygen balance, whereas yellow or rust-coloured fumes indicate an inefficient reaction, which may be due to an oxygen-positive mixture. Oxygen-negative mixture produces dark grey fumes and often leaves carbon on borehole walls. For a non-ideal mixture, even for higher VOD explosive products, the blasthole may not have sufficient energy to break the rock properly in front of the hole. Rather, such a situation may lead to airblast, flyrock or undesired rock breakage.

In limestone mining there are situations where bulk blasting agents, that have strengths less than bulk ANFO, can be substituted to achieve better blasting performance and cost effectiveness. This can be achieved by using smaller-diameter charge with a smaller degree of confinement than that for the fully charged blast hole (in other words "decoupling"), or this can be achieved by using some additives in the conventional ANFO to form a stable homogeneous mixture.

12.4.3 Detonation Velocity and Pressure of ANFO/Sawdust Mixture

Wilson and Moxon (1988) conducted both laboratory and field tests in which low-density materials, such as polystyrene, bagasse, sawdust, perlite, were mixed with ANFO and detonated. Their study basically concentrated on the process of reduction of explosion pressure and, thereby, the amount of undesirable damage to the surrounding rock mass. A practical and field usable recommendation was not available in their work.

Detonation pressure of explosive can best be estimated by—

$$P_{DET} = 0.25(10^{-3})\rho\,(VOD)^2 \qquad (12.14)$$

where

P_{DET} is the detonation pressure (MPa)
ρ is the explosive density (g/cc) and
VOD is the detonation velocity (m/s).

The explosion pressure which would be applied on the hole wall is approximately half of the detonation pressure. As the explosion gases expand to come into contact with the borehole wall, the pressure drops. Assuming adiabatic expansion, one can quote (Hustrulid, 1999)—

$$P_h = P_e\,[D_e/D_h]^{2\gamma} \qquad (12.15)$$

where

P_h is the effective pressure at the hole wall (MPa),
P_e is the explosion pressure for ANFO (MPa),
D_e is the diameter of the explosive (mm),
D_h is the diameter of the hole (mm) and
γ is the equivalent adiabatic expansion factor for the pressure range, $P_h \to P_e$.

Figure 12.12 shows the plot of some of the data generated by Wilson and Moxon (1988) in association with the data generated by CMRI. This figure shows the variations of detonation velocity (m/s) and consequent pressure (MPa) at various percentage of sawdust mixture. Field experiments showed that 30% volumetric mixture of sawdust could be taken as optimum from both economic and performance points of view. Figure 12.13 shows that the velocity of detonation (VOD) of ANFO is dependent on the diameter of the hole. The trend of increasing VOD with drillhole diameter is stabilized at a drillhole diameter of around 250 mm. ANFO (density 850 kg/m^3) when mixed with sawdust (density 190 kg/m^3) in the ratio of 10% (by weight) (or 68%:32% by volume), the mixture had a density of 645.40 kg/m^3. The loading density (kg/m) of such a mixture was reduced by nearly 13% to 15% with respect to ANFO for 152 to 165 mm drillhole diameters. It indicated that better charge column could be obtained using lesser amount of explosives.

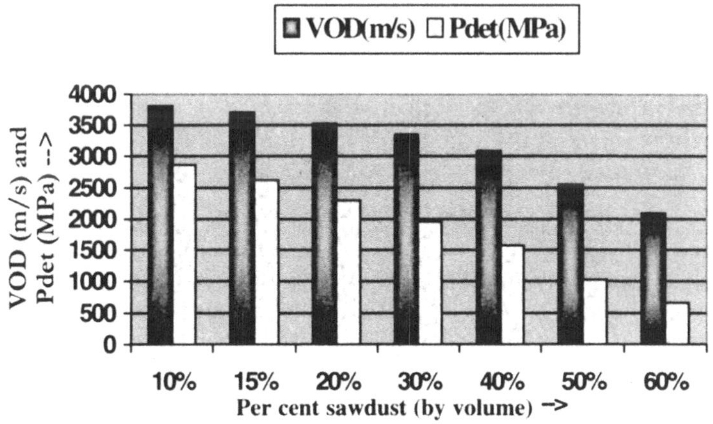

Fig. 12.12 Effect of sawdust mixture on detonation velocity (m/s) and pressure (MPa) of ANFO

12.4.4 Properties and Chemical Reaction of ANFO/Sawdust Mixture

Cellulose is the principal constituent of the cell walls of all higher plants. In terms of absolute amount, it is probably the most abundant organic

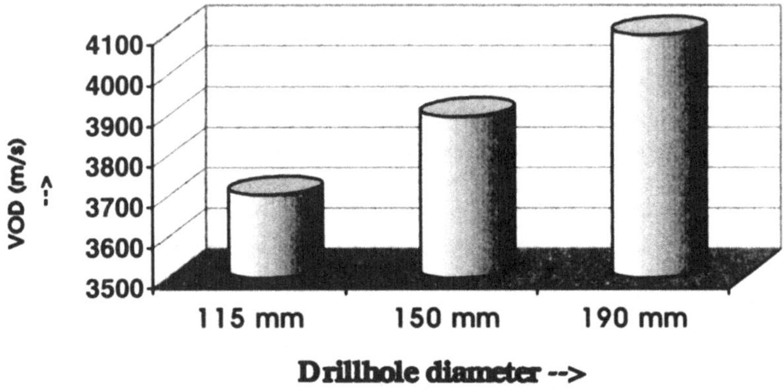

Fig. 12.13 Effect of drillhole diameter on the VOD of ANFO

compound present on the earth. Most of the cellulose in these plants occurs with or is encrusted by other materials. Some fibres however, such as those of the cotton plant, are practically pure cellulose. Molecular formula of cellulose is $C_{18}H_{29}O_{16}$, with a molecular weight 501.4222 (Foster, 1990).

Most ingredients of explosives are composed of the elements—oxygen, nitrogen and carbon, plus certain metallic elements (aluminium, magnesium, sodium, calcium etc.). The composition of an explosive is varied and balanced chemically to produce the desired effects on blasting. Because an explosion is essentially a combustion process, the chief criterion of efficient energy-release is the oxygen balance; that is, an oxygen-balanced explosive is one that has optimal energy release. Zero oxygen balance is the point at which explosive mixture has sufficient oxygen to completely oxidize all the contained fuels but no excess oxygen to react with the contained nitrogen (Table 12.2) (Hartman, 1987).

Table 12.2 Oxygen balance of ANFO/sawdust mixtures (Pal Roy and Mohanty, 2002)

Ammonium nitrate, (%) $N_2H_4O_3$	*Fuel oil, (%) CH_2*	*Cellulose, (%) $C_{18}H_{29}O_{16}$*	*C, gram-atom /100 g*	*H, gram-atom /100 g*	*N, gram atom /100 g*	*O, gram-atom /100 g*	*Oxygen balance, gram-atom /100 g*
94.48	5.52	x	0.3935	5.5084	2.3607	3.5411	0.0
85.89	5.02	9.09	0.6842	5.5336	2.1461	3.5092	0.626025
88.66	2.25	9.09	0.4870	5.2775	2.2152	3.6128	0.0

12.4.5 Procedure of Mixing Sawdust with ANFO

Sawdust can be mixed easily with ANFO either in a bowl truck or manually. In the case of manual mixing, the sawdust is spread onto a thick polythene sheet. Ammonium nitrate prills are then placed on top of the sawdust to form a heap. Then, approximately 6% fuel oil (by weight of ANFO) is dropped on the prills, and the ingredients are uniformly mixed.

12.4.6 Experimental Blasts

Experiments were conducted in Paraswani Limestone Mine at Hirmi and Naubasta Limestone Mines at Statna, located in the Chhattisgarh and Madhya Pradesh Sates of India, respectively. The geological setting of the Paraswani Limestone Mine consists of overburden soil, upper green shale, grey limestone, pale pink limestone and chocolate brown limestone. The average compressive strength is 655 kg/cm^2 (9,300 psi), the tensile strength is 32.1 kg/cm^2 (456 psi), and the density is 2.50 t/m^3 (0.071 ton per cu ft).

Seven experimental blasts were conducted at the site using 5 to 10% (by weight) sawdust with the ammonium nitrate prills and fuel oil mixture. One intermediate deck of 0.75 to 1.0 m (2.5 to 3.3 ft) was used while charging explosive in the holes. The depth of the 152.4 mm (6 in.) diameter holes varied from 5.5 to 8.5 m (18 to 28 ft). Shock tubes as well as detonating cord systems were used in such blasts. The burden varied from 3.0 to 4.5 m (9.8 to 14.8 ft), while the spacing varied from 5.0 to 6.0 m (16.4 to 19.7 ft). In general, the top stemming was kept between 2 and 2.25 m (6.6 and 7.4 ft) to eliminate boulder formation from the top column. An overall charge factor of 0.37 kg/m^3 (0.023 lb per cu ft) was used during charging of holes.

The geology of the Naubasta Limestone Mines of M/s Jaypee Rewa Cement mainly consists of top 3 m (9.8 ft) of overburden soil, 6 m (20 ft) nodular limestone, 5.5 m (18 ft) of lower-grade limestone and 5 m (16 ft) high-grade limestone. Five experimental blasts using ANFO/sawdust mixture (10% sawdust by weight of ANFO) were conducted in low-grade

Plate 12.2 Fine fragmentation produced in an experimental blast using ANFO-sawdust mixture

Plate 12.3 View of light grey coloured fumes appeared in ANFO-sawdust mixture

limestone and high-grade limestone benches. The depth of the 152.4 mm (6 in.) diameter holes varied from 5.5 to 8.25 m (18 to 27 ft). Burden and spacing varied from 3 to 4.5 m (9.8 to 14.8 ft) and from 4.0 to 6.0 m (13.1 to 19.7 ft), depending on the hole depth and location of the blasts. Out of these five blasts, one blast in the bottom bench was conducted where all the holes were watery. In which case, two polythene bags were used to pour ANFO/sawdust mixture in the form of cartridge and pushed them inside the hole by stemming rod and stemmed them properly.

12.4.7 Observation

All the experimental blasts that were performed, produced very good results in respect of fragmentation and wall control. The size of fragments in such blasts was less than 0.65 m^3 (23 cu ft). It was interesting to note that the post-blast fumes in conventional blasting with ANFO (added with booster) produced dark yellow and rust-coloured fumes, indicating an inefficient reaction, which could be due to an oxygen positive mixture. As sawdust acts as a low-grade fuel that reacts or burns more slowly in the presence of ammonium nitrate than fuel oil, the oxygen positive mixture (usually obtained in conventional ANFO shooting) gets balanced in the ANFO/sawdust compound. This could be the reason that detonation of ANFO/sawdust mixture produced light-grey post-detonation fumes obtained in experimental trials. It is a known fact that the maximum energy is released when light-grey fume is seen on detonation.

12.4.8 Conclusions

The use of sawdust can save substantially on explosive cost without hampering blast performance as measured in terms of fragmentation and wall control. At present, many limestone mines and Kota stone quarries in India are now using this additive to save the explosive cost by nearly 10%. The main aim of this section is to establish the scientific reasoning of the usage of sawdust in ANFO for performing economic and efficient production blasting. The following points are raised to make the best use of the system:

1. Sawdust can form a stable homogeneous mixture with ANFO.
2. A sawdust/ANFO mixture would have a lower detonation pressure than ANFO alone. The magnitude of the detonation pressure can be controlled by the proportion of the sawdust.
3. The mixture does not segregate when loaded into a borehole.
4. The procedure can reduce explosive costs by nearly 10% in any blast without compromising fragmentation and productivity.

5. The mixture was found effective even in water-filled holes. In water filled holes, two polythene bags can be used to pour the mixture inside the hole in the form of cartridges.
6. 10% mixture (by weight) of sawdust with ANFO (or 68%:32% volume) produced excellent results in all the experimental blasts conducted by CMRI during experimental trials.
7. In place of sawdust, rice-husk can also be used as another economic additive. Its success would also help the neighbouring farmers by providing a market for what is now a waste material.

12.5 INDUCED CAVING BY BLASTING

12.5.1 Introduction

Induced caving by blasting, besides its extreme importance during extraction in depillaring panels of underground coal mines, did not receive much attention from scientists and engineers the world over. As such, by and large, no valued literature is available on this particular topic. To avoid the occurrence of airblast, this technique has become an integrated part of mining operation during depillaring of developed coal pillars by Blasting Gallery (BG) method. On experimental basis, systematically organized deep hole blasting was successfully carried out from underground split galleries of BG panels. A constructive package study which included optimization of drilling and blasting parameters, gas hazards, strata behaviour and ground vibration were undertaken as part of one S&T Project of the Ministry of Coal and Mines, Government of India. Strata behaviours during blasting as well as ground movements were critically investigated before implementation of blasting and charge loading parameters. All experiments were conducted using newly developed explosive and d-cord systems (Pal Roy et al., 2003).

12.5.2 What is Induced Blasting?

Induced caving by blasting (in short Induced Blasting) is essential to bring down the hanging roof in order to prevent air blast when the roof strata are difficult to cave in by its own weight. In Blasting Gallery (BG) method of extraction, induced blasting is generally carried out at regular intervals when the exposed roof span is about 120-190 sq. metres. Constant monitoring of roof convergence and load are very useful to get indications for carrying out the induced blasting operation. Jumbo drill is used for drilling of holes. Generally 5 to 8 number of holes (depending on the exposed roof span) with 15-22 m depths may be drilled through the roof of the developed gallery keeping 40-45° inclination to the horizontal at a distance of 2.5-4.2 m from the goaf edge. Such parametric values are entirely dependent on the seam thickness, height of gallery and inclination of the roof face. Powergel-101 (P_1-Group, Emulsion explosive) is the approved explosive while Super-cord or G-cord is the approved detonating fuse system for carrying out induced blasting. Explosive cartridges are placed only in the stone portion of the shot holes, leaving approximately 1.0 to 1.5 m from the stone and coal junction. The remaining portion of the hole is stemmed with bamboo spacers (or plastic spacers). All the holes are fired in a single round with instantaneous detonator. Before firing, five rows of girders are reinforced with extra props.

12.5.3 Results of Strata and Gas Monitoring

Field investigations, on experimental basis, were conducted at VK-7, GDK-10 and GDK-8 Incline mines of M/s Singareni Collieries Company Ltd. (SCCL), Andra Pradesh and Chora-10 Pit Colliery of M/s Eastern Coalfields Ltd. (ECL), West Bengal.

During field investigations, it was found that the sandstone roof was very massive in nature and therefore difficult to cave by its own weight. The main fall generally occurred after 1600 sq. metres area of exposure in the worked out panels, whereas local fall did not occur after 120 m^2 area of exposure. These were posing the danger of air blast as well as overriding of the pillars. Therefore, induced caving by blasting was essentially required to bring down the overhanging roof.

Comprehensive gas survey was conducted at the site of blasts and district returns before and after ringhole and induced blasting with a view to assessing the safe underground working environment during blasting operations. The results are given in Table 9.2 of Chapter 9. Strata monitoring of roof rock were conducted using load cells (mechanical type) and convergence indicators (Rheostat type). Initially load cells were installed in the level galleries at 20 and 25 m from the working faces (goaf edge). Leak-proof hydraulic props were used for installation of load cells. When there was water in the level gallery, instead of installing near the floor, load cell was installed near the roof of the gallery. In order to compare the roof load with that of convergence; a pair of instruments was installed adjacent to each other. The results of strata monitoring are shown in Table 9.3 of Chapter 9. Critical convergence, which is defined as the convergence beyond which the immediate roof in developed galleries may pose a stability problem in normal conditions, is given in Section 9.12 of Chapter 9. Blast Damage Index (BDI), which is denoted by 'D_{ib}', is discussed, in detail, in Section 9.14 of Chapter 9.

12.5.4 Design Patterns

During experiments, two different types of patterns were tried. Thorough descriptions of the patterns are given in figures 12.14 to 12.17 (Pal Roy et al., 1996 and 2003).

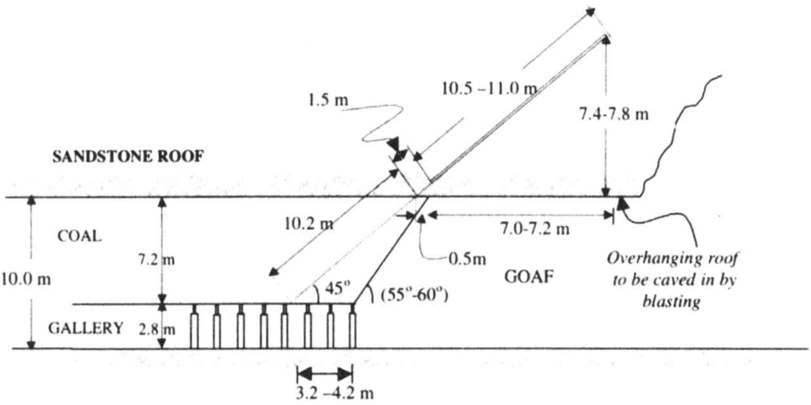

Fig. 12.14 Sectional view of single-row induced blasting pattern

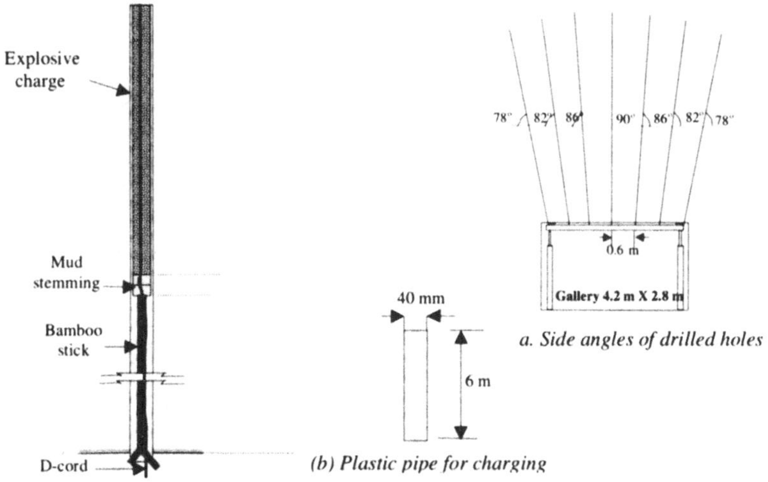

Fig. 12.15 Charging pattern of holes followed on experimental basis

12.5.5 Fragment Size Analysis

Fragment size analysis of the broken rocks in one of the experimental blasts (following single-row design pattern) was carried out using FRAGALYST image analysis software developed by CMRI. The contour

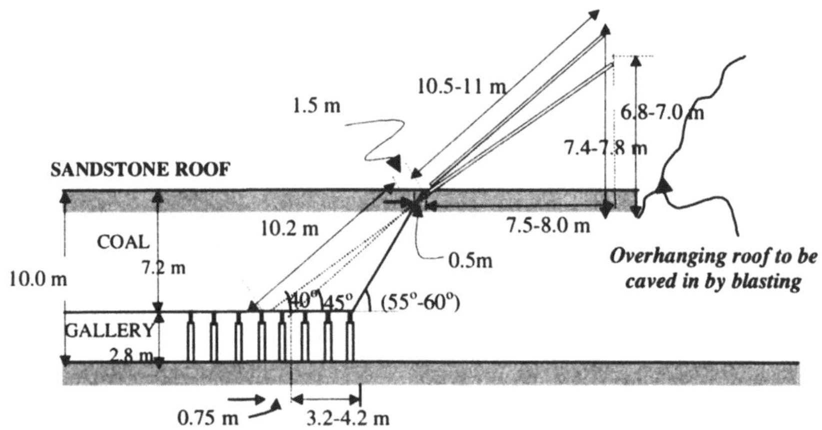

Fig 12.16 Sectional view of double-row staggered induced blasting pattern

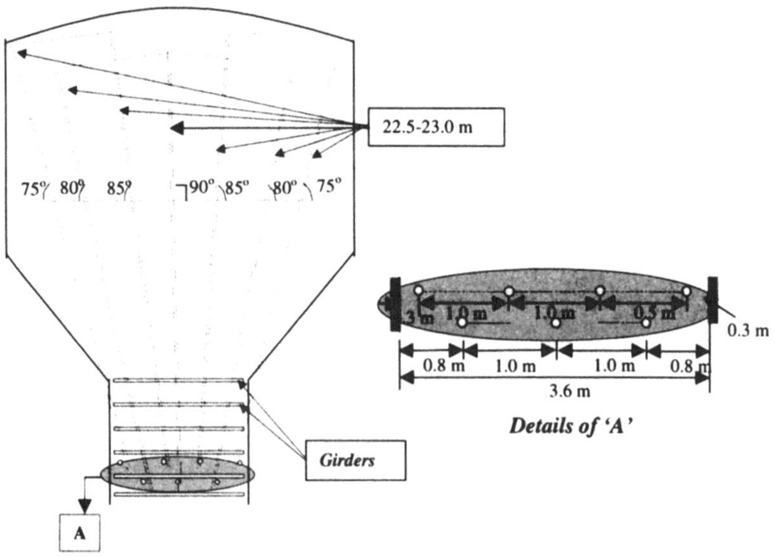

Fig 12.17 Plan view of induced blasting pattern described in figure 12.16

of the fragments after image processing of the broken rock is shown in figure 12.18. These images were then calculated using Rosin-Rammler Distribution (Figure 12.19) to get the actual size distributions. Uniformity index, characteristics size and mean fragment size were obtained as 2.633, 0.2577 m and 0.2242 m (K_{50}) respectively. Plate 12.4 shows the broken roof rock when blasting was conducted using the double-row staggered pattern as depicted in figure 12.16.

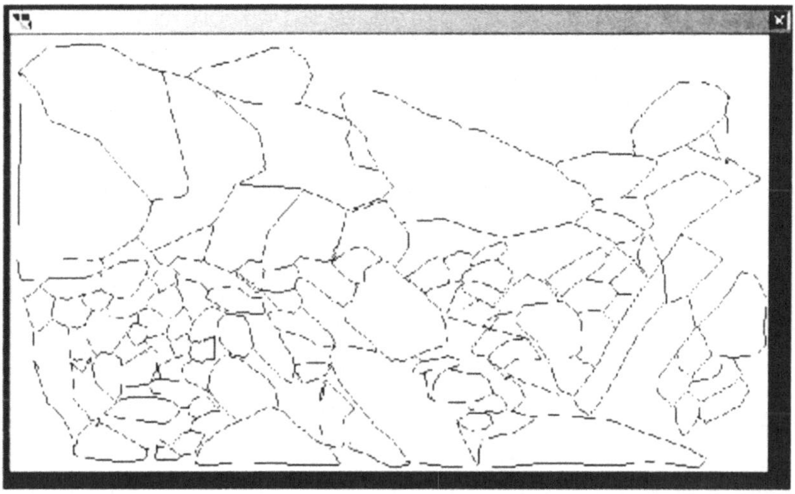

Fig. 12.18 Object contours of the fragments obtained by image processing

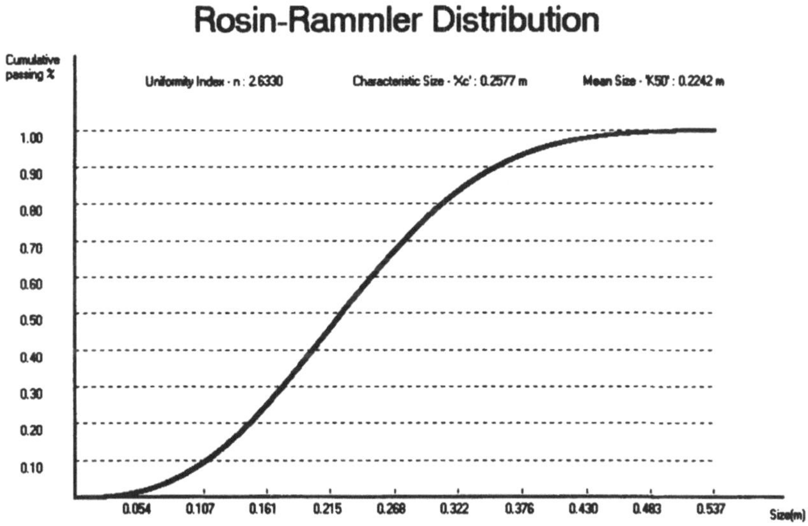

Fig. 12.19 Fragment size distribution of the blast shown in figure 12.18

12.5.6 Software Developed

A comprehensive software named 'INCAB' (INduced CAving by Blasting) was developed for designing of drilling and blasting patterns as well as performance evaluation of the blast. 'INCAB' has the following features—

- A comprehensive, menu-driven, GUI software package in JAVA (Swing) language under windows environment. Operative in any standard Pentium-based IBM compatible system with VGA display unit.

Plate 12.4 View of one fully packed level gallery after induced caving by blasting

- Provides perfect methodology and directives for proper implementation of blasting and charge loading parameters related to induced caving by blasting in Blasting Gallery panels of underground coal mines.
- Provides analysis of blast design, optimization of design parameters, sectional view of drillholes and geometry (in scale), charge loading parameters, Blast Damage Index (BDI) and consequent safety zones, histogram of gas survey, blast performance parameters like area and volume of rock fall as well as charge factor.

12.5.7 Conclusion

CMRI-experimented induced blasting trials led to the generation of vast amount of base data for future research in this particular area. The trials showed good breakage of overhanging roof rock and ensured safe underground environment. However, before conducting any such induced blasting operation, special permission of the Directorate General of Mines Safety (DGMS) is essentially needed. The description and results given in this section can only provide some sort of scientific guidance for any further work in this particular area.

12.6 BLAST CLOSURE

12.6.1 Introduction

The objective of mine closure is to leave sites in a condition which is safe, stable, and limits further environmental impacts so that mining tenements can be relinquished for alternative land use. The concept of mine closure was mentioned in the policy statement under 'National Conservation Strategy and Policy Statement—Environment and Development, 1992', which directs that the mined out area should be reclaimed and rehabilitated concurrently with the ongoing mining operations to ensure ecological restoration of the affected areas and that the abandoned areas should be rehabilitated in a phased manner so that the land resources can be brought back under productive use (Bhatia and Hasan, 2003). As per policy, the government is to be informed before the mine is closed. Adequate attention is also required on the issue of environmental protection, safety and socio-economic aspects.

In India, there are many examples of abandoned mines, both in coal and non-coal sectors. It is essentially required to see that harmful gases are not released from any abandoned mine to the atmosphere, thus posing an environmental problem. In such cases, these workings and mine openings are to be properly sealed to avoid any untoward incident. If the mines do not have any economic or historic value but present a significant hazard, then they may be permanently closed by using controlled blasting method to prevent access and release of harmful gases to the atmosphere, so that in future, they do not require any inspection and maintenance.

"Mine closure" basically signifies the steps taken for reclamation measures of a mine or part thereof, commencing from cessation of mining or processing operations. And "abandonment of mine" means final closure of a mine either whole or part thereof when the mineral deposits within the mine or part thereof, have been fully extracted or when the mining operations have become uneconomic.

The concepts, definitions and issues related to mine closure revolve around the rehabilitation of disturbed lands to safe, stable and post-mining landform, which is suitable and/or acceptable to the community. The closure of a mine will not compromise environmental quality in the future and will limit the extent of any prospective liabilities for the operator, the government and the community (Sassoon, 1996).

Mine closure is a continuous series of activities that begins with pre-planning prior to the project's design and construction and ends with the achievement of long-term site stability and the establishment of a self-sustaining ecosystem (WMI, 1994).

In India, many mines are temporarily closed to become abandoned mines. Specific hazards in such mines include unstable rocks at the portal and within the underground workings, rotting timber, rusting metal, low oxygen levels, stagnant air, deep water, and other unknown factors. They also pose special problems to the government, industry as well as the health and safety of local communities.

Practically, these abandoned mine openings are dangerous for the public health and safety and pose an obstacle in further reclamation work.

However, to eliminate the hazards associated with abandoned mine openings and having no economical or historic value, the openings should be:

- Fenced to keep people away from the hazard;
- Barricaded to restrict access; or
- Sealed to prevent entry.

12.6.2 Reasons for Closure

Some compelling reasons for closure of a mine:

- Uneconomical (prices of minerals/metals are not satisfactory, grade of the deposit is too low, workings are too deep, nature of the ore poses problems in efficient metal extraction)
- Technical
- Environmental (acid mine drainage, reclamation and rehabilitation)
- Privatization
- Mineral deposit gets depleted
- Unsafe workings
- Mine management loses interest in further operations
- The mining lease expires, and no further renewal is granted by the government

12.6.3 Closure Planning

Mine closure essentially requires planning to ensure the following:

- Future public health and safety are not compromised;
- Environmental resources are not subject to further physical and chemical deterioration;
- The after use of site is beneficial and sustainable in the long term;
- A post mining land use compatible with the surrounding land use; and
- Adverse socio-economic impacts are minimized.

12.6.4 Methods of Closure

Before selecting any method to secure a mine opening, it is important to assess the level of hazard, the potential future economic or historic value of the mine, the associated cost and effective life of the selected closure method. Typically closure methods are classified as temporary, long term and permanent in nature.

(A) Temporary Closure Methods: Includes fencing, wire screening and steel grating. This method can remain effective for 20 years.

A.1: Fencing: Perimeter of openings fenced in conjunction with warning signs to restrict entry but required inspection and maintenance.

A.2: Steel Wire Screen: Steel wire screening placed across an abandoned mine opening restricts unauthorized entry, while still providing limited access to small wildlife. Entry is also possible for any important purpose in the future.

A.3: Steel grating: Steel grate closure methods restrict entry into abandoned mine openings while still providing access for small wildlife and airflow. Entry is possible by removing the barrier.

(B) Long-Term Closure: It includes pre-cast concrete caps, cast-in-place concrete caps, monolithic concrete caps, and native rock or concrete block bulkhead. Life of this method can remain effective for 30 to 50 years.

B.1 Concrete Cap Closures: It completely seals-off vertical shafts or raises, and permit reclamation and revegetation of the area over the mine openings. A pre-cast or cast-in-place concrete cap must be designed for each site-specific shaft or raise and should be constructed only where the shaft collar or raise openings lie in competent rock. Monolithic concrete caps are typically used to secure shaft or raise opening over which the surface soil or fill has begun to subside into the opening, although the mine opening is still covered and not visible. Revegetation is possible after completely sealing the abandoned mine openings.

(C) Permanent Closure Method: It includes backfilling and blasting.

C.1 Backfill closure: Backfill closure consists of completely sealing-off the mine openings with on-site imported fill. This will prevent access for exploration for archaeological purposes, and obviously block any natural airflow and wildlife. Backfill is suitable for vertical shafts, raises, adits, inclines, slopes and subsidence features. After backfilling, the opening is covered with soil or overburden followed by the revegetation of the area.

C.2 Blast Closures: The openings of abandoned mines sometime require filling with broken rock through the use of controlled blasting. The broken rock creates a permanent seal which completely restricts access into the abandoned mine opening. Before blasting work, conditions above and around the openings should be inspected for the knowledge of strata, opening structures, nearby residential or commercial sites for the designing of safe controlled blast pattern to contain vibration and flyrock within safe limits. All the blasting activities must be conducted in compliance with the blasting norms.

A specified design is required for each site depending upon the structure and geology of the adjacent strata. Based on an approved blasting plan, blast holes are drilled above and around the entrance of openings.

The blastholes are usually situated in points of difficult access. For this reason, the most suitable drills are jackhammers with a drill diameter of 38 mm. If conditions permit, blast holes should be drilled inside the mine opening to ensure more complete closure. For this purpose, safe commercial explosive (diameter 25-32 mm) with high density and detonation velocity is required. Near mine openings, generally foundations are built for the safety of openings and erection of surface structures. At the time of closure, these foundations need to be blasted. For blasting, the foundation is drilled with vertical blastholes and with lengths that depend upon its size. Burden (B) and spacing (S) are made equal. The specific charges and patterns recommended by Gustafsson (1973) in function with the material of the foundation are indicated in Table 12.3. The explosive charges are prepared by uniting pieces of about 50 g to a detonating cord and with a spacing that is in function with the density of the calculated charge. For controlling the flyrock and vibration, an initiation sequence with millisecond delay detonator is recommended. If the walls of the opening are made of bricks or concrete, the recommended patterns in function with the thickness of the brick walls are indicated in Table 12.4. Depending upon the quality of the material, consistency and risk of flyrock, the specific charge lies between 0.5 and 1.0 kg/m^3. Drilling lengths depend upon the thickness of the wall and mostly equal to the 2/3 of the thickness. Special attention is required to check the deviation of holes to prevent flyrock due to high concentration of charge at the bottom of holes. For concrete walls, the recommended patterns in function with thickness are indicated in Table 12.5. As stated earlier, the length should be 2/3 the thickness of the wall, but the specific charge should be in between 0.9-1.5 kg/m^3. In such type of blasting, some experimental tests should be made on part of the wall to determine the ideal patterns and charges. Table 12.6 lists a few globally cited abandoned mines, which were closed by blasting.

Table 12.3 Recommended specific charges and blast patterns (Gustafsson, 1973)

Material	*Powder factor* (kg/m^3)	*Drilling B=S (m)*
Non-reinforced concrete of poor quality	0.25-0.30	0.70-0.80
Non-reinforced concrete of good quality and material strength	0.30-0.40	0.60-0.70
Reinforced concrete with only surface reinforcement	0.60-0.75	0.55-0.60
Reinforced concrete with heavy reinforcement	0.80-1.0	0.50-0.55
Extra powerful reinforced concrete of military type	1.5-2.0	0.40-0.50

Table 12.4 Recommended patterns in function with the thickness of wall of bricks (Jimeno et al., 1995)

Thickness of wall (m)	*Pattern: B x S (m x m)*	*No. of rows*
0.35	0.30 × 0.30	2
0.45	0.35 × 0.35	2
0.60	0.45 × 0.45	2
0.70	0.55 × 0.55	3
>1.0	0.55 × 0.55	3

Table 12.5 Recommended patterns for non-reinforced concrete walls in function with thickness

Thickness of wall (m)	*Pattern: B x S (m x m)*	*No. of rows*
0.20	0.25 × 0.30	2
0.30	0.30 × 0.30	2
0.40	0.30 × 0.30	3
0.50	0.35 × 0.40	3

Table 12.6 Name of a few abandoned mines which were closed by blasting (Global citation)

Name of mine	*Location*
Black Eagle	T. 8S. 4E., Sec. 24. , 1/2 mile west of the gate on the Forest Road 2209
Ruth # Discovery Cut	T. 8S., R. 5E., Sec. 27., 1.5 mile east of Jawest Flats along Forest Road 2209

12.6.5 Conclusion

In India, there are a number of abandoned mine openings which require specific attention during closure so that public health and safety remain unaffected. It should not, at the same time, pose any obstacle in further reclamation work. As a task for the future, a few mining sites may be selected, for blast closures of mine openings which may be carried out in collaboration with DGMS and other regulatory authorities. Planning for such closure is the way to ensure that the environmental hazards are minimized and at the same time additional benefits of reducing the financial costs are achieved. It should be remembered that 'prevention is always better than cure'.

12.7 DEMOLITION BLASTING

12.7.1 Introduction

Demolition or dismantling of a structure by blasting in a controlled fashion is essential only when the progress of work has to be smooth and fast and/or when there are other structures in close proximity to be protected.

The controlled demolition technique was developed in Europe during the reconstruction of the cities destroyed in World War II and, owing to its advantages; it was later extended to the rest of the world (Jimeno et al., 1995).

Jimeno et al. (1995) classified the demolition job into three important aspects like a) the breakage of the construction elements by eliminating the unions and sectioning rigid parts, b) division and distribution of charges and c) proper choice and application of the initiation sequence for provoking the directional fall of the structure.

12.7.2 Basic Design Patterns

Extensive site-specific studies on controlled demolition, covering both fundamental and technical aspects, are available in global citation (Internet). In Sweden, USA and China, a large group of scientists are specially involved in demolition jobs in high-risk areas. A good summary of fundamental and technical aspects of demolition blasting is available in Jimeno et al. (1995) and Guftafsson (1973). Their recommendations, in brief, are summarized in Tables 12.3 and 12.7.

Table 12.7 Recommended pattern in function with the thickness of reinforced concrete walls (after Jimeno et al., 1995)

Thickness of wall, m	*Pattern: B x S (m × m)*	*Number of rows*
0.35	0.25 × 0.25	2
0.45	0.30 × 0.30	2
0.60	0.45 × 0.45	3
0.70	0.50 × 0.50	3

12.7.3 Demolition of Piers

CMRI conducted one demolition job wherein, piers of an old bridge, each 11 m tall, located within 4.5 to 14.5 m from the existing Nonia Bridge on G. T. Road near Asansol Bye-pass junction, were successfully demolished (Singh et al., 1998). Two factories were located within 40 m distance to the structures. There were in total seven piers out of which five were of plain cement concrete and the rest were made with stone blocks and cement mortar. Since, the plain cement concrete blocks were difficult to be demolished manually, attempts were made for its safe, quick and efficient demolition by blasting. These piers were to be demolished for construction of a new bridge without disturbing the existing bridge and at the same time, avoiding any untoward accident due to flyrock in the two nearby factories (one glass and the other of gypsum), located within 40 m distance from the site (Plate 12.5).

Plate 12.5 View of piers of an old bridge before demolition

In view of directional fall by displacing the centre of gravity, experimental blasts were conducted using 47 number of holes and 17.55 kg of total charge to demolish a single pier. The distribution of delay numbers and associated charges are shown in Table 12.8 and figure 12.20. All the holes were drilled horizontally for a depth of 1.5 m with burden and spacing of 0.7 m and 0.8 m respectively. Prior to the demolition of the structures, ground vibrations due to movement of heavy vehicles were measured on the floor of the bridge. It ranged between 5 and 9.25 mm/s. While the level of vibration measured during blasting varied between 5.98 and 10.29 mm/s, and the bridge was completely safe without any adverse effect (Plate 12.6).

Table 12.8 Delay numbers and associated charges used in the demolition of piers

Short Delay Nos.	*No. of holes*	*Charge per delay (kg)*
Z (Short delay detonator)	12	3.90
I (-do-)	6	2.34
II (-do-)	5	1.95
III (-do-)	5	1.95
IV (-do-)	5	1.95
V (-do-)	4	1.56
VI (-do-)	4	1.56
1 (Long Delay)	3	1.17
2 (Long Delay)	3	1.17
Total	47	17.55

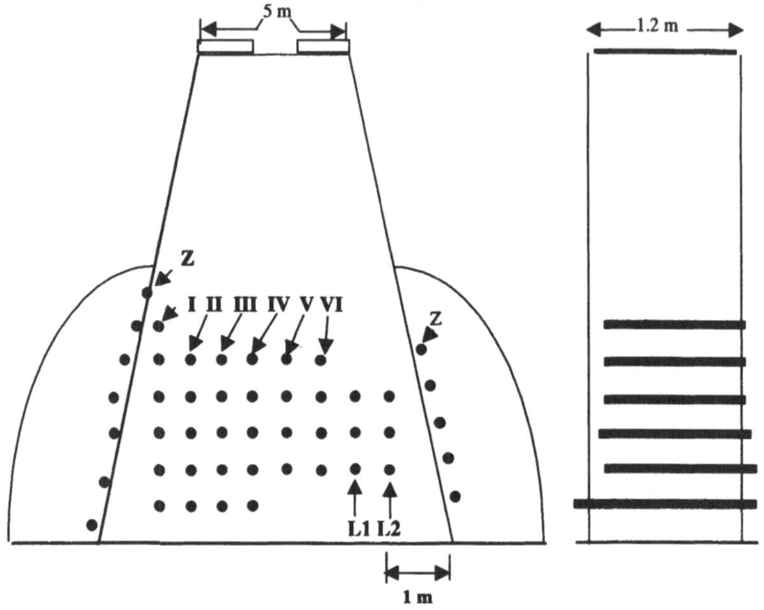

Fig. 12.20 Blast design and charging patterns adopted during demolition of piers (after Mandal et al., 2001)

Plate 12.6 Controlled demolition of piers without affecting Nonia Bridge

12.7.4 Conclusion

Demolition blasting has the advantage of lower global cost of dismantling structures with elevated heights. It can be executed within a short time and is safe even when carried out in residential areas or near traffic arteries. However, proper planning, delay arrangements and safety measures like control of flyrock, vibration, noise, dust and air overpressure are the prime factors to be considered with utmost care. The direction of fall, may be vertical or angular, would solely depend on the initiation sequence which should be designed as per the need of the job.

REFERENCES

Bhatia, R. L. and Hasan, M. 2003. Mine closure plan-legislative aspects, Proc. National Seminar on Strategies for Mine Closure (SMC), A. K. Ghosh, B. K. Tewary and M. K. Chakraborty (eds.), 11-12 October, CMRI, Dhanbad, India, pp. 1-6.

Borovikov, V. A. and Vanyagin, I. F. 1995. Modelling the effect of blasting on rock breakage, Oxford & IBH Publishing Co. Pvt. Ltd., New Delhi, India, p. 238.

Busey, John and Borg, Davis, G. 1995. Pre-splitting with the new air-deck technique, Proc. 21st Int. Conference on Explosives and Blasting Technique, February 5-9, Nashville, Tennesse, USA, Vol. 1, pp. 197-213.

Chernigovskii, A. 1985. Application of directional blasting in mining and civil engineering, Oxonian Press Pvt. Ltd., New Delhi, pp. 98-109.

Chiappetta, R. F. and Postupack, C. 1995. An update on causes and recommendations for controlling coal damage when blasting overburden, Proc. EXPLO'95 Conference, Brisbane, Australia, pp. 345-360.

CMRI Report of Investigations, 1998. Techno-economic assessment of drilling and blasting parameters vis-à-vis establishment of optimum blast design parameters for opencast mines, (Pal Roy et al.), GAP/2/Blasting/MOC/96, February, p. 86.

CMRI Interim Project Reports, 2000. GC/MT/122/99-2000 (March, 2000) and GC/MT/65/2000-2001 (October, 2000), p. 17, 21.

Chironis, N. P. 1987. Air-shock blasting, smoother blasts for your buck, Coal Age, McGraw-Hill Publishing Company, November, USA, pp. 48-51.

Foster Dee Snell (Editor-in-Chief), 1990. Encyclopaedia of industrial chemical analysis, Interscience Publishers— a division of John Wiley and Sons, Inc., New York, Vol. 9, pp. 27-94.

Gustafsson, R. 1973. Swedish blasting technique, SPI Publication, Gothenburg, Sweden, p. 328.

Hartman, Howard L. 1987. Introductory mining engineering, John Wiley & Sons, New York, USA, pp. 111-148.

Hustrulid, William, 1999. Blasting principles for open pit mining. Vol. 1(Chapter-7), A. A. Balkema, Rotterdam, Netherlands, p. 382.

Jimeno, Carlos Lopez, Jimeno, Lopez Emilio, Carsedo, F. J. A. 1995. Drilling and blasting of rocks, A. A. Balkema, Rotterdam, Netherlands, p. 391.

Kalam, A. and Singh, B. 2000. Deep hole blasting using NONEL—an experiment at Jayant Project of Northern Coalfields Limited, Proc. National Seminar on Drilling and Blasting for the Next Millennium, Bhubaneshwar, January 17-18, A. K. Ghose, G. K. Pradhan and J. K. Hota (eds.), pp. 17-19.

Mandal, S. K., Singh, M. M. and Singh, R. B. 2001. Segregation of coal by single-shot blasting of composite strata in opencast mines, Transactions of the Institution of Mining and Metallurgy, Section-A (Mining Technology), Vol. 110, January-April, pp. A69-A73.

Mandal, S. K., Singh, M. M. and Singh, R. B. 2001. Demolition of old bridge piers by blasting, Proc. Int. Soc. of Explosives Engineers (ISEE), January 28-31, Orlando, Florida, USA, Vol. I, pp. 215-233.

Pal Roy, P., Singh, R. B. and Mandal, S. K. 1995. Air-deck blasting in opencast mines using low cost wooden spacers for efficient utilization of explosive energy, Journal of Mines, Metals and Fuels, Explosives and Blasting Special Number, August, pp. 240-243, 248.

Pal Roy, P., Singh, M. M. and Singh, R. B. 1996. "Induced caving by long-hole underground blasting — A case study in BG Panel at GDK-10 Incline Mine of SCCL", Proc. 2nd National Conference on Ground Control in Mining, Oct. 14 and 15th, pp. 265-276.

Pal Roy, P. and Singh, R. B. 2001. Air-decking principle and its applications in production and presplit blasting, CIM Bulletin, 94(1047): 60-64.

Pal Roy, P. and Mohanty, B. 2002. A study on the usage of sawdust in ANFO, Proc. 7th International Symposium on Rock Fragmentation by Blasting, 11-15 August, Beijing, China, p. 73-77.

Pal Roy, P., Sawmliana, C., Bhagat, N. K. and Madhu, M. 2003. Induced caving by blasting: Innovative experiments in blasting gallery panels of underground coal mines of India, Mining Technology (Trans. Ins. Min. Metall. A), April, Vol. 112, pp. A57-A63.

Pal Roy, P. and Bhagat, N. K. 2003. A review on the strategic planning for mine closure with particular reference to blast-closure, Proc. National Seminar on Strategies for Mine Closure (SMC), A. K. Ghosh, B. K. Tewary and M. K. Chakraborty (eds.), 11-12 October, CMRI, Dhanbad, India, pp. 104-108.

Patrick, R.B., Bailey, M.L. and Yang, R.L. 1988. Minimizing coal edge losses in dragline blasting operations, Proc. The AusIMM, Explosives in Mining Workshop, Melbourne, Australia, pp. 69-74.

Sarma, Kanchibotla, S. and Scott, Andrew, 1999. Application of baby deck initiation to reduce coal damage during cast blasting, Inter. J. of Surface Mining, Reclamation and Environment, Vol. 14, pp. 75-85.

Singh, M. M., Mandal, S K., Pal Roy, P., Singh, R. B., Singh, P. K. and Pingua, B. M. P. 1998. Controlled blasting pattern for demolition of abandoned bridge piers near the existing Nonia bridge on G. T. Road, Asansol, CMRI Report of investigation, GC/MT/3/1998-99, p. 21.

Wedmair, R. and Scott, A. 1995. The assessment and control of coal damage and loss, Final Report ACARP Project C 3017, Australia, p. 204.

Wilson, J. M. and Moxon, N. T. 1988. The development of a low shock energy ammonium nitrate based explosive, Proc. Explosives in Mining Workshop, AusIMM, Nov. Melbourne, Australia, pp. 27-31.

EXTRA READING

Mineral Conservation and Development Rules (MCD), Government of India, 1992.

Sassoon, M.: Closure or Abandoned. Mining Magazine, August 1996.

Abandoned Mines Closure Project- Decision Memo (Through Internet Search).

Remediation and Closure Options: (Through Internet Search).

Overview of Mining and Mineral Industries in India: (Through Internet Search).

WMI (Whitehouse Mining Initiative): Environment issues group final report, October 1994, (Through Internet Search).

13

Rock Excavation by Non-explosive Methods

13.1 INTRODUCTION

In today's environmentally sensitive times, more and more emphasis is being placed on the negative effects of blasting in the extraction and construction industries. In addition, the legislative control of the use of explosives is becoming more vigorous, making their use more difficult in many civil engineering operations and in quarries where blasting requirements are infrequent or are deemed to cause nuisance. For a century or more, the universal utilization of commercial explosives for exploitation of medium to hard rock in mines, quarry, civil construction and tunnelling, established the place of explosives as an important tool of engineering endeavour. The main field of use is in breaking hard rock to minimize the arduous nature of the work, and to increase the productivity of such operations.

However, the growth of urbanization along with increased human sensitivity to blasting operations especially in close proximity of mining and quarrying areas, safer excavation by non-explosive methods is sometimes more important than conventional activities.

Since the development of many upcoming techniques and hard metals like tungsten carbide, excavation of the less tenacious rocks is being increasingly accomplished by non-explosive methods. There are obviously certain merits and demerits for all such methods, which cannot obviously be ascertained unless one goes into the depth of technology.

In this chapter, different non-explosive methods of rock excavation are scientifically summarized in the context of environment friendly controlled blasting activities in sensitive and conscious areas. All such methods may not be feasible or cost-effective at a particular site when considering deeper aspects of technology implementation, manpower, productivity and cost-effectiveness. However, in a non-conducive working atmosphere under repeated threats of the closure of operations, the management can give a second thought in a particular patchwork, using non-explosive methods of excavation. After reading thischapter, one can surmise which method is most suitable for the desired job.

13.2 ACCONEX: NON-EXPLOSIVE DEMOLISHING COMPOUND

The Associated Cement Company Limited has developed "Acconex" for breakage of rock, PCC or RCC material without causing any damage to the environment due to flyrock, ground vibration or noise. It functions solely due to expansive forces it generates when mixed with water. The right quantity of water to be used is 30% by weight of Acconex or 600 ml of water in every 2 kg packet. Generally, 4 kg of Acconex is needed for breaking one tonne of rock.

When Acconex is mixed with water, chemical reaction starts immediately after and the effects begin to manifest normally in 10 to 12 h as the cracks appear linking the holes or otherwise. It may take 24 to 48 h to obtain the full effects depending upon the ambient temperature, the characteristics of the material to be demolished, and the quantity of material used in a particular case.

However, there are some basic precautions and limitations of using the Acconex. They are listed below:

— Continuous stirring is required while mixing water
— Acconex slurry is to be used within 10 minutes of mixing
— In case of rock cleavages or cracks, plastic sleeves are to be used
— Holes with Acconex should be covered with gunny bags or tarpaulins
— Very careful operation is required to avoid any accident.

13.2.1 Limitations

— It is a costly process and requires a lot of preventive measures
— Can not be used in rainy season
— Can not be used in production faces, as it will hamper other mining activities
— It will not work without proper drilling.

13.2.2 Results of Laboratory Tests

During 1989-1990, a group of researchers at the Indian School of Mines conducted trials with Acconex, both in the laboratory and field with a view to dealing with massive longwall roof rocks (Gupta et al., 1989, 1990).

In the CMRI laboratory, a number of concrete blocks (both PCC and RCC) were constructed to see the effect of Acconex in different conditions (Pal Roy and Singh, 1997). There were two different kinds of blocks in which the ratio of cement, sand and stone chips were respectively [1:2:4] and [1:1.5:3].

The strengths properties of such blocks (in smaller cubical and cylindrical configurations) were determined in the laboratory. The stone chips, used in the blocks, were actually 0.5" (small) and 0.75" (medium) sizes. The test results are as follows:

Bulk density 2.14 g/cc

Young's modulus: 0.5 GPa

Uniaxial compressive strength: 8.26 MPa

Tensile strength: 0. 95 MPa

The loading density of Acconex was 2.31 kg/m of hole length in 40 mm hole diameter. It was observed that the charge factor of 4.25 kg/t of Acconex broke the P.C.C. blocks into pieces (Plate 13.1). However, in reinforced (R.C.C.) concrete slab as shown in plate 13.2, the chare factor of 6.14 kg/t could produce only deep cracks, joining the holes in the slab. The hole spacing was between 26 and 33 cm and the hole depths were up to 80% of the height of the slab. It was experimentally found that only 610 gm (1.48 kg/t charge factor) of conventional explosive was able to demolish the slab sufficiently, but it created too much vibration and noise near the surrounding (PPV=111 mm/s at 1.8 m distance from the source on the slab itself, PPV=70.24 mm/s on ground, at a distance of 3.6 m from the slab and air-overpressure level=134.4 dB-L at 12.5 m distance). It was found that the temperature inside the hole at the time of pouring was 34° centigrade and it rose up to 42° centigrade within 10 hours, and then started decreasing slowly and reached to the normal temperature after 27 hours of pouring. Thus, it could be concluded that the Acconex-water slurry did not produce exothermic reaction inside the hole. It was also observed that rectangular pattern of holes worked better in concrete slabs than the staggered pattern.

Another interesting point to note was that after filling Acconex of 50 % of the hole length (40 mm diameter), only 3 cm of Acconex level rose up after 30 h of pouring. This indicated that Acconex would work even if the holes were not covered. However, covering seemed to be essential from the point of view of safety. It was also noticed that the cost of using Acconex was nearly 4 to 5 times to that of the cost of using conventional explosives for the same job.

13.2.3 Acconex in Sandstone Boulders

For laboratory testing, boulders were collected from the nearby Chasnalla Opencast Mine of IISCO and transported to the site of experiments at CMRI. The strength properties of such boulders are given below:

Plate 13.1 Types of breakage with Acconex in square (4 nos.), Line (3 nos.), corner (2 nos.) and central drilling of PCC blocks of different sizes

Plate 13.2 A view of RCC structure with four different drilling patterns designed for testing

Density: 2.51 g/cc
Young's modulus: 9.4 GPa
Compressive strength: 55 MPa
Tensile strength: 4 MPa

The depth of hole in each boulder was nearly 40% of the boulder width. Depending upon the size of the boulders, one, two or three holes were drilled in them. In experiments, approximately 4-4.5 kg/t of Acconex was used for breaking the boulders.

Action of Acconex started after 18 h of pouring and it took 48 h for complete expansive action.

13.2.4 Observation

When the size of the boulder was too big and massive (more than 2 m^3), then Acconex took a lot of time to act. Charge factor (in kg/m^3) was also very high. However, Acconex worked better when the size of the boulder was less than 1 m^3 and the holes were drilled along any line of weakness or joints.

After pouring of Acconex-water slurry inside the hole, no upward thrust was observed. In a separate experiment, two holes of equal sizes were made side by side in a block. In one hole, Acconex was added almost to the top of the hole length whereas the other was filled only 25% of the hole length and both the holes were covered. After 48 h, it was seen that the Acconex level in the second hole was only 35% of the hole length. Figure 13.1 shows the diagrammatic view of breakage formations in different test-blocks after 24 and 48 hours of pouring of Acconex. While, Table 13.1 gives an overview of the type of action which Acconex created on such blocks.

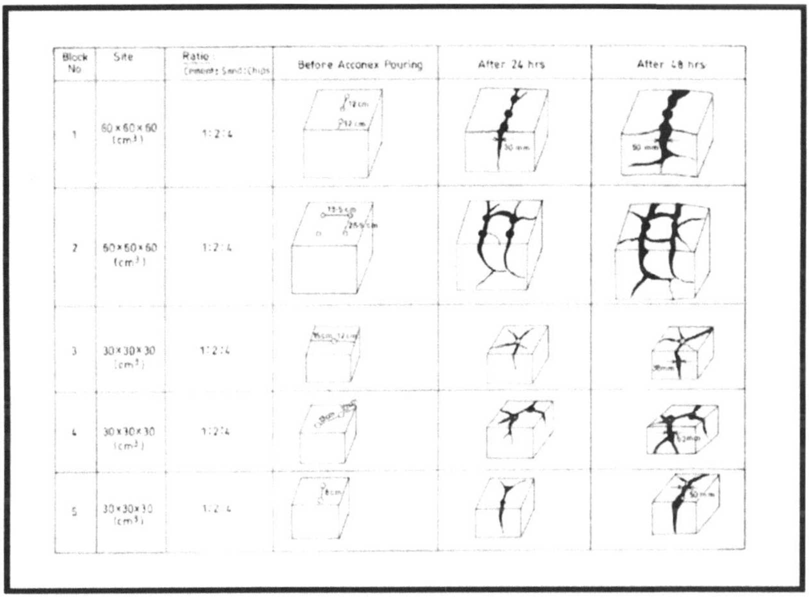

Fig. 13.1 Diagrammatic view of breakage formations in different test-blocks

13.3 CAPSULED COMPRESSED GAS CARTRIDGES

Two general types of non-explosive compressed gas systems have been developed in the UK and USA, primarily for 'soft ground' applications.

To enhance safety, these are used in underground coal mining for loosening compacted bulk materials in storages, and for dislodging accretions from hot furnaces. The impact depends upon the build-up of gas pressure in a cylinder, which, following a trigger action, is suddenly expanded into the material to be broken or loosened. In case of coal mining, the cylinders are located in regular boreholes. For loosening bulk

Table 13.1 Overview of the action of Acconex on test-blocks (Loading density of Acconex: 2.32 kg/m)

Block No.	*Remarks*
1	Depth of holes: 44-47 cm Spacing between holes: 12 cm Hole diameter: 4.0 cm Distance between extreme hole and block corner: 12 cm ***Observation:*** Vertical splitting occurred along the axis of the holes.
2	Depth of holes: 44-47 cm ***Observation:*** Shearing occurred in one portion, which could be due to the placement of heavy sandbags on top of the holes. The block was fractured into pieces.
3	Depth of holes: 20-25 cm ***Observation:*** Block was broken into 4 parts.
4	Depth of holes: 20-25 cm ***Observation***: The block was broken into two major parts of which the one had a minor breakage. The breakage was along the hole axis.
5	Depth of holes: 20-25 cm ***Observation:*** Block was broken into almost two equal parts.

materials, or breaking up furnace accretions, the cylinder is inserted into similar holes; or (for Airdox), located permanently in a recess in the wall of a storage bin (Gregory, 1984).

For coal mining, a mobile machine is available to drill holes and place cartridges of Airdox; or a specially designed self-propelled 'shooting car' is used for multiple sequence shooting.

Airdox cylinders are inserted in drill holes in the face and charged to 10,000-12,000 psi (69-83 MPa) by a nearby high pressure six-stage compressor through a feeder hose. Cylinders (reusable) are fired remotely through shooting hoses. This equipment is available from the manufacturers on lease arrangements.

For loosening bulk materials, Cardox cylinders are also available; but especially for intermittent *ad hoc* duty at infrequent intervals.

This unit consists of an alloy steel cartridge filled with liquid carbon dioxide and a chemical igniting agent. When required, the reaction is initiated by an electric current, controlled from a safe distance. The expanding action of the CO_2 gas breaks a soft steel disc at the discharge port, and the escaping expanding gas 'pressurizes' and heaves the enclosing material. Cardox cylinders may be recharged at the factory. Standard discharge pressures range from 10,000 to 19,000 psi (69-131 MPa), depending upon the application.

Where there is a continuing need for loosening stored bulk materials, this system is recommended.

Cardox consists of a high strength reusable steel tube filled with liquid carbon dioxide, a chemical heater and a rapture disc. When energized by the application of a small electrical charge, the chemical heater instantly converts the liquid carbon dioxide to a gas. This conversion expands the CO_2 volume and builds up pressure inside the tube until it causes the rapture disc at the end of the tube to burst. This releases the CO_2—now 600 times the original volume—through a special discharge nozzle to create a powerful heaving force, at pressure up to 3000 bar. Cardox can dislodge over 3 ton of blockage in a single blast. This all takes place in milliseconds.

Carbon dioxide is an inert gas that is commonly used in fire extinguishers, thus it is safe to use without the fear of generating secondary reactions with gases in the processing line. In addition, the quick release of the gas refrigerates the discharge, bringing it to a temperature low enough to avoid ignition of any air-gas mixtures inside the blocked vessel.

Simple secure coupling systems are mounted on the equipment in areas of known build up, which allows to set the Cardox tube to a predetermined depth and discharge direction. This control of the discharge minimizes any risk to refractory in the equipment.

After use, the Cardox tube is easily removed and replaced with a sealing plug that returns the equipment back to normal operation until the next time Cardox is needed.

13.4 MECHANICAL METHODS

Tungsten carbide tipped steel (rotating and traversing) abrading elements are now used to excavate rock of soft to medium hardness.

Specialized 'continuous miners', tunnelling moles, road-headers, and large diameter shaft/raise borers have been developed recently to employ these principles in soft to medium rock without the use of explosives. However, for hard rock, where the compressive strength generally exceeds about 35,000 psi, and where such mechanical methods are uneconomical, explosives must be used in the conventional drill/blast/muck cycle of operations.

Secondary breaking of large hard rock boulders by thermo-mechanical methods has recently been developed, wherein, the rock is first heated and then fragmented by mechanical impact.

13.5 ELECTRICAL METHODS

Various methods of breaking rock by electrical energy have been investigated, including heating by induction, high frequency, microwave, electron beam, plasma jet, arc, and condenser discharge techniques.

The 'Electrofrac' method uses electrode contact and resistance heating to stress the rock and to create fractures for subsequent separation by mechanical handling. Electrical fragmentation for secondary breaking of magnetic iron ore has been demonstrated in a field trial. It is claimed that it is feasible to use this method for primary breaking of iron ore (Gregory, 1984).

An electrothermal loosening (ETL) technique was tested in the erstwhile Soviet Union, aimed at saving explosives and providing greater fragmentation of iron ore. The technique involved the normal drilling of blast holes followed by a pair of electrodes (using AC at normal frequency) in adjacent holes; then a breakdown voltage (0.4 to 10 k V) was applied as a succession of vigorous electrothermal jolts, at the pre-blasting stage.

13.6 LIQUID JET SYSTEMS

A panoply of high-pressure jets, both continuous and discontinuous systems, with or without abrasive materials have found interesting applications worldwide (Res et al., 2003). Pulsed water jets at pressures of 100,000 psi (6896 kg/cm^2) can fracture rock and concrete. By incorporating a battery of high-pressure water jets on a tunnelling machine, it is conceivable that conventional tunnelling practices may be revolutionized.

13.7 EXPANSIVE MORTAR

Daigh Chemical Rock Splitting of USA supplies "Fract.AG" expansive mortar which can fracture rock and concrete with expansive force in excess of 15,000 psi (1034 kg/cm^2). The system eliminates problems caused by blasting and increases productivity of mechanical breakers. It can also be used for pre-splitting rock in sensitive conditions and has the capability to function properly even in larger hole diameters up to 3 inches (76 mm), both in dry and watery conditions (Plates 13.3 and 13.4).

13.8 SURFACE MINER

Surface miner, which is the outcome of an adaptation from road milling machines, can be used for cutting coal and soft to medium-hard rock (< 90 MPa) with mechanically driven cutting drum and two-part loading conveyor. There are essentially two principal types of surface miners, the milling type machines (Wirtgen, Bitelli, Huron and Man Takraf) and bucket type machines represented by Krupp surface miner. Wirtgen GmbH, Germany has developed a number of models viz. 2100 SM, 2600 SM, 3500 SM, 4200 SM etc. for economically cutting rocks of varying strengths. In India, a number of Surface miners are successfully operating in the limestone mining sectors. However, in the coal mining sector, it has been implemented, on recently an experimental basis, in Mahanadi Coalfields Ltd., a subsidiary of Coal India Limited.

Plate 13.3 A view of "Fract.AG" in operation to excavate a large pit (Courtesy: Daigh Company, Inc., USA)

Plate 13.4 A view of "Fract.AG" in operation in a hard strata to accelerate removal time (Courtesy: Daigh Company, Inc., USA)

The Surface miner has rigid, welded honeycomb design with integrated tanks for diesel, hydraulic fluid and water. The four crawler tracks are free-floating and connected to supporting arms which are linked to the frame via a moving parallelogram made upto two strut constructions in each case. The trucks are fitted with double grouser track shoes.

The height is adjusted via hydraulic cylinders. The rear crawler suspensions are controlled by an automatic cutting depth control unit. Each crawler unit is driven by a planetary gearing with hydraulic motor. Each of the front and rear travel drive motors is fed by a hydraulic variable displacement pump. The advance speed can be infinitely adjusted from zero to maximum in both travel gear and operating gear. The operating brake works via a closed-loop hydrostatic system. Parking brake for each crawler unit correlated with spring-loaded multiple disc brakes. The machine is steered by two hydraulic cylinders on each of the front crawler tracks. It is equipped with a central lubrication system which lubricates all the most important bearing points with grease. The operator's cab is enclosed, air-conditioned, soundproofed and mounted on vibration-damping elements. Different operational aspects of surface miners are depicted in Plates 13.5 to 13.9.

Plate 13.5 Surface Miner in operation in a limestone mine (Courtesy: Wirtgen GmbH)

Plate 13.6 Side discharge of mined mineral on an intermediate stockpile (Courtesy: Wirtgen GmbH)

13.8.1 Functionalities

The cutting drum rotates in an up-cutting direction. The cutting tools are mounted in tool-holders welded onto the body of the drum. The use of the tools depends on the machine's momentary use and on the properties of the material being cut. The cutting drum is driven via a mechanical

Plate 13.7 Widening of road foundation while upholding flow of traffic by removing the rock formation layer by layer (Courtesy: Wirtgen GmbH)

Plate 13.8 View of cutting and loading in one pass (Courtesy: Wirtgen GmbH)

Plate 13.9 Surface Miner can provide precise cutting depth and slope adjustment by fully automatic controls (Courtesy: Wirtgen GmbH)

clutch on the flywheel side of the diesel engine, as well as transmission belts and a planetary gearing installed in the cutting drum. The cutting speed can be adjusted in line with the momentary application by replacing the pulleys for transmission belts. The drum chamber is sealed by a hydraulically actuated scraper blade behind the cutting drum, thus ensuring a clean cutting surface.

The water spraying system largely prevents the generation of dust while cutting material and reduces the degree of wear on the cutting tools. The spray nozzles are additionally installed for the loading conveyors.

The conveyor system consists of a wide primary belt which picks up the cut and transfers the materials onto trucks. The loading conveyor can be adjusted in height and slewed to both sides. The conveyor speed can be varied.

13.8.2 Applications in Coal Mining

In 1999, the first Continuous Surface Miner (CSM), Wirtgen 2100 SM Model (CSM), was introduced at Lakhanpur Opencast Project of Mahanadi Coalfields Limited (a subsidiary of Coal India Limited), on an experimental basis for excavating coal. On the basis of the preliminary experiments conducted in this mine, it was decided to implement this system on a larger scale and to prepare a feasibility study for its applicability in the neighbouring Lingaraj Opencast Project of the same company (Sen et al., 2000). Lakhanpur OCP selected a category III Wirtgen 2100 manufactured by Wirtgen GmbH. This model is already working satisfactorily in the limestone sector of India. In India, currently 44 surface miners are successfully operating in coal and limestone mining sectors. The applications of such miners and their operating experiences have been reviewed extensively (Ghose et al., 2001).

13.8.3 Factors Affecting Selection of CSM

Surface miners are suitable for mining in fairly flat deposits, geologically undisturbed thin seams and thick seams having dirt bands of varying thickness. The following factors need to be considered before selecting a particular model of Surface miner: technical suitability, economic feasibility, environmental feasibility, quality control requirements and requirement of sized material. Technical suitability is tested subjectively by determining the application requirement of the machine vis-à-vis the geomechanical properties of the deposit. The technical suitability is also judged from the nature of deposit and its capability to accommodate the geometric configuration of the pit.

13.9 PENETRATING CONE FRACTURE

This innovative rock breaking process, called PCF (Penetrating Cone Fracture), involves energy efficient breakage of rock using high-pressure gas (Young, 1991). The PCF name is derived from the signature pattern of break. A specially formulated propellant is ignited and deflagrates producing the high-pressure gas pulse.

13.10 PLASMA BLASTING

A group of scientists at the Korea Institute of Geology, Mining and Materials (Lee et al., 1997) have established the plasma blasting method for rock fragmentation with a view of avoiding the adverse environmental effects. The plasma transformed from the electrolyte solution in a borehole reached an energy of 280 kJ by electrical input of 8.5 kV and 200 A and this rapidly released energy in the borehole developed a shock wave, which in turn produced a stress field that fractured the rock without producing excessive dust and flyrocks. The range of plasma strength varied with chemical additives. The scientists concluded that such increased energy delivery covered a wide scope of applications, from breaking a boulder to bench cutting of grounds for buildings in urban areas.

The plasma blasting system can supply 280 kJ by electrical energy for 1.5 ms, so the system may deliver 0.95 MJ additionally as a blasting energy when 30 g of aluminium completed its chemical reaction. This amount becomes 0.243 of relative weight strength of 243 g of ANFO or 320 g of Emulite.

The scientists concluded that plasma strength was improved with chemical additives and that the rock fragmentation was feasible not only for boulders but also for bench face. They also tried some numerical solution to show the practicability of multi-hole and controlled blasting in jointed rocks. It was also shown that the analysis of ground vibration data induced by plasma blasting might be predicted using the equation similar to that used in explosive-induced vibration.

Sumitomo Electric Industries Ltd. (SEI), Japan has developed compact Plasma Acoustic Blaster (PAB) for easy blasting of building structures and rocks.

In Plasma Acoustic Blaster (PAB) mechanism, a hole (or holes) is drilled into an object to be blasted and water is filled in it. Energy is discharged in it thereby generating a shock wave for breaking the rock. Compared with conventional blasting method, the PAB technology, which uses no chemical substances, is eco-friendly, because it causes less vibration, noise and dust. In PAB, a reusable blasting electrode and a power supply main body, incorporating capacitors and other devices that are connected by a cable are used.

This newly developed compact blasting equipment is suitable for blasting of platy structures and boulder stones. Holes for blasting can be opened by the hand-held tools. Therefore, the blasting procedure is easy and simple. By creating more than one hole in linear orientation and discharging at the same time, the object can be blasted as if it was cut.

This compact blasting equipment is suited for the following applications.

- Blasting of thin, platy objects (structures such as buildings and furnaces)
- Blasting of boulder stones
- Digging of riverbed
- Use in places where transportation of heavy load is difficult (such as mountains, slopes, and high places)
- Flexible and impromptu blasting under various situations (such as recovery work in a time of disaster)
- In the case where blasting like cutting is required (such as finishing of the blasted surface after main excavation work)

13.11 CONCLUSIONS

Till date, there has not been any established method that can replace the vast usage of commercial explosives for exploitation of coal and minerals. Alternate techniques discussed in this chapter can be used for secondary rock breakage, building construction and demolition purposes, low depth trenching and pit excavation requirements in supersensitive and environmentally conscious places. In brief, the following general conclusions are made—

1. Airdox can be employed to break soft rocks (e.g. coal), and it does not ignite a gassy or dusty atmosphere.
2. Cardox creates a pressure of nearly 70-130 MPa, adequate to break and undercut soft rocks and coal.
3. Acconex can be used to create fractures in concrete structures and breakage of boulders with intermittent joints. However, it cannot be used in production faces.
4. Surface miners are suitable for mining fairly flat deposits, geologically undisturbed thin seams and thick seams having dirt bands of varying thickness. Before selecting Surface miner, it is necessary to thoroughly, study, the technical suitability; economic feasibility; environmental feasibility; quality control requirements and requirement of sized material.
5. Application of plasma blasting has not yet been totally established except in a few scattered works. From the available published works, it can be said, that the method has the potential for boulder breakage

and cutting of 1-1.5 m bench faces for building constructions and other related purposes.

6. All mechanical and electrical methods are useful for secondary breakage of medium to hard boulders. They can also be used for low depth trenches and pit excavations.

REFERENCES

Airdox Shooting, Bulletin 63-1-5000, 1980. Long Airdox Company, (Company Bulletin), Oak Hill, West Virginia, USA.

Cardox International Ltd. 2000. Roe Cross International Park, Old Road, Mottram, Via Hyde, Cheshire-SK14 6NB, England (e-mail: cardox@btinternet.com).

DAIGH Company, Inc. 2393 Canton Hwy Ste 400, Cumming, GA 30040, USA, (e-mail: sales@daighcompany.com).

Energy and Environmental Technology R&D Laboratories and Sumitomo Electric Engineering Company Limited, Osaka, Japan (Contact No. +81-6-6466-6551).

Ghose, A. K., Schimm, B. and Niyogi, S. 2001. New techniques for surface mining in the 21st century—emerging role for surface miners, Proc. 10th Int. Symposium on Mine Planning and Equipment selection, New Delhi, November, A. A. Balkema, Netherlands, pp. 269-276.

Gregory, C. E. 1984. Explosives for North American Engineers, Trans. Tech. Publications, Germany, pp. 150-154.

Gupta, R. N., Dey, A. and Ghose, A. K. 1989. Evaluating a new rock breaking compound for dealing with massive longwall roof rocks—laboratory and field investigations, Journal of Mines, Metals and Fuels, December, pp. 591-597.

Gupta, R. N., Dey, A. and Ghose, A. K. 1990. Use of rock breaking compounds (RBC) in mining and civil engineering—laboratory and field investigations, Journal of Mines, Metals and Fuels, September, pp. 209-215, 227.

Lee Kyung Won, Chang Ha Ryu, Joong-Ho Synn and Chulwhan Park, 1997. Rock fragmentation with plasma blasting method, Proc. Environmental and Safety Concerns in Underground Construction, Lee, Yang and Chung (eds.), A. A. Balkema, Rotterdam, Netherlands, pp. 147-152.

Pal Roy, P. and Singh, R. B. 1997. Enhancement of safety and productivity using new blasting accessories in opencast mines, Proc. 27th International Conference of Safety in Mines Research Institutes, ICS-MRI'97, Bharat B. Dhar and B. C. Bhowmick (eds.), 20-22 February, New Delhi, India, pp. 655-664.

Res, J., Wladzielezyk, K. and Ghose, Ajoy K. 2003. Environment-friendly techniques of rock breaking, A. A. Balkema, Rotterdam, Netherlands, p. 160.

Sen, P., Pathak, K. and Dey, K. 2000. Continuous surface miner in Indian surface coal mining—a case study, Surface Mining, 52(4): 371-376.

Wirtgen Surface Miners, 1997. Technical Specifications, Wirtgen GmbH, Windhagen, Germany, p.12.

Young, C. 1991. Penetrating cone fracture for rapid excavation and mining, American Underground Space Association, 6(3): 15-18.

14

Blasting in Surface and Subsurface Hard Rocks

14.1 BLASTING IN DIMENSION STONE QUARRYING

14.1.1 Introduction

There are various methods available for recovery of dimensional stone in open-pit quarries. In India, jet flame, wedge cutting and blasting with gunpowder are specifically carried out for rock cutting in dimension stones. Amongst the mechanical means, the cutting tools for dimension stone are slot drilling and diamond wire saw (or *in situ* scale diamond saw) cutting. Although, commercial explosive is the cheapest chemical means for rock breakage, its usage in stone quarries, has so far been limited, mainly because of non-availability of its foolproof technology.

At first, the cutting of dimension stone is carried out by separating the required material from the mother rock and then subdivided into blocks for production. The first separation of the rock from its in-situ is generally carried out either by jet cutting or by slot drilling. Drilling and blasting with gunpowder along the joint planes is performed and then production is made from the separated rock. Jet-flame is generally carried out by burning kerosene oil under high pressure causing excessive noise and thereby endangering the operators and at the same time raises severe complaints from the residential buildings surrounding the quarries. Moreover, considering the jet-flame nozzle, a wide cutting of 15-25 cm is made with burning effect on the walls causing wastage of the rock.

Slot drilling is carried out with a hydraulic drilling machine, capable of drilling vertically. These machines are capable of drilling any diameter between 45 and 76 mm. With this machine, two holes of diameters between 45 and 56 mm are drilled at 65 to 75 mm distance (centre to centre) and the large diameter (76 mm) is drilled between them to create a separation line of the dimension stone from the in-situ mass. This process, though being very safe, is costly and time consuming. Wedge and hammer, is another method of breaking rock into required sizes, however not properly implemented, may result in improper breakage and wastage of the rock wealth. Controlled blasting, is an easy and the cheapest method, to be implemented judiciously for such purposes.

14.1.2 Traditional Methods

Slot drilling and jet flame cutting are very widely used methods in cutting dimension stones at gem granite quarries. These machines are generally used to separate the material to be produced, for a particular month or week, from the *in situ* mother rock. This is generally done at places where it is massive and devoid of joints. The slot drilling is carried out by hydraulically drilling with 45-56 mm drill diameters and 76 mm drill diameter. In this method, one hole of 76 mm is drilled between two 45 or 56 mm drilled holes and thereby joining the drilled holes and separating the rock. In jet flame cutting method, the rock is melted by burning it with the help of compressed air kerosene oil under high pressure. In general, 40 litres of oil is consumed per hour for cutting the rock. With this method, about 15-25 cm of rock is required to be cut with burning effect up to about 3-5 cm on either side. However, as shown in Table-14.1, all such methods produce very high levels of noise, which may cause hearing problem for the operators.

Table 14.1 Noise level from jet flame and diamond wire sawing operations (Ryu et al., 2003)

Distance from source	*Noise level (dB-A)*	
(m)	*Jet flame operation*	*Diamond wire sawing operation*
3	117	91.9
5	115	90.4
10	104.2	87.9
20	99	83.9
30	98	79.6
40	95.2	75.8
50	92.7	73.4

In the black granite quarry of India, dimension stone is recovered by drilling and blasting or wedging and hammering. In Kota stone, cutting is done with a circular diamond saw. The depth of the cut varies between 10 and 15 cm. After its separation from the mother rock, the stones are recovered with the help of crowbars and finally shaped by chiselling.

14.1.3 Presplitting Mechanism

It is a controlled blasting technique in which a series of closely spaced holes at the perimeter of the excavation are fired prior to the main blast to create a crack along the contour. The main objective of this method is to develop a tensile failure through hoop stress without any undesirable damage around the blastholes. Generally 10 to 20 times of the borehole diameter is kept as spacing between blastholes. Chiappetta (1991) reported the following relationship for calculation of spacing:

$$S \leq 2r_h\,(P_b + T)/T \tag{14.1}$$

Where, S = spacing between boreholes (cm); r_h = borehole radius (cm); P_b = borehole pressure (psi) and T = tensile strength of rock (psi).

Orientation of joint planes have a major influence or bearing on the quality of the presplitting. Joint frequency of 3 or 4 in hole spacing produces an adverse effect on presplitting. Dunn and Cocker (1995) recommended that for better wall control, hole spacing should not exceed double the joint spacing. The success of splitting reduces with the decrease of rock strength. Therefore, splitting would be best observed in rocks having UCS (Uniaxial Compressive Strength) between 100 and 150 MPa and would not be very effective in rocks having UCS between 0 and 50 MPa. It is also observed that if the presplit holes are fired with different delays, there will be a shearing effect on the wall and splitting will not be effective. A very low charge concentration varying between 0.1 to 0.2 kg/m will cause cracking to a maximum depth varying between 0.2 and 0.5 m.

In order to calculate the optimal time (T_{opt}, in ms) between two consecutive presplit holes, Rustan et al. (1996) suggested the following equation, which should be provided between two holes:

$$T_{opt} = K \times C_S \times [B + (B^2 + S^2)^{0.5}]/1000 \tag{14.2}$$

Where, B = Burden (m); S = Spacing (m); C_S = shear wave velocity (m/s) and K = safety factor (usually 1.5).

14.1.4 Dimension Stone Quarries

In the Kota stone quarry, a circular saw is used for cutting the blocks into required sizes and then removing it from the *in situ* position with the help of chiselling and crow bars. While cutting, this process sometimes creates jamming problems resulting in machine breakdown. Moreover, cutting the stone by circular disc, from its mother rock, in the hanging wall causes loss of length. Sometimes, the block cannot be separated desirably, which results in loss of production. To avoid such difficulties, CMRI conducted some presplitting blasts at a few experimental sites and observed that with spacing of holes between 15 and 20 cm produced good results with a charge concentration of 0.07 kg/m for 60-70 cm hole depths.

Air decking at the bottom of the holes was observed to reduce cracking of Kota stone and reasonably improved the overall efficiency of recovery.

In the black granite quarry of India, the stone is recovered either by drilling and blasting or by wedging and hammering. The percentage of recovery is not satisfactory due to low drilling length and very tight stemming. Due to shorter drilling length, the rock breaks in a straight line at the positions of drilled holes which takes a curvature breakage towards the free face and thereby the percentage of recovery reduces. Tight stemming prevents the ejection of gaseous energy through stemming column and hence extending the cracks, in undesired directions, along the weakest plane and lessening the percentage of recovery. Similarly, splitting by wedging and hammering for a hole depth of about 40% of the height of the block causes improper breakage and thereby reduces the percentage of recovery. Also blasting is carried out only in vertical holes and the material that is recovered by shovel has improper breakage and hence creates a loss in the percentage of recovery.

In view of such problems, CMRI conducted some trial blasts and suggested a sequence of operations for better recovery. At first, horizontal holes are to be fired to split the stone block from its base. The second firing round should be with the vertical holes to be drilled parallel to the free face. This would help in segregating the block from the mother rock. The last and final round of blast is to be carried out with the vertical holes drilled parallel to the free face. This round helps both in separating the complete block from its mother rock and at the same time helps in displacing it, depending upon the quantity of the explosive. Such a sequence of operation, if properly implemented, would result in maximum possible recovery of stone.

In India, gunpowder is mainly used for quarrying of dimension stones. Gunpowder, a low grade explosive, causes no damage to the walls of dimension stones in comparison to other commercially available explosives. Field trials carried out by CMRI (Mandal et al., 2003) showed that firing with detonating fuse (10 g PETN/m) and gunpowder of 0.1 kg in a blasthole caused no damage (or extension of exiting cracks) beyond 1 cm while 0.043 kg of NG-based explosive created damage up to 2 cm.

14.1.5 Conclusion

Besides traditional techniques, drilling and blasting with commercial explosives, if properly implemented, can play an important role for better production and productivity. Gunpowder can be used for splitting the production blocks. Trial blasts conducted by CMRI have indicated that 1.5 m of hole depth and a spacing of 0.15 cm between consecutive holes, with a charge concentration of 0.07 kg/m, produced better results. Further experiments are required for better approaches.

14.2 BLASTING IN THE HIMALAYAN ROCKS

14.2.1 Introduction

Precambrian rocks which cover a substantial part of the Himalayas are geologically presented by the following four major rock formations, listed from youngest to oldest (Bhasin et al., 2002):

(1) Everest Pelitic Formation
(2) Sikkim Group
(3) Chungthang Formation
(4) Kanchenjunga Gneiss Group

Experiments were conducted by CMRI, at the Sikkim group of rocks, which consists of schist and gneisses. East Sikkim is mainly dominated by rocks of lower metamorphic grades, mainly chlorite schist, sericite schist and quartz schist. These rocks have a phyllitic appearance.

The region of central and eastern Himalayas near the Taste river is characterized by large-scale thrust movements from the northwest to southeast. East Sikkim is mainly traversed by two thrusts called Chungthang Thrust and the Main Central Thrust (MCT), both trending in a NW-SE direction (Mehrotra et al., 1996). The MCT is a well-known tectonic boundary between sedimentary and crystalline rocks, separating the lesser and greater Himalayas. It is characterized by crushed rock and fracture zones. The Chungthang Thrust involves gneissic rocks of Chungthang formation and schistose rocks of the Sikkim group.

The Nepal-Sikkim Himalayas and adjacent fore-deep areas are dominated by conjugate strike-slip faults that face the wedges of the Indian shield. Rocks of this terrain are characterized by intense folding, metamorphism and thrusting in a number of tectono-stratigraphic units, which exhibit inversion in stratigraphic succession as well as metamorphic grading.

14.2.2 Rock-Geologic Parameters

The local geology of the investigated area, which was a road construction site of Sikkim, was very complex. Lithology was exclusively gneisses rock mainly consisting of quartz, muscovite and biotite. The geotechnical survey of the area indicated three sets of prominent joints along the excavated length of 13.5 to 16.0 m. The general strike direction of bedding plane (J_1) was N157°–358° with dip varied from 45° to 72°. Thickness of the bedding plane varied very widely from a few centimetres to 6 m.

The strike direction of the second set (J_2) was parallel to the bedding plane with its dip in opposite direction. Spacing of J_2 varied between 0.2 and 3.0 m.

The strike of the third joint set (J_3) was perpendicular to the strike of the bedding plane (with strike direction N30° – 210°). Similar to J_2, spacing of J_3 also varied from place to place. It varied between 0.5 and 7.0 m.

Joint planes were found to be generally smooth in nature. In some places, joints were filled with sandy soil of few centimeter thicknesses. These types of filling materials were more frequently present in cases of J_2 and J_3. The cohesive strength along these joints, while filled with sandy soil, was very little. However, when saturated with water, joint planes became slippery and therefore more prone to sliding and instability.

In order to design safe and optimum blasting pattern as well as to assess rock mass damage due to blasting, rock properties like density, tensile strength, compressive strength, *P* and *S*-wave velocities were determined. The samples were collected from the construction site and tested at CMRI Rock-Testing Laboratory. Table 14.2 shows different rock properties as determined in the laboratory.

Table 14.2 Laboratory determined rock characteristics parameters

Density (g/cc)	*Compressive strength (kg/cm²)*	*Tensile strength (kg/cm²)*	*P-wave velocity (m/s)*	*S-wave velocity (m/s)*
2.66	532.52	50.55	2442.15	1638.17

The tested values of compressive strength, tensile strength, primary wave velocity (*P*-wave) and shear wave velocity (*S*-wave) of the rock specimens were observed to be low, mainly due to their mineral content. The rock specimens which were tested, mainly consisted of quartz, muscovite and biotite. While quartz made the rock stronger and heavy, muscovite and biotite made them weaker in respect of compressive strength, tensile strength and elastic wave velocities.

14.2.3 Effects of Blast Vibration on Rock Masses and Slopes

Blast induced ground vibrations have two-fold actions on rock masses. On one hand they affect the integrity of the rocks or their strength parameters and, on the other, can provoke wall or slope collapses when destabilizing actions are introduced (Jimeno et al., 1995).

During the process of blasting, the energy that is not used in fragmentation and displacement of rock, reduces the structural strength of the rock mass outside the theoretical radius of action of excavation. New fractures and plane of weakness are created and joint, declasses and bedding planes that initially behaved non-critical, when open, result in an overall reduction of rock mass cohesion. This is manifested by overbreak, leaving fractured mass in a potential state of collapse.

The following modified equation can be used for evaluation of critical vibration velocity (V_{crit}) for rock mass damage based on the longitudinal wave propagation velocity, density, rock strength parameter and tensile strength of rock. The equation is a modified version of the equation described in Jimeno et al. (1995).

$$V_{crit} = 10^4 \times RT \times \alpha / (\rho_r \times V_C) \qquad (14.3)$$

Where,

V_{crit} = Critical vibration (mm / s)
RT = Tensile strength of rock (kg / cm^2)
V_C = Propagation velocity (m / s)
ρ_r = Density of the medium (g / cc)
α = Mohr's scale of hardness
[α = 1-2 for very soft: Compressive strength (MPa): <10
2-3 for soft: Compressive strength (MPa): 10-30
3-4.5 for medium soft: Compressive strength (MPa): 30-60
4.5-6 for medium hard: Compressive strength (MPa): 60-120
6-7 for hard: Compressive strength (MPa): 120-200
+7 for very hard: Compressive strength (MPa): +200]

Based on selective experiments, Oriad (1970) established the damage threshold level in rock slope as 6000 mm / s. In a separate classified study, Bauer and Calder (1978) proposed the following generalized criteria for damage level (Table 14.3).

Table 14.3 Damage level of sound rock mass based on ground vibration (Bauer & Calder, 1978)

Particle velocity (mm/s)	*Predictable damage*
< 250	No danger in sound rock
250–600	Possible sliding due to tensile breakage
600–2500	Strong tensile and some radial cracking
> 2500	Complete breakup of rock masses

14.2.4 Blast Pattern

During the field investigation, blasts of different sizes were thoroughly monitored and examined. Holes were drilled using hydraulic-leg mounted, compressed-air operated jackhammer drill machines. Drillbit used was of 32 mm diameter and the finished hole diameter was 42 mm. Depth of holes varied from 0.6 to 3.0 m. Explosive cartridges of 25 mm diameter, 200 mm length and 150 g weight of Nobelgel-80, manufactured by Nobel Explosives Chemical Limited were used. Detonating cord (PETN composition: 12 g / m), manufactured by Gulf Oil Corporation Limited (previously IDL Industries Ltd.) was used for connecting the holes. Safety fuse coupled with one plane detonator was used for firing. All the holes were fired simultaneously i.e. no delay detonator was used in any round of blast.

Overcharging of blasthole, a usual practice in road construction blasting, generally poses the following problems—

(1) Uncontrollable generation of flyrock
(2) Blown-out shoot due to insufficient stemming
(3) Overbreak and generation of cracks to the adjacent wall
(4) Increased ground vibration level etc.

These are apparent in Plates 14.1 and 14.2 where severe rock mass damage due to unscientific method of blasting and the opening of joint (nearly 4 cm) due to blasting are seen.

Plate 14.1 Opening of joints (nearly 4 cm) due to blasting

Plate 14.2 Damage of mother rock due to unscientific blasting

14.2.5 Ground Vibration Monitoring

Ground vibrations generated due to blasting, both formation cutting and boulder breaking, were thoroughly monitored. Uniaxial high frequency sensors were fixed on the rock slope using some typical non-ferric (copper) sensor attachments. Triaxial sensors were placed on the ground (newly constructed road) using plaster of paris. Vibrations were monitored at distances varying between 5.75 and 35.0 m from the blasting source. No delay detonators were used in all the blasts.

The vibration data monitored during the field investigation were grouped together for statistical analysis to obtain, in a simple form, a suitable predictor equation, for the determination of charge quantity. The following conventional predictor equation was developed.

$$V = 0.559\ (D/Q^{0.5})^{-0.707} \qquad (14.4)$$

r (correlation coefficient) = 0.61

14.2.6 Vibrations from Non-Blasting Events

Besides blasting, ground vibrations produced due to dozer operation, drilling operation and vehicle movement were also monitored for better understanding of the problem. During dozer operation, seismograph sensors were placed both on the ground level as well as on the rock slope. Distance of measurements varied from 0.75 to 6.5 m.

The maximum vibration recorded was 10.70 mm/s. Such a level of vibration could be capable of triggering and/or loosening any overhanging embedded boulder to fall. It was also noticed that the vibration during dozing time was more than that of marching time.

Drilling was carried out by compressed-air operated Jackhammer drill machine with hydraulically mounted legs. Vibrations were measured at distances between 1.75 and 3.5 m from the drilling operation. Sensors were placed on the rock slope. The measured values indicated that the distance had a great influence on the amplitude of vibration.

Monitoring of vehicular movement was conducted along the constructed road. A sensor was fixed on the road (firm ground) at a distance of 0.3-1.0 m. Due to inaccessibility, the movements of only one type of vehicle was monitored. It was observed that the vibrations generated due to vehicle movement were lower than that of dozer and drilling operations. Time history records and FFT analyses of recorded vibrations from different sources are shown in Figures 14.3 to 14.8.

14.2.7 Blast Design Patterns for Safe and Optimal Excavation

For safety and long-term stability of the excavated rock slope, controlled blasting is very essential. Such type of blasting would ensure the stability of nearby slopes during and after the excavation job. For safer and faster progress of work, Jumbo drill or wagon drill (crawler mounted) would be a better option for drilling operations. This would ensure more precise drilling pattern, faster progress and safety for the drilling operator. Jackhammer drilling was found to be unsafe for drilling in slopes. Smaller diameter blast holes always produce good fragmentation with lower intensity of vibration and air blast, and therefore creating less damage to the remaining rock. As such, recommendation of 42 mm drill diameter was good enough for depth of holes varying between 0.5 and 3.0 m.

14.2.8 Drilling Pattern

Based on hole diameter (D) and compressive strength of rock, Jimeno et al. (1995) suggested the following burden and spacing values for highway excavation works (Table 14.4). It was felt that for smooth blasting, the perimeter holes along the wall side, were required to be drilled with closer spacing. Line drilling was not recommended due to time factor.

Table 14.4 Calculation of burden and spacing from hole diameter for different rock strengths (Jimeno et al., 1995)

Design Parameter	*Uniaxial Compressive Strength (MPa)*			
	Low	*Medium*	*High*	*Very high*
	<70	*70-120*	*120-180*	*>180*
Burden	39*D*	37*D*	35*D*	33*D*
Spacing	51*D*	47*D*	43*D*	38*D*

The following three types of drilling operations are generally conducted for highway excavation works (Figures 14.1a, b)—

(1) Vertical holes, either parallel or in fan cut
(2) Both vertical and horizontal holes
(3) Horizontal holes.

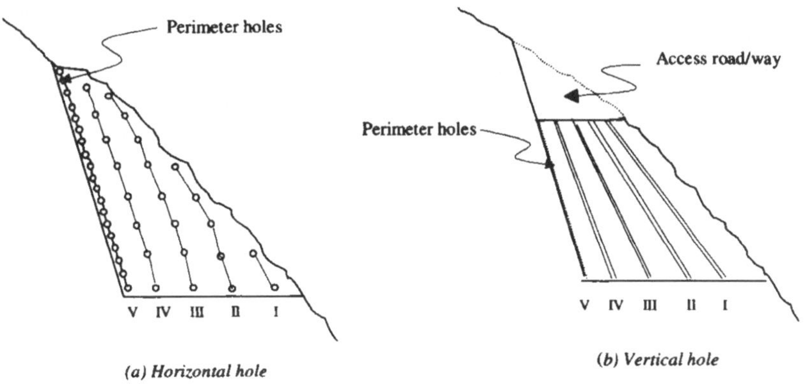

Fig. 14.1 Different types of drilling operation and consequent firing sequence (in Roman Nos.) (after Jimeno et al., 1995)

Vertical holes are difficult to drill with jumbo drill due to placement of the machine. Spacing between perimeter holes may vary from 0.25 to 0.30 m. For vertical holes, subdrilling is required which may be 0.3 times of the burden.

14.2.9 Charge Factor and Initiation Sequences

The amount of explosive charge may be calculated from the charge factor. Charge factor (kg/m^3) for different kinds of rock strength, density and joint spacing are given in Table 14.5. Since, uniaxial compressive strength of the rock at the construction site was 53.2 MPa and its density was 2.66 g/cc, charge factor between 0.38 and 0.52 kg/m^3 produced good fragmentation.

Table 14.5 Rock classification according to their facility of fragmentation by explosives in open excavation

Charge factor		*Mean distance between natural fractures in rock*	*Uniaxial compressive rock strength*	*Rock density*
Class limit (kg/m³)	*Average value (kg/m³)*	*(m)*	*(MPa)*	*(g/cc)*
0.12-0.18	0.150	<0.10	10-30	1.40-1.80
0.18-0.27	0.225	0.10-0.25	20-45	1.75-2.35
0.27-0.38	0.320	0.20-0.50	30-65	2.25-2.55
0.38-0.52	0.450	0.45-0.75	50-90	2.50-2.80
0.52-0.68	0.600	0.70-1.00	70-120	2.75-2.90
0.68-0.88	0.780	0.95-1.25	110-160	2.85-3.00
0.88-1.10	0.990	1.20-1.50	145-205	2.95-3.20
1.10-1.37	1.235	1.45-1.70	195-250	3.15-3.40
1.37-1.68	1.525	1.65-1.90	235-300	3.35-3.60
1.68-2.03	1.855	>1.85	>285	>3.55

Detonating fuse along with 25 mm diameter cartridged explosives (at the bottom or alternately) was used for charging perimeter holes. Top stemming length was equal to or more than burden to control flyrock. Suggested charging pattern of holes, using detonating fuse, are shown in figures 14.2 (a), (b) and (c).

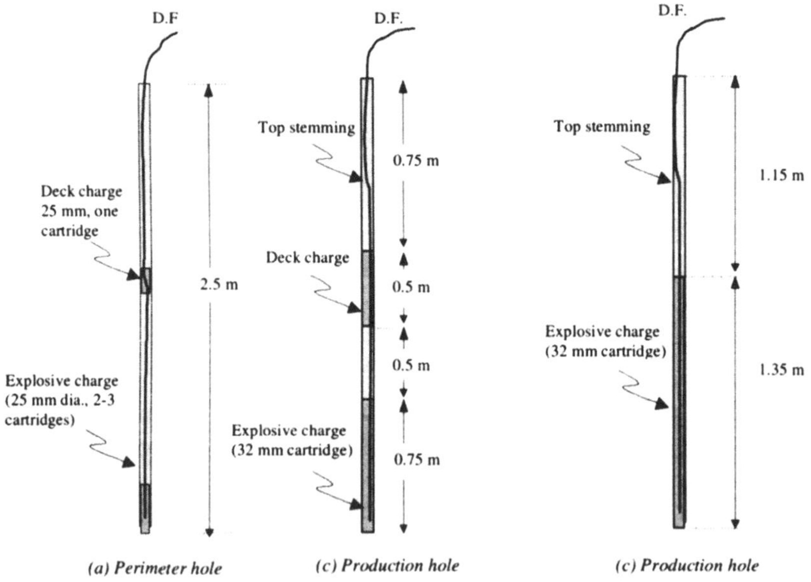

Fig. 14.2 Suggested charging pattern for 2.5 m dept, 42 mm diameter hole using detonating fuse (Burden: 0.95-1.00 m; Spacing: 1.00-1.20 m; Charge factor: 0.45 kg/m³)

14.2.10 Boulder Blasting

When detonating cord is used during blasting of boulders by pop shooting, for better control of flyrock, depth of hole may vary from 0.5 to 0.7 times of boulder thickness i.e. $(0.5 - 0.7) \times T_b$, where T_b is the thickness of the boulder. For boulders of less than two cubic metre sizes, only one drill hole at the centre is sufficient. For larger boulders, two or more holes may be drilled based on their sizes. Spacing of holes in such cases may be about $(0.5 - 0.9) \times T_b$.

During experiments, the maximum charge factor of 0.30 kg/m^3 was used i.e. for one cubic metre rock, not more than 0.3 kg (or 300 g) charge was used. The uncharged portion of the hole was stemmed properly.

While in plaster shooting, the charge factor varied between 0.7 and 1.0 kg/m^3 for gelatin explosive. The explosive charges were covered with a layer of nearly 10 cm thick of mud or sand for better utilization of explosive energy and to reduce air overpressure and also noise levels. It was observed that firing with uncovered charges produced approximately 25% more noise and air-overpressure.

14.2.11 Concluding Remarks and Guidelines

The following recommendations are made for blasting in road construction works for general guidelines:

1. It is necessary to conduct a detailed inspection/survey before drilling of holes in order to find out any overhanging loose rocks that may cause danger to the drilling operator. If any loose rock is found, it should be removed beforehand.
2. During boulder blasting, it should be ensured that the hole depth should not be more than $1/3^{rd}$ of the boulder thickness to prevent penetration of the hole.
3. Proper alignment of holes should be done specially during drilling at the wall side to prevent unwanted rock damage.
4. All the holes should be checked before charging of explosives. If any hole is found jammed, it should be cleared. Overcharging of holes with explosive must be avoided.
5. All the holes should be stemmed properly to prevent venting of premature gas. This will also protect flyrock ejection.
6. When firing is carried out by using safety fuse, the following care should be taken:
   (a) Check the burning speed of the safety fuse before using it. This will ensure the proper length to use in blasting to reach at a safe place.
   (b) Cut-off a few centimetres of the fuse to ensure a dry end. It is always better to cut a fresh end. Oblique cutting should be avoided.

(c) Visually inspect inside of detonator for foreign material or moisture. If any foreign material is present inside the detonator, remove it by gently shaking in an up-side-down position. Never use force to remove it. If the material cannot be removed, or moisture is present then avoid using that particular detonator.

(d) Insert the safety fuse gently inside the plane detonator, never use force or twist the safety fuse inside the detonator.

(e) Use crimp only for crimping the detonator with a safety fuse. Never use teeth or a knife.

7. Always give sufficient warning before firing the round. Make sure everyone is in a safe place.
8. Wait for some time after firing the round. If there is any misfire, care should be taken.
9. Do not approach the charging place hastily if blasting does not occur after the expected time of initiation. It is to be remembered that due to presence of moisture or bending of safety fuse, the burning speed may be reduced.
10. For overall slope stability and controlled blast impacts, either smooth blasting or presplitting method is strongly recommended. Delay detonators should be used during production blast (formation cutting).
11. Highly structured slopes can be affected by the low amplitude, low frequency surface waves that persist some distance from a blast. The stability of a slope is expressed in terms of its Factor of Safety (FOS). The FOS represents the balance between the forces acting to cause slope failure and those available to resist such failure. The FOS is usually obtained from a static analysis of these forces.

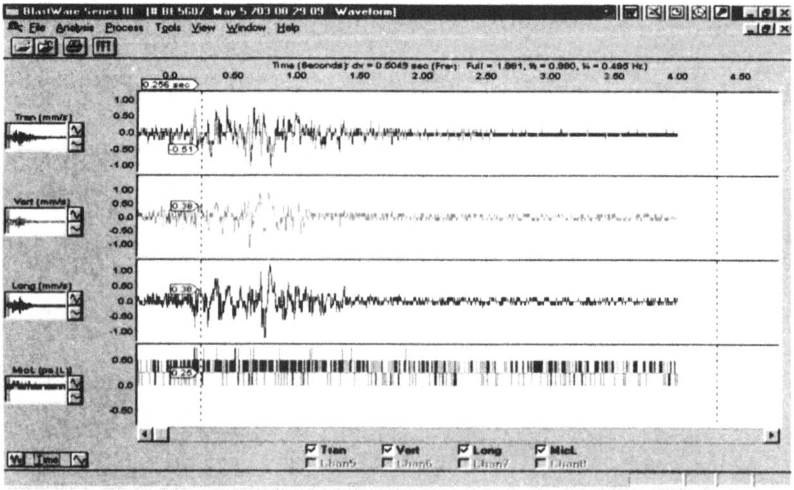

Fig. 14.3 Wave signature recorded during movement of a vehicle at 5.0 m distance

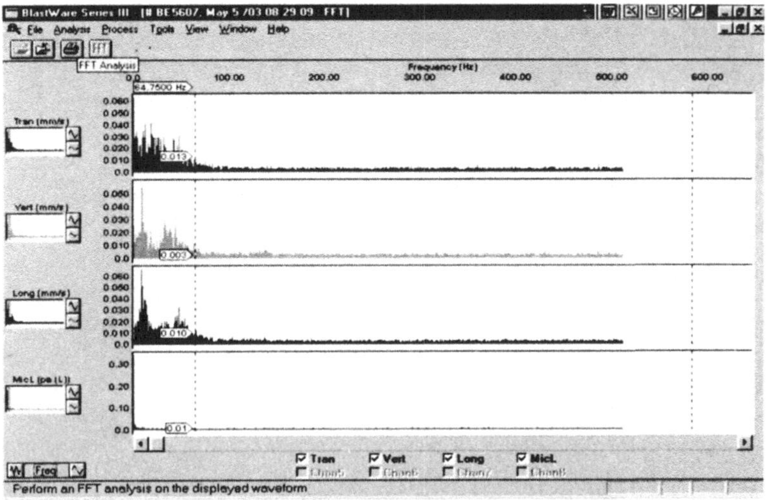

Fig. 14.4 Fast Fourier Transform (FFT) analysis of wave form of figure 14.3 showing energy concentration at different frequencies

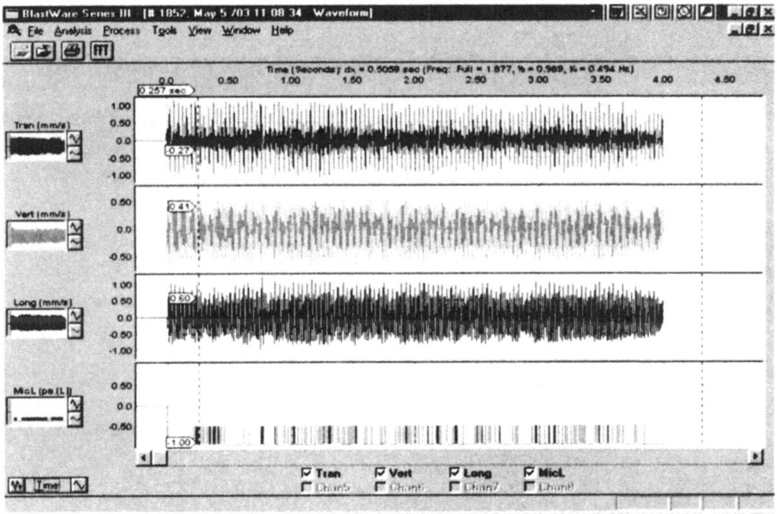

Fig. 14.5 Wave signature monitored during drilling operation at 1.75 m distance

14.3 SLOPE FAILURE DUE TO UNSCIENTIFIC QUARRYING

14.3.1 South Hlimen Quarry

South Hlimen Quarry is one of the biggest sandstone quarries in the state of Mizoram. It is located in the South Hlimen village, only 6 km from

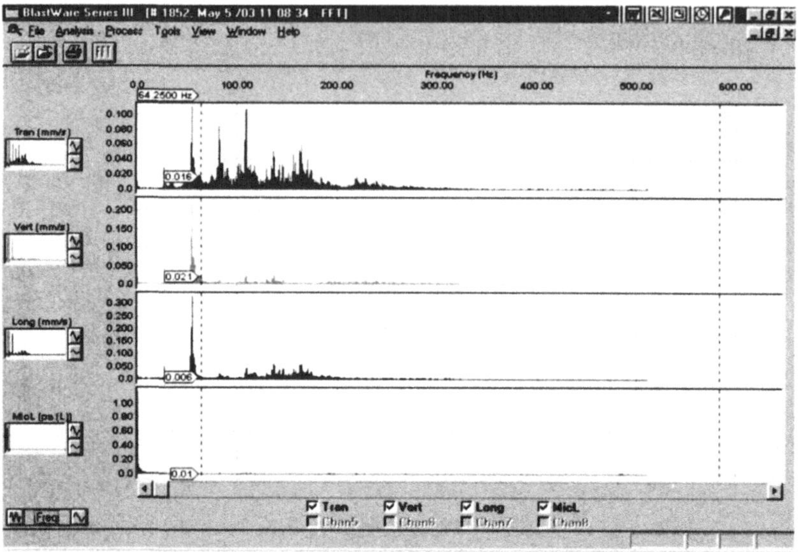

Fig. 14.6 Fast Fourier Transform analysis of wave form of figure 14.5 showing energy concentration at different frequencies

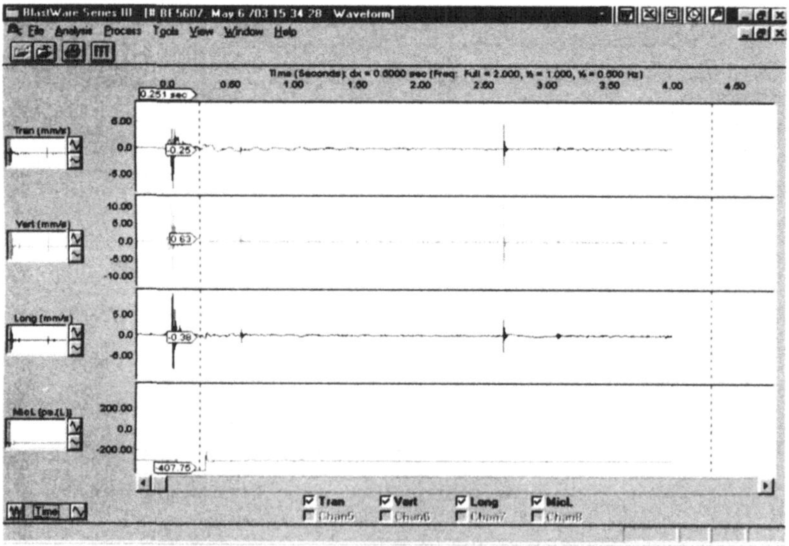

Fig. 14.7 Blast wave signature monitored during detonation of 27.50 kg of explosive at a distance of 31.30 m

Aizawl. The whole quarry is comprised of small sub-groups of quarries. The quarries are located along the length of a road which links Aizawl and Samtlang village. Before sliding took place, the rock was excavated upslope over a length of approximately 230 m along the slope with slope

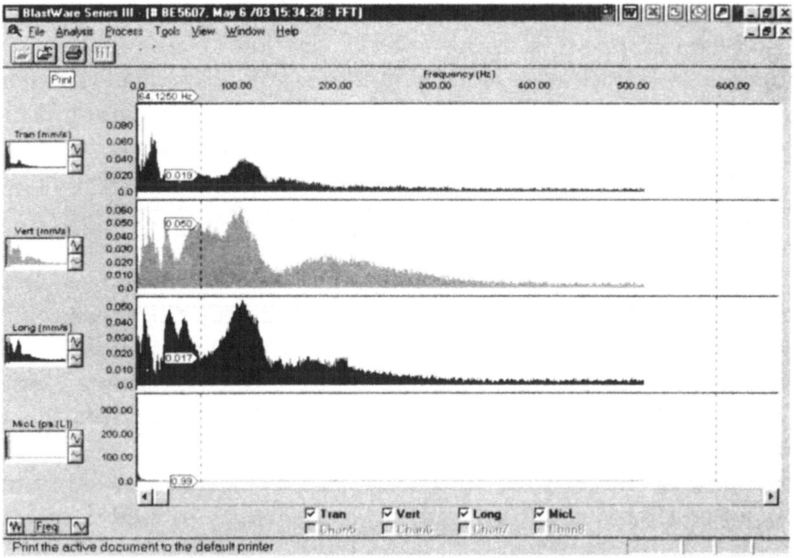

Fig. 14.8 Fast Fourier Transform analysis of wave form of figure 14.7 showing energy concentration at different frequencies

height varying between 5 and 10 m. Several houses were located along the road at the valley side of the quarries. On August 9th 1992, early in the morning (4:00 am) when the people were still asleep, the upper unexcavated portion of about 5-15 m thickness slid down along the slope of the valley killing 66 people including miners and local people (Plate 14.3).

Plate 14.3 South Hlimen quarry where an unprecedented landslide took place on September, 9, 1992, which killed 66 people and damaged several houses (Photograph was taken after 4 years of the incident)

A total number of 22 houses were damaged, alongwith 8 stone crushers and air compressors. Unfortunately, the labour camps and a few village houses, which were situated at the foothill of the quarry, suffered the wrath (Kumar et al., 1997). Due to safety and on general complaints raised by the local people, the state administration closed down the quarry for three years. It has recently started functioning again. Small scale mining and collection of rock boulders from the failed slope have started once again.

Figure 14.9 shows the block diagram of the total quarry. It is clearly depicted that the cause of the slope failure was mainly due to unscientific practice of quarrying. However, the local geology and topography of the area also favoured the slope instability. The rocks were dipping nearly 45° towards the slope direction with thickness varying between 1.0 and 5.0 m. The sandstone formation in the quarry area was also cut by a set of joints trending N75°W with vertical dip, which resulted in blocky formations. Some joints were filled with clay and thereby formed very weak joint planes when saturated with water during rainy season.

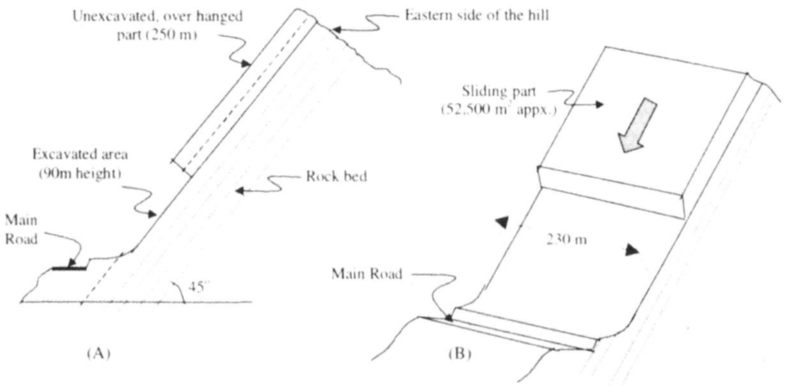

Fig. 14.9 **Block diagram of South Hlimen quarry (A) Side view, (B) Three dimensional view (Not to scale)**

14.3.2 Conclusion

From the above mentioned facts, it is clear that slope stability is very important in hilly areas which may be of natural slope or created during rock excavation. High intensity rainfall coupled with increased human activities may trigger occurrences of slope failure in the Himalayan regions (Sawmliana et al., 2000). Unscientific method of quarrying or mining activity often gives rise to disastrous slope failure in hilly terrains.

14.4 UNDERGROUND METAL MINING

14.4.1 Introduction

Theoretical and practical aspects of the optimization of drilling and blasting patterns for excavation of underground mineral deposits are discussed in this section, taking into consideration the technical and economical aspects. The case study of Khetri Copper Mine of M/s Hindustan Copper Limited, India is cited for proper understanding of the methodology.

14.4.2 Mining Methods with Vertical Crater Retreat (VCR) Stoping

In this method, a series of holes are fired in successive upward rounds with elongated spherical charges placed in optimum depth so that the excavated craters overlap each other to form as regular a roof as possible. After each round of blast, bottom of the holes are plugged before the subsequent round is charged. The length of charge may be calculated from the following relationship—

$$\text{Length of Charge, } L < 6 \times D \text{ (hole diameter in mm)} \qquad (14.5)$$

The broken ore is hauled away through cross cuts excavated from the extraction level to the draw point of the stope.

Once the explosive charge is placed inside the hole, it is stemmed to improve confinement with a length of inert material, nearly twice the charge length, by using fine sand or drill cuttings.

In this type of work, a determined initiation sequence is not necessary, as in bench blasting, owing to the characteristics of the breakage mechanism in crater blasts. However, when there are charges under the mean level of the stope ceiling, it is recommended that they be fired first. It is also convenient, whenever possible, that each charge has two free faces one at the bottom and other in the front. Sometimes, the term VCR is used although the mining method is a stoping method with end-slicing or inverted benching. Some mines are currently using a modified method where the charges are longer than 6 times the hole diameter. The term Vertical Retreat Mining (VRM) is used to indicate this modification (Persson et al., 1994).

14.4.3 Large Diameter Long Blastholes 'LBH' (115 mm and Above)

To determine the drilling pattern in the long blasthole zone, the following Langefors equation may be applied (Jimeno et al., 1995):

$$B_{max} = (D/33) \times [(\rho_c \times PRP)/(k \times f \times S/B)]^{1/2} \qquad (14.6)$$

Where, B_{max} = Maximum burden (m)
D = Hole diameter (mm)
ρ_c = Charge density (kg/dm^3)
PRP = Relative weight strength of explosive
k = Rock constant: 0.75 (For Hard rocks)
f = Fixation factor [= 0.9 Inclined holes; 3:1
= 0.85 Inclined holes; 2:1
= 1.0 Vertical holes]
S/B = Spacing / Burden

The drilling pattern influences the sizing of the drifts or cuts of the top level of drilling. The bench blasting in this method does not require toe breakage; therefore it is necessary to use the column charge. The most commonly used explosives are ANFO for hard to medium hard rock and ALANFO for very hard rock. If water is present in the blastholes, the charge can be placed in plastic covering or watergels and low-density emulsions may be used.

14.4.4 Relationship Between Length of Charge and Diameter

$$\text{Charge Length} > 20 \times D \text{ (mm)} \qquad (14.7)$$

For 165 mm drill diameter, charge length has to be greater than 3.30 m or say 3.50 m that would contain nearly 70 to 73 kg of explosives.

14.4.5 Intermediate Stemming

The volume of rock in front of the intermediate stemming tends to make fragmentation worse. If similar types of stemming and decking columns are maintained in all successive parallel holes (large diameter), then the chances of big boulders will be more pronounced. In order to place stemming in a zigzag system, it was found that in parallel holes (large diameter), alternatively 3.5 m charge column and 2.0 m stemming followed by 4.0 m charge and 2.5 m stemming columns resulted in better fragmentation. The length and type of intermediate stemming between deck charges should be such that it would prevent sympathetic detonation or desensitization of adjacent charges which are initiated at different time intervals and at the same time achieve adequate fragmentation along the length of the stemming column. For intermediate decking, the following guidelines can be used.

Explosive	*Intermediate stemming*
ANFO	12 to 15 D [1.98 - 2.48 m for 165 mm dia.]
Watergels	16 to 18 D [2.64 - 2.97 m - do -]

14.4.6 Initiation Sequence

According to Du Pont, the initiation sequence of the blast is from the bottom upward, with the following recommended delay timings:

- Charges in the same blasthole .... 50 ms
- Adjacent charges in the same row .... 10 ms
- Charges between rows .... 75 ms

During experimental trials in Khetri Copper Mine of M/s Hindustan Copper Limited, India, different initiation sequences as depicted in figures 14.10 and 14.11 were implemented. Some of these experiments were in contrast to the recommendations made by Du Pont. Such design patterns could solve the problem of big boulders and dilution of ore in open stopes of the mine and subsequently help the management to increase production and productivity under a safe working environment.

14.4.7 Case Study (Khetri Copper Mine)

Due to steep dip and lenticular nature of the deposit and the competency of wall rock and ore body, a highly productive sublevel open stoping method of mining was adopted in Khetri Copper Mine of M/s Hindustan Copper Limited, India. The thicker portion of the ore body (about 25 m or above) is worked in the transverse direction, while the other parts of it are worked by longitudinal stoping. Large diameter (165 mm) drillholes are used for widening to the full width of the stope and small diameter (57 mm) long drillholes are used in a ring pattern from the extraction level.

14.4.7.1 Method of Work

In the transverse method, 20 m wide stopes are worked from hangwall to footwall leaving 10 m wide barrier pillars. But in longitudinal stoping, stopes are worked along the strike.

Each stope is worked between two levels, vertically 60 m apart. Two sublevels are developed along the centre of the ore body dividing the vertical intervals into three equal parts. Sometimes in longitudinal stopes, the ore body varies widely in width within a single stope. In such cases, the sublevels in the wider zone are developed by a pair of sublevel drives instead of a single drive.

The stopes are worked in retreating fashion, from hangwall to footwall in the case of transverse stopes, or from one end to another along the strike in case of longitudinal stopes, as the case may be. A slot raise is first driven at the starting point of a stope connecting from bottom to top. The slot raise is driven by blasting large diameter (165 mm) blastholes in a Vertical Crater Retreat (VCR) method. It is then widened to the full width of the stope. This slot raise provides the first free face for blasting during extraction.

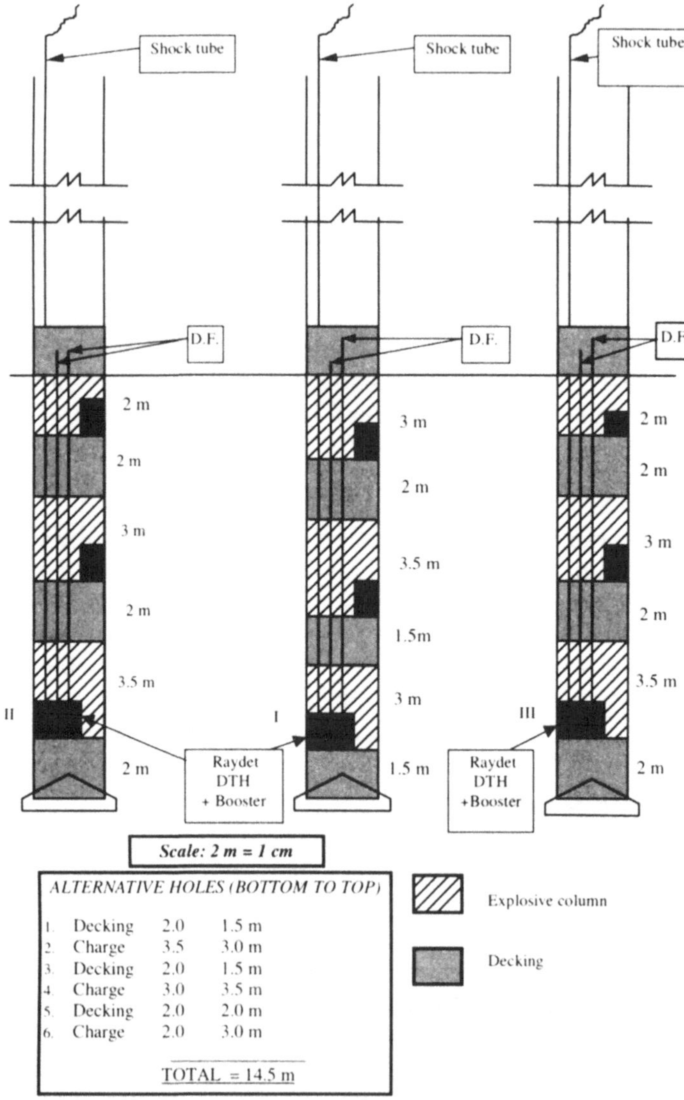

Fig. 14.10 Charging pattern of large diameter holes with DTH and D.F.

A series of 57 mm diameter long holes drilled in a ring pattern is driven from the extraction level. In the case of sublevel stoping, 57 mm diameter holes are drilled in a ring pattern from the two sublevels in the stope. In large stopes, 165 mm diameter holes are directly drilled from the crown level of the stope to the extraction level. They are blasted against the slot, which is also developed by blasting large diameter holes following the principle of "Vertical Crater Retreat".

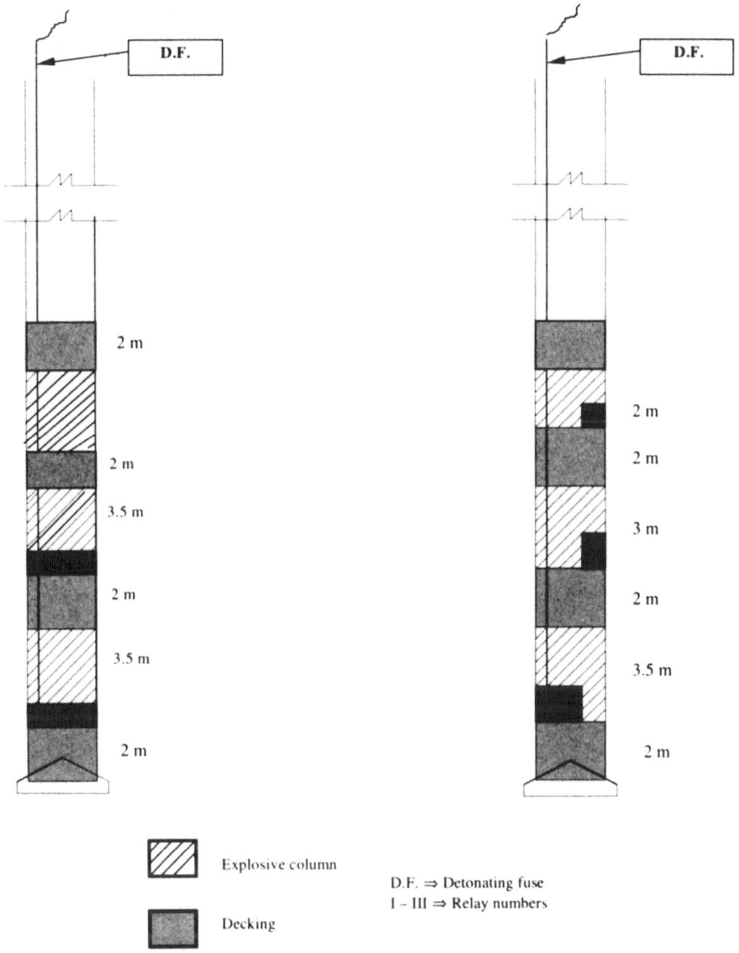

Fig. 14.11 Charging pattern of large diameter holes with detonating fuse

14.4.7.2 Geotechnical Parameters

The rock mass in this mine is jointed. The frequency and intensity of jointing vary from place to place. Joint planes are mostly straight and sharp but at a few places they are undulating. Slickensiding along the joint planes is not a rare feature. There are seven sets of joints in this area and the best-developed are trending 30° and 111°. Though, there are seven sets of joints, all of them are rarely seen in a single place. These joint systems can be conveniently classified as shown in Table 14.6. Besides these joints, there are other structural weaknesses in this area as well. Features associated with large faults indicate that they are essentially shear faults. Shears associated with folding are mainly concentrated along the bedding planes. From field observations, it is evident that

Table 14.6 Classification of joint sets (CMRI, 1989)

Joint Set	*Trend (N to N)*	*Dip*	*Remarks*
Dip Joints	110° to 125°	60° to 90° due NE	Prominent
Strike Joints (1)	25° to 35°	45° to 75° due SE	Strike joint sets (1) and (2) bear
Strike Joints (2)	25° to 35°	45° to 75° due NW	conjugate relationship. Both sets are prominent.
Oblique Joints (1)	105° to 115°	40° to 70° due E	Oblique joint sets (1) and (2)
Oblique Joints (2)	105° to 115°	40° to 70° due W	bear conjugate relationship with each other.
Bedding plane	30° to 50°	65° to 75° due W	Conspicuous, often found in quartzite.
Others	—	0° to 25°	All other joint sets mostly belong to this category.

shearing is the main mechanism for the formation of most of the faults developed in this area. Many of the faults indicated in surface plan are not traceable in underground levels.

Among the faults, longitudinal faults are not recognizable in the field; though at places they are slickensided, often with superimposed grooves of different generations indicate the presence of these faults.

There are transverse faults also and they trend oblique to the contacts of the beds or the longitudinal faults. They are in conjugate sets. Irrespective of the topography, most of the transverse faults are straight in their trend.

Besides the above structural features, localized shear zones are common in this area. They usually trend NE with varying dips and directions. There are two main shears in the mines that can be conveniently termed as the Northern Shear and the Southern Shear. They vary in size and thickness both along strike and dip. They are characterized by localised crushing, leaching along the planes, slickensiding and by a network of irregular fractures with carbonates, quartz veins and sulphide fillings. The Northern Shear is nearer to the service shaft between 1300 and 1400 departure towards west. Their strike is N50°-60°E and dip 60°-75°. The Northern Shear dips in a SE direction with an average thickness of 10 m and the Southern Shear measuring 5 to 10 m in thickness dips in a SW direction. Both the shears are dry.

Besides these two intersected shears, there are other minor shears running parallel with little or no ground problems. Borehole intersections on the south of Kudhan fault at +300 m level have indicated numerous parallel shear zones. These shears carry a good quantity of ground water. Table 14.6 shows a brief classification of joint sets.

14.4.7.3 Rock Strengths

NX core samples from 180 and 240 m levels were collected and tested using standard testing procedures in the laboratory, to determine the following average parametric values.

(1) Uniaxial compressive strength (MPa): 90.8 – 164.8
(2) Tensile strength (by Brazilian Test in MPa): 10 – 16.6
(3) Shear strength (MPa): 34 – 54.8
(4) Young's modulus (GPa): 34.3 – 42.3.

14.4.7.4 Blast Design Patterns

In the case of Khetri Copper Mine the following parametric values of equation 14.6 were used:

$D = 165$ mm; $k = 0.75$; $f = 1.0$; $S/B = 1.25$, $\rho_c = 0.90$ gm/cc; $PRP = 0.911$

Equation (14.6) determines:

$$B_{max} = 4.23 \text{ (for 165 mm) and } B_{max} = 2.95 \text{ (for 115 mm)}$$

The value of the practical burden (B_{prac}, in m) may be computed from the following equation (14.8) (Jimeno et al., 1995). This value is established from the maximum value, applying a correction factor for the deviation of the blastholes and collaring errors.

$$B_{prac} = B_{max} - 2 \times D - 0.02 \times L \tag{14.8}$$

Where L is the blasthole length and D is the diameter of hole, both in metres.

Using equation (14.8), the practical burden values are obtained as:

$B_{Prac} = 3.00$ m (165 mm dia. hole)

$= 1.82$ m (115 mm dia. hole).

Thus, for 165 mm drill diameter, $B_{max} \times S_{max} = 22.37$ m^2 and $B_{prac} \times S_{prac} = 11.25$ m^2. The average of these two values i.e. $B_{avg} \times S_{avg} = 16.81$ m^2. Considering B_{prac} which equals to 3.00 m as acceptable value of burden, the spacing becomes 5.6 m.

14.4.7.5 Study of Vibration for the Safety of U/G Structures

When a blast performs well, the available waste energy will be less in comparison to an unoptimized blast even of a smaller size. In order to see the extent of vibration levels, near-blast monitoring was carried out near a trough drive and adjoining cross-cuts while blasting small diameter (57 mm) ringholes. Special type of attachments was used to place the sensors in the roof, pillar and floor. Based on the data generated, the following predictor equation (14.9) was derived:

$$V = 590\,(D/Q^{1/2})^{-1.5} \tag{14.9}$$

$$r \text{ (correlation coefficient)} = 0.80$$

Where, V = Peak particle velocity (mm/s)
Q = Maximum charge per delay (kg)
D = Distance from the source (m).

In case of a Swedish rock, with a tensile strength of 5-11 MPa, the damage was found to occur when the peak particle velocity exceeded 700-1000 mm/s. It is interesting to note that from equation (14.9), when $V = 300$ mm/s, $D = 50$ m then Q becomes 1010 kg. With the same amount of charge, the value of '*V*' would become 100 mm/s at a distance of only 104 m from the source. Therefore, from the point of view of safety, 5 t of total charge and 1000 kg of maximum charge per delay were recommended for the mine. With such charge levels, there was no threat of danger to the underground and surface structures.

14.4.7.6 Delay Connections for Large-Diameter Ringholes

After repeated trials, the following delay patterns were established to conduct safe and efficient blasting practices in large-diameter ring pattern of drillholes (Figures 14.12 to 14.14).

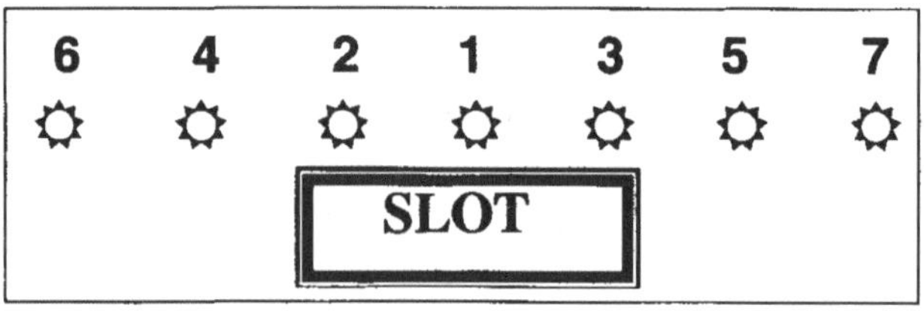

Fig. 14.12 Delay connections for single row of holes

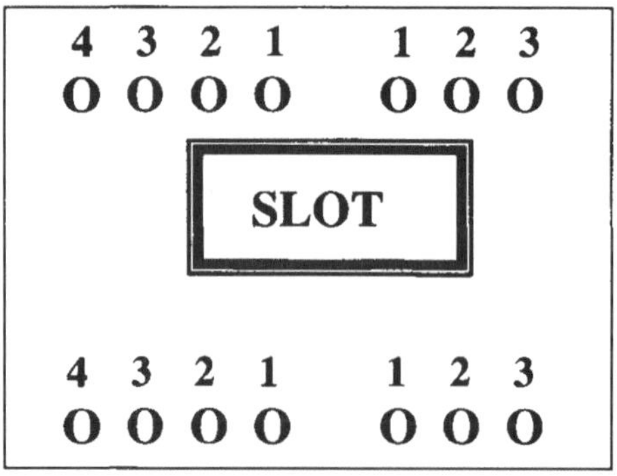

Fig. 14.13 Delay connections for two rows of holes

14.4.7.7 Calculation of Toe Spacing in Long-hole Small-diameter Design Patterns

In order to calculate the toe spacing in designing long-hole, small-diameter ring patterns in stope blasting, the following Table 14.7 was suggested for use as a tool.

14.4.7.8 Sublevel Stoping with Blastholes in Fan Pattern

Sublevel stoping with blastholes in a fan pattern is preferable for subvertical ore beds with sidewalls that have good characteristics (Figure 14.15). In this system when the ore is extracted, large open stopes are left behind. The separation between two consecutive drilled sections may be kept at around 1.2 and 1.8 m (for 51 to 64 mm drill diameter). However,

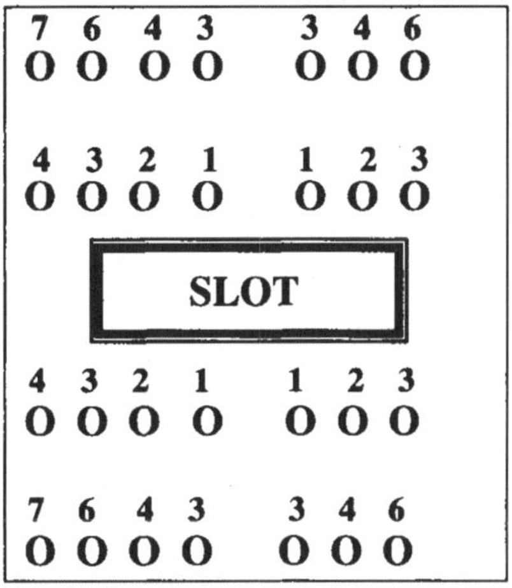

Fig. 14.14 Delay connections for four rows of holes

Table 14.7 Calculation of toe spacing

Hole length (m)	*Burden (m)*	*Spacing (m)*	*Area covered(m^2) $Ep = B \times S$*	*Toe spacing at fixed ring burden, V_{pm}*		
				V_{pm} =1.3 m	V_{pm} = 1.5 m	V_{pm} =1.8 m
				E_p/V_{pm}	E_p/V_{pm}	E_p/V_{pm}
1	1.97	2.47	4.87	3.74	3.24	2.70
2	1.94	2.44	4.73	3.64	3.16	2.63
3	1.91	2.41	4.60	3.54	3.07	2.56
4	1.88	2.38	4.47	3.44	2.98	2.49
5	1.85	2.35	4.35	3.34	2.90	2.41
6	1.82	2.32	4.22	3.25	2.81	2.35
7	1.79	2.29	4.10	3.15	2.73	2.28
8	1.76	2.26	3.98	3.06	2.65	2.21
9	1.73	2.23	3.86	2.97	2.57	2.14
10	1.70	2.20	3.74	2.88	2.48	2.08
11	1.67	2.17	3.62	2.79	2.42	2.01
12	1.64	2.14	3.51	2.70	2.34	1.95
13	1.61	2.11	3.40	2.61	2.26	1.89
14	1.58	2.08	3.29	2.53	2.19	1.83
15	1.55	2.05	3.18	2.44	2.12	1.77
16	1.52	2.02	3.07	2.36	2.05	1.71
17	1.49	1.99	2.96	2.28	1.98	1.65
18	1.46	1.96	2.86	2.20	1.91	1.59
19	1.43	1.93	2.76	2.12	1.84	1.53
20	1.40	1.90	2.66	2.05	1.77	1.40
21	1.37	1.87	2.56	1.97	1.71	1.42
22	1.34	1.84	2.47	1.90	1.64	1.37
23	1.31	1.81	2.37	1.82	1.58	1.53
24	1.28	1.78	2.23	1.75	1.52	1.27
25	1.25	1.75	2.19	1.68	1.46	1.21

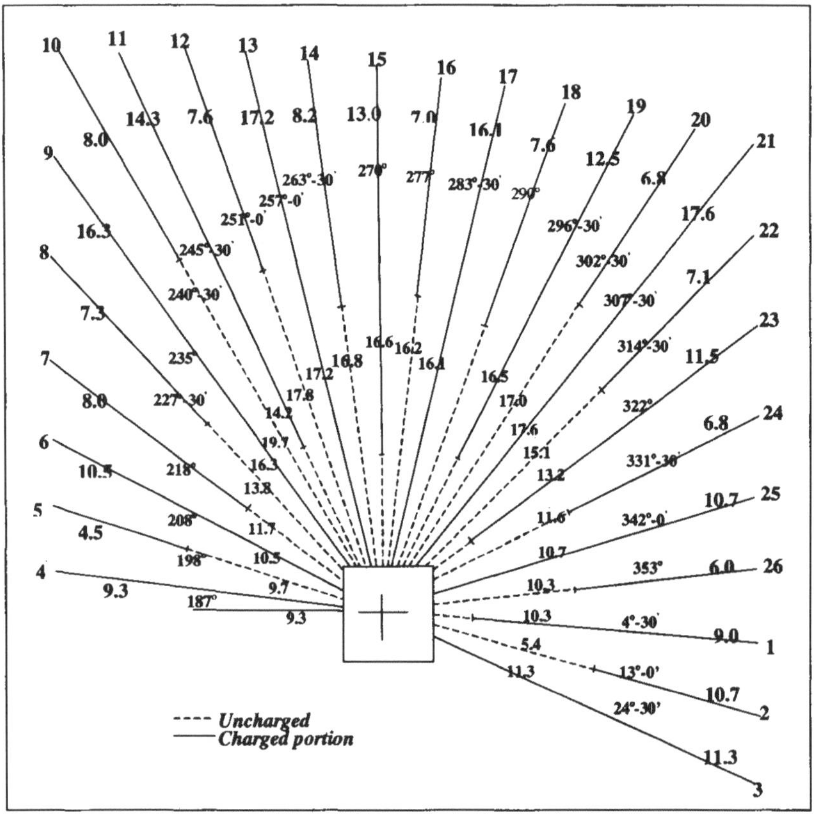

Fig. 14.15 Design of ring-hole blasting pattern used in BSL, Ring No. 11 (Lat. 4596.40, Dep. 1217.50) of Khetri Copper Mine, HCL

for 57 mm drill diameter, 1.5 m of separation is the ideal. To obtain favourable blasting results, collaring, orientation and deviation of the blastholes are the major operative conditions necessary to be checked.

The yield of broken rock per linear metre drilled is usually low because the planned spacing diminishes as it nears the collar, leaving part of the blastholes unused in the breakage.

The calculation of the drilling patterns are taken from the necessary charge factor, which is a function of the rock type, drilling length and width of the blast (Jimeno et al., 1995).

$$CE = CE_o + 0.03\,L + 0.4/A_V \tag{14.10}$$

Where, CE = designed charge factor in the bottom of the blasthole and in one-fifth the length of the same, expressed in kg/m^3 of gelatin explosive.

$$CE_o = \text{Base } P.F. = 0.4 \text{ kg/m}^3 \tag{14.11}$$

$$L = \text{Hole length (m)}$$

A_V = Width of the round (m) = 1.5 m
$A_e = S \times B = q_l$ (kg/m)/CE = 2.25/CE (for 57 mm diameter)
and $S = 2 \times B$.

Table 14.8 Theoretical determination of toe spacing and burden obtained from equation (14.10) [*width of the round = 1.5 m*]

Hole length(m)	1.0	2.0	3.0	4.0	5.0	6.0	7.0	8.0	9.0	10	11	12
Burden (m)	1.27	1.24	1.22	1.20	1.17	1.15	1.13	1.11	1.10	1.08	1.06	1.05
Toe Spacing (m)	2.54	2.49	2.44	2.39	2.35	2.31	2.27	2.23	2.19	2.16	2.12	2.09
Hole length(m)	13	14	15	16	17	18	19	20	21	22	23	24
Burden (m)	1.03	1.02	1.00	0.99	0.98	0.97	0.95	0.94	0.93	0.92	0.91	0.90
Toe Spacing (m)	2.06	2.03	2.01	1.98	1.96	1.93	1.91	1.88	1.86	1.84	1.82	1.80

14.4.8 CONCLUSIONS

The following recommendations were made for safe, economic and productive drilling and blasting operations at Khetri Copper Mine.

1. Blasting a full round of large diameter holes, either in a parallel or in a ring pattern was found to provide better fragmentation. The use of detonating fuse and relay connections was suggested as a first preference in large diameter holes. Shock tubes (down-the-hole delays) were found to be suitable in small diameter ringhole blasting.
2. Staggered charging patterns of consecutive holes as depicted in figures 14.10 and 14.11 were established for better fragmentation.
3. Charge length should be around 60 to 63% of the total length of the hole.
4. In large diameter ring patterns, if there were seven holes in the full ring, then 3.5 m charge length and 2.0 m decking might be followed alternatively. When the total number of holes became six, it was convenient to use 4.0 m charge length and 2.0 m decking.
5. Different initiation patterns for large diameter holes for 1 row, 2 rows and 4 rows are given in figures 14.12 to 14.14.
6. In view of the rock stability, it was found that a total charge of nearly 5 tonnes of explosive and 1000 kg of maximum charge per delay were safe beyond a distance of 50 m.

14.5 UNDERWATER BLASTING

Underwater blasting is generally carried out for the purpose of construction of berths/jetties, making of channels and trenches for placing underwater pipes, deepening dock basins in ports and canals. In this kind of blasting, the drillhole openings are located under water and thereby complicate the charging work. It is also very difficult to check drilling precision with respect to collaring and direction. There are several operational constraints like positioning of drillhole, charging of holes and delay arrangements that make underwater blasting more cumbersome and time consuming. These are added with detrimental effects like ground vibrations, water-borne shock waves and noise. These side effects may cause damage to vessels, human beings, jetties, sensitive installations around, and aquatic life (Sastry, 2000). Gustafsson (1981) proposed the following design patterns for underwater blasting—

1. Charge factor (C): 0.90 kg/m^3 (which is almost double than the normal bench blasting)
2. In case of vertical holes, increment in charge factor: 10% of C
3. In case of water pressure, increment in charge factor: 0.01 × water depth (m)
4. In case, where the rock is covered with clay, increment in charge factor: 0.02 × thickness of clay (m)
5. For rock sections, increment in charge factor: 0.03 × bench height (m)
6. Under drilling (m): Usually same as burden (m) but minimum 0.8 m
7. Hole depth (m): Bench height (m) + under drilling (m)
8. Spacing (m) = Burden (m); square pattern
9. Uncharged section (m) = 1/3rd of burden (m) but at least 0.5 m.

Any slurry (like Kelvex-P) or emulsion explosive with specific gravity of 1.2 may be used for underwater blasting. Initiation of explosive charges is done by using electric delay detonators with long lead wires.

REFERENCES

Bauer, A. and Calder, P. 1978. Open pit blast seminar, Course note, Queen's University, Kingston, Ontario, Canada.

Bhasin, R., Grimstad, E., Larsen, J.O., Dhawan, A.K., Singh, R., Verma, S.K. and Venkatachalam, K. 2002. Landslide hazards and mitigation measures at Gangtok, Sikkim, Himalaya, Engineering Geology, Elsevier, Vol. 64, UK, pp. 351-368.

Ryu, Chang-Ha, Choi, Byung-Hee, and Sunwoo, Choon, 2003. A case study of dimension stone blasting in Korea, Proc. 29th Annual Conference on Explosives and Blasting Technique, Feb. 2-5, Nashville, Tennessee, USA, pp. 231-248.

Chiappetta, R. F. 1991. Pre-splitting and controlled blasting techniques including air- decks and dimension stone criteria, Proc. Seminar on Blasting Techniques, Instrumentation and Explosives Application, R. F. Chiappetta (ed.), San Diego, USA: Section 25. Allentown, PA: Blasting Analysis International.

CMRI Report of Investigations, 1989. Ground control and rock mechanics studies for evolving new mining methods to promote productivity, safety and conservation in non-coal mines, GAP/VI/1 (MOM), Catholic Press Publ., India, pp. 33-72.

Dunn, P. and Cocker, A. 1995. Analysis of mining conditions by integration of high-wall mapping and geological modeling techniques. Proc. Bowen Basin Geologists Symposium, Geological Society of Australia, Brisbane, Australia.

Gustafsson, Rune, 1981. Blasting technique, Dynamit Nobel Wien Gesellschaft m.b.H., Vienna, Austria, p. 327.

Jimeno, Lopez Carlos, Jimeno, Emilio Lopez and Carcedo, F. J. A. 1995. Drilling and blasting of rocks, A. A. Balkema, Rotterdam, Netherlands, Chapter-24, p. 391.

Kumar, S., Singh, T. N. and Sawmliana, C. 1997. On the instability problems of South Hlimen Quarry, Mizoram, Indian Journal of Engg. and Mat. Sciences, Vol. 4, pp. 78-80.

Langefors, U. and Kihlstrom, B. 1978. The modern technology of rock blasting, John Wiley and Sons Inc, New York, USA, p. 438.

Mandal, S. K., Singh, M. M. and Pal Roy, P. 2003. Pre-split blasting vis-à-vis dimension stone quarrying, Proc. New Equipment, New Technology Management and Safety in Mining and Mineral based Industries, May, 11-12 Bhubaneswar, Part-I, pp. 7-9.

Mehtrotra, G.S., Sarkar, S., Kanungo and D.P., Mahadevaiah, K. 1996. Terrain analysis and spatial assessment of landslide hazards in parts of Sikkim Himalaya, Journal of Geol. Soc. Soc. India, Vol. 47 (April), pp. 491-498.

Oriard, Lewis L. 2000. Explosives engineering, construction vibrations and geotechnology, International Society of Explosives Engineers (ISEE), Cleveland, Ohio, USA, p. 680.

Persson P.A., Holmberg R. and Lee J., 1994. Rock blasting and explosives engineering, CRC Press, Inc., USA, p. 380.

Rustan, P. A. 1996. Micro-sequential contour blasting-theoretical and empirical approaches, Proc. 5th Int. Symposium on Rock Fragmentation by Blasting, Fragblast'5, Montreal, Canada, 25-29 August, pp. 157-165.

Sastry, V. R. 2000. KREC, Mangalore, Karnataka (Personal Communication).

Sawmliana, C., Singh, R. B., Singh, P. K., and Singh, T. N. 2000. Slope reinforcement techniques to mitigate landslide in Aizawl town, Mizoram. Proc. National Symposium on Rock Mass Reinforcement (NSRR), 28-29 January, IT, BHU, Varanasi, India, pp. 107-124.

15

Performance Evaluation of Explosives and Accessories

15.1 PERFORMANCE TEST OF EXPLOSIVES

15.1.1 Introduction

When charges are detonated in boreholes, the shock energy transmitted to the rock depends on the detonation pressure of the explosive. It is a form of kinetic energy. The detonation pressure is maximum in the direction of shock-wave travel. After the shock energy passes, gas energy is released. The expanding gas energy causes the majority of the rocks to break. The gas pressure is the pressure exerted on the borehole walls by the expanding gases after the completion of the chemical reaction. The amount of explosion pressure is related to the volume of gas liberated per unit weight of explosive and the amount of heat liberated during the reaction. If more volume of gas is liberated at a constant temperature, the pressure would also increase. The gas energy produces a force that is constantly maintained until borehole ruptures. The action of gas pressure is important in loosening, expanding and throwing the mineral.

15.1.2 Explosive Ratings

Several parameters such as energy, strength, brisance, impulse and bubble energy are often in use to rate explosives in terms of blasting performance. However, multiplicity of terms has led to their unrealistic rating. Though in absolute terms, capability of an explosive is decided by the energy available from the mass, some factors are important to choose the best explosive for a job. They are:

- Velocity of detonation
- Detonation pressure

- Energy output/strength
- Safety characteristics
- Post-blast fumes
- Density
- Sensitivity
- Water resistance
- Temperature stability
- Shelf life

RBS (Relative Bulk Strength), RWS (Relative Weight Strength) compared to ANFO, Shock and Bubble energies in underwater tests are used to assess the performance characteristics of commercial explosives. VOD (Velocity of Detonation) is an excellent means of checking the performance of explosives and initiation system in a blast. A high VOD creates a high frequency of fine cracks in the vicinity of drillholes. Measurement procedures of VOD under varying conditions and their implications were discussed, in detail, in Section 12.1.2.1 of Chapter 12 (also available in Pal Roy, 1995).

The results indicated that when the decking was made of drill cuttings, the VOD in the decking portion abruptly reduced to 1/10 to 1/13 of the explosive's VOD. When low strength explosives were used for firing, then material decking posed almost a non-effective zone. It was seen that the shock tube alone (i.e. without using detonator) could not detonate the explosive because the amount of energy liberated instantaneously through shock tube was too little to detonate a commercial explosive. Sharp attenuation of VOD was also marked in air gaps under confined condition. Tables 15.1, 15.2, 15.3 and 15.4 show some of the measured values in laboratory set-up (Pal Roy et al., 1998).

15.1.3 Mixture of Ammonium Nitrate and Fuel Oil

In the mixture of ammonium nitrate and fuel oil, it has already been shown in Table 4.6 of Chapter 4 that approximately 6% fuel oil mixture makes the optimum balance. The calculation was based on "Energy" Software developed by the Precision Blasting Services, USA. It was observed that as the oil content decreases from the desired 6%, there was a rapid energy loss. The extent and the rapidity of energy loss would be more in less fuel oil mixture than that in the excess fuel oil mixture. The strength of the explosive and the energy delivered during blasting process are not linearly related with heat released and therefore, using the heat released chart alone, as an indicator of strength or energy loss, would be misleading.

15.1.4 Conclusion

The strength of the explosive and the energy delivered during blasting process do not have any linear relationship with heat released and

Table 15.1 Results of VOD measurements carried out in laboratory set-up (Permitted explosives)

Type and brand of explosive	*Test No.*	*Dimension of explosive*	*Velocity of detonation*
Powergel GE-222 Emulsion explosive (P_5 group)	[A]	21.5 cm: length 32 mm: diameter 150 g: weight	Explosive = 3438 m/s (3250 m/s was determined through Dautriche method) EXEL shock tube = 2215 m/s
Belgex Coal-M (P_5 group)	[B]	22 cm: length 32 mm: diameter 200 g: weight	Explosive = 1805 m/s (1825 m/s was determined through Dautriche method)
Powergel GE-222 Emulsion explosive (P_5 group) *Firing was done through shock tube initiation. The end of the shock tube was inserted in the explosive cartridge.*	[C]	21.5 cm: length 32 mm: diameter 150 g: weight	*Remarks*: No firing took place in the explosive.
Powergel GE-222 Emulsion explosive (P_5 group) in one end and Prachar (P_5 group: NG) explosive at the other end. Detonator was tagged with the GE-222 explosive and fired.	[D]	GE-222: 32 mm: diameter 150 g: weight 21.5 cm: length and Prachar: 22 cm: length 32 mm: diameter	1. The explosive Prachar detonated 2. V_{12} = 3000 m/s, V_{23} = 2824 m/s V_{34} = 2616 m/s; V_{ij} is the VOD between sensor positions i and j. VOD (explosive) = 3438 m/s. 3. Reduction of VOD with respect to the distance from the source point. 4. 70.5 cm was the gap in which explosive was detonated under confined condition.
Prachar (P_5 group: NG) explosive was tagged with the detonator and fired in confined condition.	[E]	22 cm: length 32 mm: diameter 200 gm: weight	At a distance of 45 cm, the measured VOD was 2378 m/s. This value was comparatively lesser than the value posed by emulsion explosive.

Table 15.2 Air-gap sensitivity in unconfined conditions

Type and brand of explosive	*Dimension of explosive cartridge*	*Air gap*
Powergel GE-222 Emulsion explosive (P_5 group)	21.5 cm length, 32 mm diameter, 150 g weight. It was fired with a No. 6 strength copper detonator.	2 cm
Prachar (P_5 group: NG)	20.0 cm length, 32 mm diameter, 200 g weight. It was fired with a No. 6 strength copper detonator.	5 cm

Table 15.3 Measurement of strengths by Ballistic Mortar Test

Type and brand of explosive	*Explosive quantity*	*Strength*
Powergel P-301 (GE-221) Emulsion explosive (P_3 group)	10 g cartridge in tinfoil and 22 mm diameter	51% (Considering BG as 100)
Powergel GE-222 Emulsion explosive (P_5 group)	- do -	Inconclusive—because of partial detonation.
Powergel GE-223 Emulsion explosive (P_1 group)	- do -	68.5%

Table 15.4 Test results of a few permitted (P) and non-permitted (NP) commercial explosives [Relative strengths are w.r. t. B.G.]#

Explosive brand	*Type of explosive*	*RWS (%)*	*RBS (%)*	*Specific gravity*	*VOD (m/s)*
Permaflo - 1	Slurry (permitted)	61	47	1.10	3200
Permaflo-3	-do-	58	44	1.10	3200
Powergel-I	Emulsion (permitted)	69	53	1.20	4800
Powergel-C	-do-	70	70	1.20	5000
Kelvex-100	Slurry (non-permitted)	-	117	1.10	3600
Kelvex-220	-do-	-	171	1.22	4500
Dynex-C	-do-	71	54	1.20	3500-4000
Dynex-3	-do-	60	46	1.10	3500-4000

Source: Jr. Mines, Metals and Fuels, November 1986, pp. 519-528.

therefore, using the 'heat released chart' alone as an indicator of strength or energy loss is misleading. The velocity of detonation (VOD) has an important role in the breakage process. Lower velocity of detonation may not result in coarser fragmentation because the geology seems to be the dominant factor in the fragmentation process. Using an explosive with a high VOD will lead to more fines after blasting as compared with an explosive with a lower VOD, keeping the same powder factor. Visualization of post blast fume is also a way to check the performance of explosives in field conditions. An air-column followed by explosive columns poses significant detonation pressure, which can detonate subsequent explosive columns. However, the value of VOD attenuates rapidly while passing through the air columns. This is the reason why one should select optimum gap sensitiveness while using air columns.

15.2 SCATTERING IN DELAY TIMING OF DELAY DETONATOR

15.2.1 Introduction

Explosive and detonators are widely used for rock blasting in open-pit mines, tunnelling and dam construction works. In such works, there are three major areas of concern to a shot blaster. They are productivity including good fragment size, few boulders and low overbreak. Environmental problems such as ground vibration, air blast, flyrock, dust and fumes can affect neighbours. To overcome several blast hazards in practical blasting, it has been proved that delay detonators (short delay and half-second delay) play a key role in controlled blasting. In India, delay detonators of the following specifications are commercially produced.

1. Short delay (aluminium or copper tube of delay series 0 to 10) with delay periods varying between 25 and 75 ms. These are used in quarrying, open-pit mines, tunnelling and shaft sinking.
2. Short delay (copper tube and also copper coated steel non-incendive of delay series 0-6). These have a delay interval of 25 ms and are used in blasting of coal off the solid and drifting in gassy coal mines.
3. Long delay (aluminium tube of delay series 0-10 and delay interval of 300 and 500 ms). These are used in quarrying, tunnelling, shaft sinking, drifting etc.

15.2.2 Winzer Index

Winzer et al. (1979) derived a non-dimensional critical parameter, which would estimate a satisfactory probability of success in delay blasting. According to him, a value for the Winzer Index, '*S*' of less than 3 indicates a strong probability of overlapping between adjacent delay periods.
The Winzer Index '*S*' is computed using the following relationship:

$$S = \frac{T_{n+1} - T_n - r}{\left[S_n^2 + S_{n+1}^2\right]^{1/2}} \quad (15.1)$$

where

T_{n+1} = average delay time period for (n+1)th delay number
T_n = average delay time period for nth delay number
r = required firing time interval between adjacent numbers
S_{n+1} = standard deviation of periods of (n+1)th delay number
S_n = standard deviation of periods of nth delay number

The probability that two boreholes, primed with adjacent periods of detonators, would fire within a given time interval relative to each other, can be estimated from the measurement of firing times and standard deviations of the firing times for each of the two periods. However, with the time interval, the time requirement for boulders to move out sufficiently so that the subsequent shots do not fire in tight formation is not precisely known. The most liberal assumption of this time is zero.

15.2.2.1 Short Delay Series 0 to 10

Nabiullah et al. (1989) carried out extensive experiments on the performance analyses of commercially available delay detonators. They carried out experiments on the following series—

(a) Al tube short delay series 0 to 10
(b) Copper-coated steel body non-incendive short delay detonators
(c) Short delay non-incendive type of copper tube detonators
(d) Al-tube long delay detonators.

The conclusions made by the investigators, in brief, are as follows (Nabiullah et al., 1989)—

(a) Al-tube short delay detonators had probability of overlapping in their delay timings. The percentage of misfires was 2.8 to 14.2.
(b) The probability of overlapping of delay timings and misfires of copper-coated steel detonators was negligible.
(c) Delay interval of copper tube non-incendive short delay detonators had low probabilities of overlapping.
(d) Delay interval of detonating relays had no probability of overlapping and the average delay interval was equal to nominal values.
(e) Post-blast observations in several open-pit, underground mines and tunnelling indicated that the single factor responsible for variation in results was overlapping of delay periods of most commonly used Al-tube short and long delay detonators. It is therefore suggested that the probability of overlapping in delay periods of delay detonators should be checked by random sampling before their use in rock blasting.

15.2.3 Estimation of Initiation Gap

Olsen proposed the following relationship between the air gap and quantity of the donor explosive (Fedoroff and Sheffield, 1969).

$$K = 0.1522\ (W/L^{2.25}) \qquad (15.2)$$

where W is the charge of the donor explosive (kg)
L is the air gap (m)
K is a constant.

In order to use equation (15.2) to estimate the maximum gap for sympathetic detonation to take place, it is essential to determine the value of the constant K. This can be done experimentally by determining the maximum value of L (air gap) for three to four weights of the donor.

Baum et al. (1959) suggested the following relationship—

$$K = R_{50}/W^{0.5} \qquad (15.3)$$

Where, R_{50} is the air gap in metres when 50% detonation takes place and W is the weight of the donor explosive in kg.

It is appropriate to mention that equations (15.2) and (15.3) may be used for approximate estimation of the air gap for a given quantity and type of explosive provided the value of K is known. Investigations by CMRI indicated that for precise estimation of air gap for an explosive, the relationship between R_{50} and W should be established by regression analysis which then should be used only for that type of explosive. The power of W was found to vary with the type of explosive. For example, in case of nitroglycerine (NG-based) explosives, the power of W varied between 0.355 and 0.53 and the value of K between 0.34 and 1.28.

Sympathetic detonation tests for different types of explosives were conducted by CMRI, and the air gap for 5 kg of explosive was estimated from the test results. Observations indicated that the air gap for NG-based explosives was greater compared to slurry explosives (Nabiullah et al., 1989). More detailed results are available in the book authored by Singh and Pal Roy (1993).

The spacing to be kept for sympathetic detonation to take place between columns of explosives in a hole (or for estimation of safe distance between two heaps of explosives, or for estimation of minimum spacing between two rows of holes) is not only dependent on the type and strength of explosives, but also on the total quantity. A strong relationship between the air gap and the quantity of explosive was observed, and it was possible to estimate the air gap (R_{50}) for any given quantity of explosive with the help of this statistical relationships.

15.2.4 Thermal Behaviour of Commercial Explosives

CMRI studied the thermal behaviour of some commercial explosives including detonating fuse at different temperatures to measure ignition and explosion temperatures (Nabiullah et al., 1988). It was observed that detonating fuse burnt in unconfinement at 130°C but at the same temperature it exploded in steel pipe confinement. Aluminium sensitized watergel explosives lost their gel properties at 80°C and began to decompose at 150°C. Scanning calorimetry curves of watergel and ANFO explosives showed an exothermal reaction at 135°C, 150°C and 275°C respectively.

In order to examine drop test of explosives with detonating fuse on hot surface, explosive cartridges weighing 6.25 kg each were tied with a detonating fuse 1.5 to 3.0 m above the hot surface, then dropped on the hot sand bed through the cement pipe using a pulley and rope. The dropped cartridges were allowed to remain in contact with the hot sand bed for periods ranging between 3 minutes to 4 hours in order to study the behaviour of explosives under simulated borehole conditions. The experiment was repeated several times at different bed temperatures like

70°, 100°, 120° and 150°C. It was observed that when the temperature was 100°C or more and the height of fall was 3.0 m, NG-based explosives burnt without explosion after 3 min. While no burning and explosion occurred in slurry explosives, except in a case where the temperature was 150°C and the height of fall was 3.0 m. In that case, fumes were observed without flame after 30 minutes but the explosive slowly burnt after 1 hour.

15.2.5 Conclusions

The experiments conducted by CMRI (Nabiullah et al., 1988, 1989) led to the following conclusions—

(a) In NG-based explosives, three endothermic peaks were observed at 37°, 80° and 120°C, due to phase transitions of ammonium nitrate, and a large exotherm at 135°C corresponding to the thermal decomposition of the material leading to release of gaseous products.

(b) Slurry explosives showed a slightly higher temperature of ignition ranging between 155° to 175°C. Slurry explosives normally have a tendency to lose water and their gel structure above 80°C. Hence, many of the compositions showed large endothermic peaks and then the final decomposition peak.

(c) AN (ammonium nitrate) and ANFO (ammonium nitrate and fuel oil mixture) were similar in nature but the decomposition temperature of AN was greater than that of ANFO because the fuel oil used to sensitize AN lowered the decomposition temperature of ANFO.

(d) PETN explosive showed one endothermic peak at 135°C due to phase transition (and melting) and exothermic decomposition at 171°C.

15.3 NON-ELECTRIC INITIATING DEVICES

15.3.1 Introduction

Due to high velocity of detonation and detonation pressure, the detonating fuse partly desensitizes the explosive charge in contact with it, which in turn affects the available energy of the explosive. This phenomenon is more pronounced in water gels (slurries and emulsions) which are sensitized by gas bubbles, micro-balloons etc. In case of ANFO, it causes crushing and consolidation of ammonium nitrate prills, rendering a portion of the charge as useless.

Moreover, the conventionally used cap and fuse system often cause air blast, erratic time sequence, toe problem and restricted blast geometry, resulting in unoptimized blast performance. Such restrictions with the conventional methods led to the development of non-electric initiation systems. At present, in India, two non-electric initiation systems are widely available. 'EXEL' shock tube systems are manufactured by IES India Ltd. and marketed by Indian Explosives Ltd. (IEL) and 'Raydet' non-electric initiation systems are manufactured and marketed by M/s Gulf Oil Corporation Ltd. (previously known as M/s IDL Industries Ltd.). These devices are non-disruptive, immune to extraneous electricity sources, facilitate multiple delay intervals, bottom hole initiation and insensitive to friction and impact effects.

Eco-friendly and productive blasting operations can be achieved using these systems when implemented judiciously. Blasting can also be done in sensitive regions with greater ease and confidence.

15.3.2 Raydet DTH and TLD

M/s Gulf Oil Corporation Ltd. has developed Raydet DTH (Down-the-hole) and TLD (Trunk Line Delay) shock tube delay blast initiation system for surface blasting operations. Shock tube helps in achieving either bottom hole initiation or initiation at any desired location of the charge column. The shock tube is initiated by the high order shock imparted either by a detonator, detonating cord or special surface delay initiation units, specifically made compatible for use with Raydet shock tube. While initiated, a low order shock wave, with a velocity of detonation of around 2200 m/s travels inside the tube to initiate the delay detonator.

Around 20 different delay periods in milliseconds and 12 periods in the half-second (long delay) range are currently available for field use. Every shock tube DTH contains a plastic connector on its free end for hook-up with detonating cord trunk-line. This has made the system easier when down-the-hole Raydet non-electric initiators combined with detonating cord on surface trunkline are initiated either instantaneously or delayed using cord relays (Sarathy, 1995).

15.3.2.1 Raydet DTH (Down-the-hole)

Raydet DTH consists of a non-electric detonator, a length of Raydet shock tube and a connector. The non-electric detonator is a precise full strength unit with a predetermined delay element inside. The delay timing usually has 450, 475 and 500 milliseconds. Raydet DTH is used in-hole in conjunction with surface delay unit Raydet TLD. Other delay intervals are also available on requirement. The basic idea of using a relatively high delay interval DTH is to ensure that the detonation of the surface hook-up would pass to several drillholes ahead before the first drillhole gets initiated and ground movement starts subsequently and thereby minimising the risk of misfires due to ground movement cut-off. The available lengths of Raydet shock tubes varying between 3 to 45 metres.

15.3.2.2 Raydet TLD (Trunk-Line Delay)

Raydet TLD is a surface delay initiation unit which assists Raydet DTH to detonate in varying desired firing times. Raydet TLD consists of a Raydet millisecond delay detonator with varying colours and lengths assembled with delay intervals of 0 (Green), 17 (Yellow), 25 (Red), 42 (White), 50 (White), 67 (Blue) and 100 (Black) milliseconds. Other delay intervals are also available against specific indent. A specially designed bunch block connector for hooking-up up to 6 shock tubes at a time is fitted on the detonator shell.

15.3.3 EXEL System

EXEL shock tube is a plastic tube of 3.2 mm outer diameter and 1.1 mm inner diameter. A fine dust of HMX and aluminium powder is sprayed during extrusion, thus forming a fine coating of explosive in the inner walls of the tube. During preparation of shock tube it uses commonly available plastics and suitable additives which make it sticky during molten stage so that the coated explosive powder sticks properly to its inner walls. The tensile strength of this tube is little more than 45 kg. An EXEL shock tube detonator consists of a length of shock tube which is sealed at one end and a delay detonator at the other end. Usually 175, 200, 250, 400 and 450 ms delay detonators are available in the market. However, on demand, others may also be available. EXEL DTH systems are generally available in lengths between 3 and 35 metres (Bhushan and Srihari 1995; Singh et al., 1996).

EXEL DTHs are commonly used with EXEL NTD (Noiseless Trunkline Delay) units or the EXEL MSC (Millisecond Connectors) units. It can also be initiated by detonating cord using the cobra connector at the sealed end. It is safe against initiation by static electricity, stray electrical currents and radio frequency transmissions. Technological information of Raydet and EXEL initiating devices are given in Table 15.5.

15.3.4 Study of Component Scatter

When there is any scattering of the delay detonators used in the non-electric initiation system, then there may be some probability of overlap for adjacent charges by firing sequence. In scatter calculation, the following three general assumptions are made—

(1) Actual firing times of delay initiators follow normal distributions.
(2) Nominal (specified) time represents the mean for the actual firing time for each component.

Table 15.5 Technological information of Raydet and EXEL systems

	Raydet (GOCL)	*EXEL (IEL)*
Base Charge		
Material	PETN	PETN
Charge (g)	0.56	0.80
Colour	Dull Powder	White Powder
Initiating Charge (Delay Powder)		
Material	LA Powder	Powder
Charge (g)	0.23	0.11
Height of the PETN+LA bed (mm)	15	18
Plastic Tube (Shock Tube)		
Colour	Transparent	Pink
Rheology	Slightly stiff	Soft
OD (mm)	3.0	3.2
ID (mm)	1.3	1.1
Material inside the plastic tube	PETN + Al Powder	HMX / Al Powder
Charge wt. mg/m	15	16
Plastic tube inside the Al. cap (mm)	41	On the top of the antistatic cap (variable)
Gap between the plastic end inside the shell and the element (mm)	8	No gap. Signal tube rests on the top of the antistatic cap.
Down-the-hole Detonator		
Thickness (mm)	0. 50	0.50
OD (mm)	7.2	7.5
ID (mm)	6.2	6.5
Shape	Flat	Flat
Delay Elements		
Material	Lead	Lead
No. of elements	Single	Double; Main and Sealer
Core	Double Core	Single core each for main and sealer
Shape of the core	Rectangular	Round
Crimping	Single crimping	Double crimping
Closure Bush		
Colour	Black; Moulded ring on top	Black; No ring
OD of the closure bush (mm)	5.5	6.1
ID of the closure bush (mm)	4.3	3.6
Length of the closure bush (mm)	74.2	35.5

(3) Scatter represents a limit above and below the mean. This means that if there is a 5% scatter on a 100 ms component, one standard deviation of scatter is assumed to be 100 ms plus or minus 5% of a period (95 – 105 ms). If we assume one S.D. to be 5% of the period, which means that 68% of the caps should fire within that range. For more accuracy we can set the S.D. to 2 which will cover 95% of the caps.

15.3.5 Probability Calculations

The overlap can be calculated in two different ways. One can either use nominal times and a specified number of milliseconds between nominal times as a criterion or can use a more accurate method of calculation by considering the tolerance in the firing time of the components or "Cap Scatter".

Analyses were carried out for 7 different cases as shown in Table 15.6. Out of these cases, firing times with three different component scatters are depicted in figures 15.1, 15.2 and 15.3.

Table 15.6 Scattering test of non-electric initiation systems

Case	*Name of the mine*	*Total no. of holes*	*Component Scatter (%)*	*Overlap criteria*	*Minimum delay time allowed (ms)*	*95% confidence level (±2 S.D.)*
1	Parej East OCP, CCL	30	1	Charge per delay	8	- do -
2	Parej East OCP, CCL	20	3	Charge per delay	15	- do -
3	Sonepur Bazari, OCP, ECL	40	3	Charge per delay	15	- do -
4	Sonepur Bazari, OCP, ECL	40	1	Charge per delay	8	- do -
5	Chasnalla OCP, IISCO	20	3	Charge per delay	8	- do -
6	Chasnalla OCP, IISCO	20	1	Charge per delay	15	- do -
7	Chasnalla OCP, IISCO	20	3	Charge per delay	15	68% confidence level (±1 S.D.)

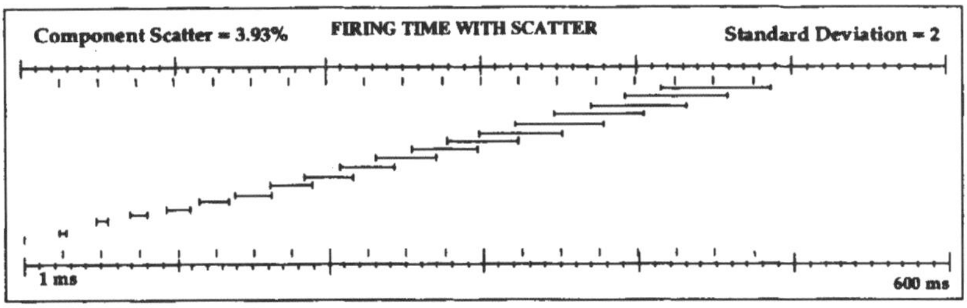

Fig. 15.1 Pictorial presentation of the firing times with component scatter 3.93% of cord relays

15.3.6 Observations

Based on experiments conducted by CMRI, the following observations

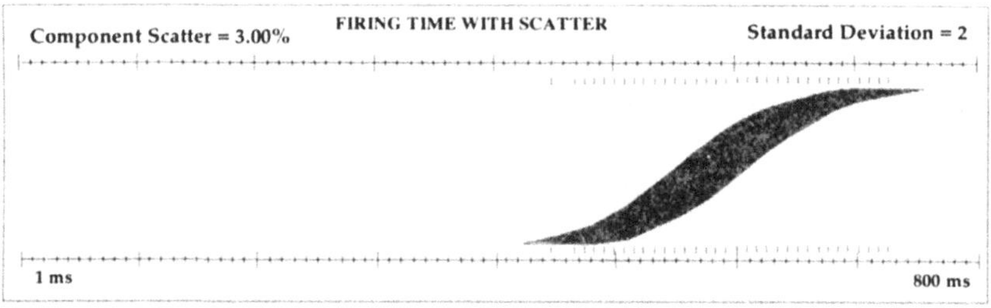

Fig. 15.2 Scattering test of non-electric initiation systems (component scatter-3.00%) [Theoretical evaluation]

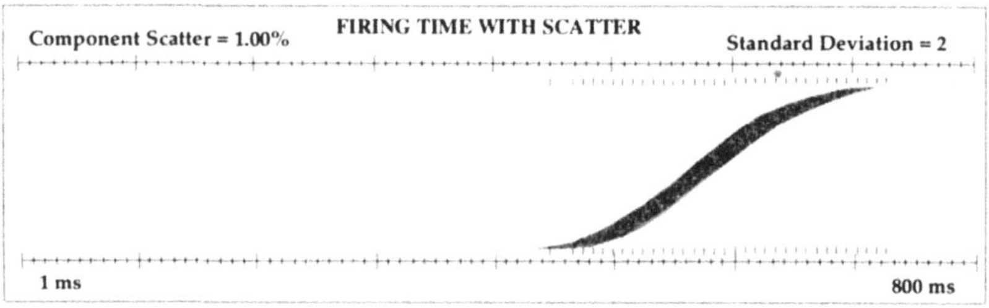

Fig. 15.3 Scattering test of non-electric initiation systems (component scatter-1.00%)

are made. They may be used as referred guidelines for the potential usage of the system.

(a) *Noise level due to blasting* - Significantly controlled

(b) *Air-overpressure level* - Reduced marginally

(c) *Ground vibration level* - Reduced in cases where muck profile was good, throw was near optimum (nearly 1.8 times of the bench height) and fragmentation was optimum.

(d) *Control of toe problem* - Could be achieved with proper blast pattern.

(e) *Control of flyrock* - Could be achieved desirably when the stemming was perfect, rock movement was proper and the hook-up system was correct.

(f) *Attenuation of vibration* - It was measured that the ground vibration generated in bottom hole initiation had the tendency of sharper attenuation in comparison with the vibration generated due to conventional detonating fuse and relay system blasting.

15.3.7 Electronic Detonator

The electronic detonator is based on microchip technology. A miniature timing circuit replaced the chemical delay element in conventional electric detonators. It is basically a device, which stores electrical energy for a certain time and then delivers that energy as a sharp pulse at a precise time to a conventional fuse head. The electronic blasting system consists of components namely programmable digital detonator, logger and blaster.

Each electronic detonator has a unique identification number (ID No.) and it contains a microchip, capacitor and safety components as well as conventional explosive components. The strength is equivalent to detonator no. 8. The microchip circuitry includes an oscillator for timing, memory for retaining its programmed delay and communication functions to receive and deliver messages to and from the control unit. The capacitor can store sufficient energy to run the microchip independent of external power for 14 seconds with enough energy remaining after this time period has elapsed to fire the fuse head. Protection against electrostatic discharges, stray current and radio frequencies are provided through detonator design, chip design and protective components on the input side of the detonator. Shock protection is provied both internally and externally. Each detonator has a connector which allows easy connection to the harness wire when hooking up.

The logger is used to communicate and assign the desired delay times to the individual detonators. When connected to the twin wire (bus wire), the logger will read the equivalent ID No. of the respective detonator and the user can then assign the intended delay time to that particular detonator. First of all, the logger will note the position of the detonator in the firing circuit and then check its ID No. During the check of the ID No. of the detonator to be logged, the logger also checks the correct contact between detonator and bus wire as well as the functioning of the detonator and check the wire of short-circuit or current leakages. Only after this examination, the user is asked to set the delay time. One logger can manage maximum of 200 detonators.

The blaster is a unit which controls the programming and triggering of blast. It contains the required voltages and codes capable of firing the detonators. The blaster communicates to each detonator in turn via the logger. The function of the blaster is to check the firing line and harness circuitry of the blast, calibrate and program each detonator with its delay time, arm all detonators, then fire the blast by providing a unique signal to all detonators. One blaster can manage up to 8 loggers. Thus, the firing system can operate a maximum of 1600 detonators in one go. The maximum bus wire length is 2000 m per logger and 1000 m firing line between loggers and blaster. A time period of less than 5 minutes is required for programming the maximum of 1600 detonators (Bartley et al., 2000; Tucker and Kay, 2000).

In India, the first blast in opencast mine using electronic delay detonator was carried out at West Bokaro Collieries, TISCO on 21st July 2004. There were 33 number of holes each of 14 m depth (154 mm diameter). Charge per hole was 225.4 kg and a total of 7438.20 kg explosive was fired in the round. The recorded vibration was 25 mm/s (vector sum) at a distance of 75 m at the backside of the blast and the level of air overpressure (PSPL) was 94 dB (L). It was successful blast casting with a throw of 45 m with desired fragmentation.

15.3.8 Conclusion

Judicious implementation of non-electric initiation systems like EXEL and Raydet may be beneficial from production and safety points of view. Implementation of such system would enhance the cost of blasting in comparison to the conventional detonating fuse and delay/relay systems. The techno-economics of these systems were thoroughly studied by CMRI in an S&T Project sponsored by the Ministry of Coal, Govt. of India (1998). Such cost enhancement can be bypassed if the blasting performance is satisfactorily improved with reduced toe problem and reduced blasting hazards especially in environmentally sensitive places. Hook-up system and design pattern are extremely important when this system is practiced. It is generally seen that the manufacturers, while implementing their product, calculate only the surface timing of delay connections without going much into the details of firing time of each deck charge and thereby the actual firing sequence does not come into the picture. It is very essential to check them before establishing the pattern, otherwise misfire may occur in any round of the blast. The accuracy, simplicity and safety of electronic delay detonators will lead to total control of fragmentation, vibration and flyrock in blasting operation. It is expected that these products would help the mining industry to maintain its sustainable growth keeping in view the increased human sensitivity and safety consciousness.

REFERENCES

Baum, F. A., Stanyukovich, K. P. and Shekhter, B. I. 1959. Safe distances for propagation of detonation between some explosive charges, Balastic Research Laboratory Report, Aberdeen, UK, pp. 754-771.

Bartley, D.A., McClure, R., Wingfield, B. and Trousselle, R. 2000. Electronic detonator technology: Field application and safety approach, Proc. 1st World Conference on Explosives and Blasting Technique, Munich, Germany, September 6-8, pp. 149-158.

Bhushan, V. and Srihari, H. N. 1995. Evolution of shock tube detonators technology, Jr. Mines, Metals and Fuels (Special Number), August, India, pp. 234-239.

CMRI S&T Report of Investigations, 1998: GAP/2/Blasting/MOC/96, Techno-economic assessment of drilling and blasting parameters vis-à-vis establishment of cost effective optimum blast pattern for opencast mines, February, p. 88.

Fedoroff, B. T. and Sheffield, O. 1969. Encyclopedia of Explosive and related items, pt. 2700, Vol. 4, Picatinny Arsenal, Dover, New Jersey, USA, pp. 395-399.

IDL Training Programme Manual on Special Initiation Techniques at BCCL, 1994. Technical Services Cell, IDL Chemicals Limited, August, India, p. 66.

Nabiullah, Md., Gupta, R. N., Singh, B. and Gangopadhya, P. K. 1988. A study on thermal behaviour of some commercial explosives, Thermal Analysis, Proc. 9th ICTA Congress, Jerusalem, Israel.

Nabiullah, Md., Gupta, R. N. and Singh, B. 1989. Investigation of delay periods for some commercial delay detonators, The Journal of Explosives Engineering, USA, pp. 12-15.

Pal Roy, P. 1995. Mathematical modelling to evolve explicit relationships in the study of energy balance in rock blasting, R&D Project Report, EMR/PPR/90-91, December, India, p. 117.

Pal Roy, P. and Singh, R. B. 1997. Enhancement of safety and productivity using new blasting accessories in opencast mines, Proc. 27th Int. Conf. of Safety in Mines Research Institute, New Delhi, India, pp. 655-664.

Pal Roy, P., Bhattacharya, M. M. and Pal, R. K. 1998. Performance test of explosives for better energy utilization, Visfotak'98, Proc. National Seminar on Explosives safety and Technology, September 12-13, Nagpur, India, pp. 59-69.

Sarathy, M. O. 1995. Raydet non-electric blast initiation system for efficient and environment-friendly surface blasts, Jr. Mines, Metals and Fuels (Special Number), August, India, pp. 205-211.

Singh, B. and Pal Roy, P. 1993. Blasting in ground excavations and Mines, Oxford and IBH Publ. Co. Pvt. Ltd., New Delhi, p. 177.

Singh, S. K., Jana, S., Singh, D. P. and Bhushan, V. 1996. Exel initiating system—a study at Noamundi iron ore mine, Proc. National Seminar on Drilling and Blasting, Drill Blast'96, September 20-21, India, pp. 193-199.

Tucker, G.B. and Kay, D.B. 2000. The use of electronic blasting system at the M.I.M. George Fisher Mine, Proc. 1st World Conference on Explosives and Balsting Technique, Munich, Germany, September 6-8, pp. 117-122.

Winzer, S. R., Furth, W. and Ritar, A. 1979. Initiator firing times and their relationship to blasting performance, Proc. 20th US Symposium on Rock Mechanics, Austin, June 4-6, Texas, USA, pp. 461-470.

EXTRA READING

Nielsen, Kai and Kristiansen, Jan, 1996. Blasting-crushing-grinding: Optimization of an integrated comminution system, Proc. Rock Fragmentation by Blasting, Mohanty (ed.), A. A. Balkema, August 25-29, Montreal, Canada, pp. 269-277.

Gynnemo, Magnus, 1996. The importance of velocity of detonation in rock blasting-a full scale study, Proc. Rock Fragmentation by Blasting, Fragblast'5, (ed. B. Mohanty), A. A. Balkema, 25-29 August, Montreal, Canada, pp. 297-300.

CMRI S&T Report, 2000. Evaluation of damage to underground coal mines caused by surface blasting vis-à-vis establishment of blast vibration threshold, GAP/3/Blasting/MOC/96 (Project leader: P. K. Singh), August, p. 51.

CONVERSION FACTORS

The following conversion factors of various quantities are given for practical usefulness.

Quantity	*Conversion factors*	*Quantity*	*Conversion factors*
Length	1 ft = 0.3048 m	Mass	1 ton = 1.016 tonnes
	1 in = 25.4 mm		1 lb = 0.4536 kg
	1 micron = 10^{-3} mm		
Area	1 ft^2 = 0.0929 m^2	Density	1 g/cc = 1 t/m^3
	1 in^2 = 645.2 mm^2		
Volume	1 yd^3 = 0.7646 m^3	Force	1 lb f = 4.448 N
	1 ft^3 = 0.02832 m^3		1 kg f = 9.807 N
	1 gallon = 4.546 litres		1 lb f = 0.004448 KN
	1 ft^3 = 28.32 litres		1 ton f = 9.964 KN
	1 cc = 1 ml		
	1 in^3 = 16.39 cm^3		
Elastic moduli	1000 lb f/ft^2 = 0.04788 MN/m^2	Pressure and stress	1 lb f/in^2 = 6.895 KN/m^2
	1 ton f/in^2 = 15.44 MN/m^2		1 lb f/ft^2 = 0.0478 KN/m^2
			1 ton f/ft^2 = 107.3 KN/m^2
			1 kg f/cm^2 = 98.07 KN/m^2
			1 bar = 100 KN/m^2
			1 Pascal = 1 N/m^2
			1 MPa = 145 psi
Plane angle	No change	Temperature	nF = 0.5556 (n-32)C; Celsius: preferred name for centigrade.

Index